Enzyme-Prodrug Strategies for Cancer Therapy

Enzyme-Prodrug Strategies for Cancer Therapy

Edited by

Roger G. Melton

Enzacta Ltd.
Porton Down
Wiltshire, United Kingdom

and

Richard J. Knox

Imperial College School of Medicine
Charing Cross Campus
London, United Kingdom

KLUWER ACADEMIC / PLENUM PUBLISHERS
NEW YORK, BOSTON, DORDRECHT, LONDON, MOSCOW

Library of Congress Cataloging-in-Publication Data

Enzyme-prodrug strategies for cancer therapy / edited by Roger G. Melton and Richard J. Knox.
p. cm.
Includes bibliographical references and index.
ISBN 0-306-45895-0
1. Antibody-directed enzyme prodrug therapy. 2. Cancer--Chemotherapy. 3. Prodrugs. 4. Antibody-enzyme conjugates--Therapeutic use. I. Melton, Roger G. II. Knox, Richard J.
[DNLM: 1. Antineoplastic Agents--therapeutic use. 2. Prodrugs--therapeutic use. 3. Neoplasms--drug therapy. 4. Antibodies--therapeutic use. 5. Enzymes--therapeutic use. 6. Antigens--drug effects. QZ 267 E61 1998]
RC271.A66E56 1998
616.99'4061--dc21
DNLM/DLC
for Library of Congress 98-37682
CIP

ISBN 0-306-45895-0

233 Spring Street, New York, N.Y. 10013

10 9 8 7 6 5 4 3 2 1

A C.I.P. record for this book is available from the Library of Congress.

Printed in the United States of America

Contributors

K. D. Bagshawe • Charing Cross & Westminster Medical School, London W6 8RP, United Kingdom

P. J. Burke • Department of Medical Oncology, Charing Cross Hospital, London W6 8RF, United Kingdom

K. A. Chester • Department of Clinical Oncology, Royal Free Hospital School of Medicine, London NW3 2PF, United Kingdom

Tom Connors • Centre for Polymer Therapeutics, The School of Pharmacy, London WC1N 1AX, United Kingdom

R. E. Hawkins • Department of Oncology, MRC Centre, Cambridge CB2 2HQ, United Kingdom

Richard J. Knox • Department of Medical Oncology, Charing Cross Hospital, London W6 8RP, United Kingdom

Roger G. Melton • Enzacta Ltd., Salisbury SP4 0JQ, United Kingdom

Mark Napier • The Meyerstein Institute of Oncology, The Middlesex Hospital, London W1N 8AA, United Kingdom

M. A. Sims • Centre for Applied Microbiology & Research, Salisbury, SP4 0JG, United Kingdom

Preface

Since the concept was first described in 1987, the use of antibody-enzyme conjugates directed at tumor-associated antigens to achieve site-specific activation of prodrugs to potent cytotoxic species, termed antibody-directed enzyme prodrug therapy (ADEPT), has attracted considerable interest. ADEPT attempts to overcome a major problem associated with the administration of cytotoxic drugs for treating cancer—the intrinsic lack of tumor selectivity. Thus, chemotherapeutic agents have the potency to kill tumor cells *in vitro* but, because of host toxicity, the treatment of human patients has to be discontinued at dose levels well below those which are expected to kill all of their tumor cells. In ADEPT a cytotoxic agent is generated selectively at the site of a tumor by an antibody-targeted enzyme. The antibody delivery combined with the amplification provided by the enzymatic activation of prodrugs enables adequate selection to be made between tumor and normal tissue.

A rich diversity of potential enzyme and prodrug combinations exists for use in ADEPT applications, and ADEPT offers new research ideas for the chemist, enzymologist, molecular biologist, protein engineer, and the cancer chemotherapist. It offers opportunities for the therapy of systemic cancer and may be a major advance for treating solid tumors. ADEPT has shown very encouraging results in animal models and in a preliminary clinical trial. The results of more extensive trials are awaited with interest, and ADEPT systems may soon form an important element in the clinical oncologist's armament.

In compiling this book we have been fortunate to have authors participate who have been involved with ADEPT since it was conceived, and we are grateful for their expert contributions. We hope this book will strengthen the understanding of this field of research and perhaps stimulate some new ideas in the area.

ROGER MELTON
RICHARD KNOX

Contents

Chapter 3

Factors Influencing Tumor-Selective Localization of Antibody Conjugates

M. A. Sims and Roger G. Melton

Chapter 4

Enzymes and Prodrugs Used for ADEPT

Richard J. Knox

1

Introduction

Richard J. Knox and Roger G. Melton

1. INTRODUCTION

In the 1970s it was thought that the advent of monoclonal antibody technology heralded the fruition of the search for the "magic bullet" first proposed by Ehrlich in the early years of this century (Ehrlich, 1960). In practice, however, this goal has remained elusive as a range of obstacles have presented themselves and have proved very difficult to overcome. To date, there have been numerous attempts to improve the cytotoxicity of the antibody "missile" by attaching a variety of "warheads" to them, for example, cytotoxic drugs, such as adriamycin or methotrexate, toxins, such as ricin A-chain or *Pseudomonas* exotoxin, and radioisotopes (Pietersz and McKenzie, 1992).

Prodrugs, defined as inert precursors of potent drugs that can be activated at the intended site of action to potent drugs, have long been sought for use in cancer therapy as described in Chapter 2. For many years, the focus of attention lay with trying to produce agents which could be activated by endogenous enzymes. Thus, it was not until the late 1980s that the possibility of using antibodies to deliver suitable enzymes to tumors was successfully explored. In part, the success of antibody targeting depends on the ability of the antibody conjugate to escape from the tumor vasculature and actually reach the target cells. Yet there is ample evidence that antibodies penetrate tumors poorly (Cobb *et al.*, 1987; Moshakis *et al.*, 1981). The implications of tumor physiology for antibody targeting strategies are discussed in the first half of Chapter 3. The best access is at the tumor

---

Richard J. Knox • Department of Medical Oncology, Charing Cross Hospital, London W6 8RP, United Kingdom. *Roger G. Melton* • Enzacta Ltd., Salisbury SP4 0JQ, United Kingdom.

Enzyme-Prodrug Strategies for Cancer Therapy, edited by Melton and Knox. Kluwer Academic/Plenum Publishers, New York, 1999.

periphery, and cells located near the center of the tumor mass may be inaccessible to immunoconjugates, even though they may express antigen. Cells in such inaccessible anoxic zones may be quiescent but possess the ability to regrow if exposed to a supply of oxygen and nutrients after more vulnerable cells in oxic zones are killed. For this reason radio-immunoconjugates have been extensively studied because the isotopes used clinically penetrate several cell diameters, killing cells distal from where the antibody is localized. Although this approach is attractive, in practice the relatively low tumor uptake of the radioimmunoconjugate may give rise to dose-limiting myelosuppression, necessitating the use of autologous bone marrow grafts (Goldenberg *et al.*, 1991).

For any form of antibody-mediated targeting to be successful, it is axiomatic that there must be selective expression of the target antigen by the tumor cells. This provides the first barrier to overcome, for the only well-characterized tumor specific antigens described to date are the idiotypic determinants on the surface immunoglobulins of B-cell lymphomas (Stevenson and Stevenson, 1975). Many antigens and other potential targets, such as growth factor receptors, are present in elevated levels in tumor tissue, but most, if not all, are found to a greater or lesser extent in other tissue. Furthermore, it is well documented that tumor cells exhibit considerable antigenic heterogeneity (Fargion *et al.*, 1986; Natali *et al.*, 1983), and some are devoid of antigen expression. So a single monoclonal antibody conjugate targeted at a tumor mass does not bind to all of the cells present. This is a disadvantage for immunoconjugates constructed by using cytotoxic moieties which need to be bound to the target cells and be internalized to exert their effect. However, the problem may be overcome to some extent by using antibody "cocktails" in which a mixture of antibodies aimed at two or more discrete antigens are used in noncompetitive combination (Tagliabue *et al.*, 1986). The subject of tumor physiology and tumor-associated antigens is covered in some detail in the second half of Chapter 3.

The use of antibody-targeted enzymes to activate prodrugs, an approach which has been termed "Antibody Directed Prodrug Therapy" (ADEPT) (Bagshawe, 1989), or "Antibody Directed Catalysis" (ADC) (Shepherd *et al.*, 1991) offers the potential of overcoming most of the problems associated with the use of drug or toxin immunoconjugates described previously. The system is shown by the diagrammatic representation in Figure 1. An antibody-enzyme conjugate is injected and allowed to localize at the tumor while clearing from the rest of the tissues. Therapy can proceed as either a two- or three-phase system. In the two-phase system the nonlocalized, unbound antibody is simply allowed to clear naturally from the body. Then the prodrug is administered. For the three-phase system this unbound antibody in hastened on its way by a "clearance" antibody, and then the prodrug is administered. The prodrug is innocuous to the normal tissues but is converted to a cytotoxic form by enzyme localized within the tumor tissue. The active low molecular weight drug can diffuse to adjacent tissues,

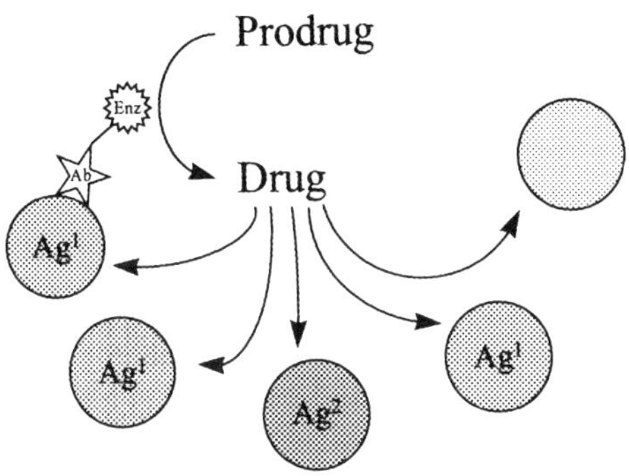

Figure 1. The generation of a cytotoxic drug by ADEPT. Antibody-enzyme conjugate is administered and allowed to bind to the target tumor cells. After unbound conjugate has been cleared by natural or accelerated processes, a prodrug is administered. This is converted to an active drug by the bound enzyme. Importantly, the active drug can migrate and have cytotoxic effects on cells that do not have conjugate bound to them.

including cells that express the target antigen but which have not bound conjugate, cells that may express alternative tumor-associated antigens not recognized by the vector, or cells that are antigen-negative. This is the so-called bystander effect. It is implicit in this approach that the antigen to which the antibody-enzyme conjugate is targeted must remain extracellular. Thus secreted antigen that accumulates in the tumor interstitial spaces may assist in achieving high levels of conjugate at the tumor site. Internalized conjugate is expected to be rapidly degraded within the lysosomal compartment.

A further advantage of ADEPT is that a single enzyme molecule has the potential to cleave many prodrug molecules—up to 800 molecules per mole of enzyme per second in the case of the benzoic acid mustard substrates of carboxypeptidase G2 (Springer *et al.*, 1991). This amplifies the effect, giving high levels of drug localized at the tumor that may be an important advantage in the clinic in view of the typically low localization of immunoconjugate in humans (Begent, 1985). Some evidence also indicates that high levels of drug generated at the surface of tumor cells are more effective than equivalent concentrations of free drug (Haenseler *et al.*, 1992). In selecting an enzyme, one looks for activity under physiological conditions, low immunogenicity, and little or no equivalent endogenous enzyme in humans. The enzyme must localize efficiently at the tumor when coupled to an antibody and the development of immune responses to non-human enzyme is undesirable because localization of the conjugate is likely to be drastically reduced on readministration after the onset of an immune response (Meredith *et al.*, 1993). Immunosuppressive drug regimes designed to extend the time window before the onset of the human immune response, however, have been developed to counteract this problem (Bagshawe and Sharma, 1996; Sharma *et al.*, 1996). A wide range of enzymes have been used for experimental systems and are described in detail in Chapter 4.

The prodrug, an integral component of ADEPT systems, requires careful design in its own right. An ideal prodrug is one with a large differential in cytotoxicity between drug and prodrug, which is a good substrate for the enzyme under physiological conditions, and for which there is no mammalian homologue capable of performing the same reaction. Equal cytotoxicity of the released active drug toward proliferating and quiescent cells is also desirable if residual deposits of viable but nonproliferating cells with the potential for outgrowth are to be eradicated. Once formed, it is desirable for the drug to have a very short half-life, limiting the possibility that active drug escapes back into circulation and has access to healthy tissue. Development of drug resistance may limit the effects of conventional active drugs produced from prodrugs. Thus, the nitrogen mustard group of compounds have been the most commonly used active drugs for ADEPT because they are not cell-cycle-specific, can kill both well-oxygenated and hypoxic cells, and cells develop only low levels of resistance to them (Frei *et al.*, 1988). The chemistry of prodrug development is reviewed in Chapter 5.

The manufacture of antibody-enzyme conjugates that retain both enzymatic and antigen-binding activities and are stable *in vivo* is an important facet of the development of ADEPT systems, and a range of conjugation chemistries is available for use. The appropriateness of these and their relative advantages and disadvantages are discussed in Chapter 6. Second generation systems are likely to be based on conjugates produced by protein engineering techniques, and a number of such constructs have been produced. The approaches currently available are described in Chapter 7.

Experimental studies have been followed by a limited clinical study in patients being treated for advanced metastatic colon or rectal cancer, as described in Chapter 8. In common with all types of therapy, ADEPT has certain limitations that were highlighted in the clinic.

1.1. The Activity of the Activated Drug

An advantage of the ADEPT concept is the catalytic formation of the active drug that can then migrate to parts of the tumor that have not bound the conjugate. However, the active drug must not migrate so far that normal tissue is cytotoxically affected. Further work is needed to determine the optimal characteristics (in particular the half-life) for the active drugs when they are generated by ADEPT so that bystander effects are maximized in the tumor but damage to normal tissue is kept to a minimum. These considerations are discussed further in Chapter 5.

1.2. Activation of the Prodrug by Unbound, Circulating Conjugate

After administration, an antibody-enzyme conjugate can remain in the vascular compartment for several days. Although the concentration is much less than that bound to the tumor, the volume of plasma is much greater, and thus the actual enzyme activity may be greater in the plasma than the tumor. In this case, administration of a prodrug may be no more selective than systemic administration of the active form. Thus it is necessary to wait for plasma levels of the conjugate to fall to very low levels (during which time the enzyme activity at the tumor may also be decreasing). Alternatively, the plasma may be *cleared* of enzyme activity, and this can be achieved by use of a second, clearance antibody directed toward either the enzyme or antibody. Much high tumor to blood levels of the conjugate have been reported (38:1 compared to 1.3:1 without clearance) (Sharma *et al.*, 1994; Sharma *et al.*, 1990). Inhibiting the active site of the enzyme enhances clearance of enzyme activity, but it is important that the clearance antibody has little effect on bound conjugate. This has been achieved by the rapid clearance of the second antibody through the carbohydrate receptors in the liver by introducing galactose residues into the antibody (Rogers *et al.*, 1995; Sharma *et al.*, 1994; Sharma *et al.*, 1990). Alternatively the conjugate itself is galactosylated and the conjugate localized whereas the liver receptors are blocked by administering a competitive binder, such as asialo-bovine submaxillary gland mucin. Using this type of approach, a tumor to blood ratio of 45:1 for an anti-CEA-carboxypeptidase G_2 conjugate was achieved 24 hr after injection of the conjugate. This increased to 100:1 at 72 hr (Sharma *et al.*, 1994). Tumor to blood ratios of a conjugate (but not other tissue to blood ratios) were also significantly increased by co-administering tumor necrosis factor alpha that enhances the tumor uptake of antibody-enzyme conjugates (Melton *et al.*, 1993). Co-administration of 1.5μg TNF with conjugate led to a twofold increase in tumor uptake with only transient increases in normal tissue localization. It should be possible to use this system combined with a clearing antibody because the two approaches are complementary.

1.3. Immunogenicity of the Antibody-Enzyme Conjugate

An immune response was observed in patients treated with ADEPT therapy using an antibody-carboxypeptidase G2 conjugate (Bagshawe *et al.*, 1991; Sharma *et al.*, 1992). Raising an immune response would preclude extended ADEPT therapy. Several strategies have been proposed to overcome the host immune response. These include co-administering immunosuppressive agents, such as cyclosporin A (Bagshawe *et al.*, 1991; Sharma *et al.*, 1992), or it may be possible to administer a second ADEPT conjugate consisting of a different antibody-enzyme combination. There is also the possibility of masking the immunogenicity of conjugates by modification with polyethylene glycol or other similar polymers (Sehon, 1991). Immunogenicity of the antibody moiety is reduced by using chimeric (e.g., rat antigen-binding domains coupled to human constant regions), humanized or fully human-derived antibody fragments (Desutter and Fiers, 1994). Efforts to reduce the immunogenicity of the enzyme can be attempted by reducing the size of the active fragment; modifying the bacterial enzyme at key immunogenic residues by site-directed mutagenesis; engineering the enzymes active site into the antibody or engineering a human enzyme, such that its activity and specificity more closely match that of the bacterial enzyme. Model compounds consisting of a mouse Fab fused with nucleases or polymerases were reported some time ago (Neuberger *et al.*, 1984). Such constructs could be produced in relatively high yield and with a greater homogeneity of conjugate compared with current chemical linking techniques, where yields are relatively low and the product's homogeneity is ill defined (Melton, 1996). The use of antibody engineering techniques to produce conjugates suitable for use in second generation ADEPT systems is described in Chapter 7. The first fusion protein developed specifically for an ADEPT application consisted of human β-glucuronidase fused to the CH domain of BW431, a highly avid anti-CEA antibody, humanized by complementarity determining region (CDR) grafting. The protein thus produced has antigen affinity very similar to that of the parent antibody and retains enzymatic activity (Bosslet *et al.*, 1994). A *B. cereus* β-lactamase II, single-chain antibody fusion protein has also been constructed. The antibody used, L6, recognizes an antigen expressed on breast, colon, lung, and ovarian carcinomas. The fusion protein is produced by *E. coli* in an active form that does not require refolding and retains immunological and enzymatic properties similar to an equivalent chemically linked conjugate (Goshorn *et al.*, 1993). A further refinement made possible by protein engineering techniques is to make a bifunctional antibody in which one arm possesses an enzyme-mimicking catalytic function. Using combinatorial libraries of human antibodies as sources of the components offers the possibility of producing virtually nonimmunogenic constructs, although it is likely that this will not avoid the development of anti-idiotype responses. A catalytic antibody capable of mimicking cytosine deaminase by converting 5-FC to 5-FU has been described (Campbell *et al.*, 1994). Although turnover capacity of this catalytic antibody was very low, it seems probable that systems capable of matching current chemically linked conjugates will be developed in due course as the technology matures.

Finally, Chapter 9 describes the advances being made with systems that are similar in principle to ADEPT, but use other methods, such as viral vectors, to deliver genetic material encoding the activating enzyme to the target cells. This approach is in its infancy, and at present many problems have yet to be overcome, in particular the issue of ensuring that only the target cells are transfected with the desired gene. Failure to achieve this goal is the equivalent of nonspecific binding in ADEPT and, just as the relatively low percentage of administered antibody that localizes in tumors remains disappointing, it is likely that early *in vivo* transfections will show poor specificity. Early gene therapy experiments have also shown that duration of expression of introduced genes is also quite transient. So this is another issue that must be addressed.

In summary, apparently there is a rich diversity of potential enzyme and prodrug combinations for use in ADEPT applications, and it is highly probable that additional examples will materialize. As yet, there is relatively little *in vivo* data available, and although what does exist provides exciting grounds for encouragement, it is already becoming clear that a number of crucial issues must be resolved. The specificity of conjugate uptake needs to be improved by either selectively increasing tumor uptake using vascular permeability-enhancing drugs or by constructing conjugates with improved pharmacokinetic properties. In the longer term, the latter approach will probably be the more important because the ultimate goal of all workers will be minimizing the need for additional clearing steps or treatments to modify tumor vasculature so as to simplify as far as possible what is a relatively complex system. At the same time, antibody engineering techniques also hold the promise of developing nonimmunogenic conjugates derived from human antibodies. In the shorter term there is the alternative possibility of masking the immunogenicity of conjugates by modification with polyethylene glycol or other similar polymers. Further in the future, the use of antibody vectors and enzymes may be superseded by using vectors capable of delivering genetic material that encode the enzyme of interest. There is a long way to go, but the pace of progress is rapid and ever quickening. Hopefully, it will not be too long before we see that ADEPT systems will form an important element of the clinical oncologist's armament.

References

Bagshawe, K. D., 1989, Towards generating cytotoxic agents at cancer sites, *Br. J. Cancer* **60:**275–281.

Bagshawe, K. D., and Sharma, S. K., 1996, Cyclosporine delays host immune response to antibody enzyme conjugate in ADEPT, *Transplant. Proc.* **28:**3156–3158.

Bagshawe, K. D., Sharma, S. K., Springer, C. J., Antoniw, P., Boden, J. A., Rogers, G. T., Burke, P. J., Melton, R. G., and Sherwood, R. F., 1991, Antibody Directed Enzyme Prodrug Therapy (ADEPT)—clinical report, *Disease Marker* **9:**233–238.

Begent, R. H. J., 1985, Recent advances in tumour imaging: Use of radiolabelled monoclonal antibodies, *Biochim. Biophys. Acta* **780:**151–166.

Bosslet, K., Czech, J., and Hoffmann, D., 1994, Tumor-selective prodrug activation by fusion protein-mediated catalysis, *Cancer Res.* **54:**2151–2159.

Campbell, D. A., Gong, B., Kochersperger, L. M., Yonkovich, S., Gallop, M. A., and Schultz, P. G., 1994, Antibody-catalyzed prodrug activation, *J. Am. Chem. Soc.* **116:**2165–2166.

Cobb, L. M., Humphreys, J. A., and Harrison, A., 1987, The diffusion of a tumour-specific monoclonal antibody in lymphoma infiltrated spleen, *Br. J. Cancer* **55:**53–55.

Desutter, K. and Fiers, W. A., 1994, Bifunctional murine-human chimeric antibody with one antigen-binding arm replaced by bacterial beta-lactamase. *Mol. Immunol.* **31:**261–267.

Ehrlich, P. *The Collected Papers of P. Ehrlich*, Pergamon Press, London, 1960, pp. 505–518.

Fargion, S., Carney, D., Mulshine, J., Rosen, S., Bunn, P., Jewitt, P., Cuttitta, F., Gazdar, A., and Minna, J., 1986, Heterogeneity of cell surface antigen expression of human small cell lung cancer detected by monoclonal antibodies, *Cancer Res.* **46:**2633–2638.

Frei, E. III, Teicher, B. A., Holden, S. A., Cathcart, K. N. S., and Wang, Y., 1988, Preclinical studies and clinical correlation of the effect of alkylating dose, *Cancer Res.* **48:**6417–6423.

Goldenberg, D. M., Sharkey, R. M., Hall, T. C., Murthy, S., Siegal, J. A., Izon, D. O., Swayne, L. C., Lake, D., Hansen, H. J., and Pinsky, M., 1991, Radioimmunotherapy of B-cell lymphomas with ^{131}I-labeled LL2 (EPB-2) monoclonal antibody, *Antibody Immunoconj. Radiopharm.* **4:** 763–769.

Goshorn, S. C., Svensson, H. P., Kerr, D. E., Somerville, J. E., Senter, P. D., and Fell, H. P., 1993, Genetic construction, expression, and characterization of a single chain anticarcinoma antibody fused to β-lactamase, *Cancer Res.* **53:**2123–2127.

Haenseler, E., Esswein, A., Vitols, K. S., Montejano, Y., Mueller, B. M., Reisfeld, R. A., and Huennekens, F. M., 1992, Activation of methotrexate-alpha-alanine by carboxypeptidase-A monoclonal antibody conjugate, *Biochemistry* **31:**891–897.

Melton, R. G., Rowland, J. A., Pietersz, G. A., Sherwood, R. F., and McKenzie, I. F. C., 1993, Tumour necrosis factor increases tumour uptake of co-administered antibody-carboxypeptidase-G_2 conjugate, *Eur. J. Cancer* **29A:**1177–1183.

Melton, R. G., 1996, Preparation and purification of antibody-enzyme conjugates for therapeutic applications, *Adv. Drug Delivery Rev.* **22:**289–301.

Meredith, R. F., Khazaeli, M. B., Plott, W. E., Liu, T., Russell, C. D., Wheeler, R. H., and Lobuglio, A. F., 1993, Effect of human immune response on repeat courses of ^{131}I-chimeric B72.3 antibody therapy, *Antibody Immunoconj. Radiopharm.* **6:**39–46.

Moshakis, V., McIlhinney, R. A. J., and Neville, A. M., 1981, Cellular distribution of monoclonal antibody in human tumours after I.V. administration, *Br. J. Cancer* **44:**663–669.

Natali, P. G., Cavaliere, R., Bigotti, A., Nicotra, M. R., Russo, C., Ng, A. K., Giacomini, P., and Ferrone, S. (1983), Antigenic heterogeneity of surgically removed primary and autologous metastatic human melanoma lesions, *J. Immunol.* **130:**1462–1466.

Neuberger, M. S., Williams, G. T., and Fox, R. O., 1984, Recombinant antibodies possessing novel effector functions, *Nature* **312:**604–608.

Pietersz, G. A. and McKenzie, I. F. C., 1992, Antibody conjugates for the treatment of cancer, *Immunol. Rev.* **129:**57–80.

Rogers, G. T., Burke, P. J., Sharma, S. K., Koodie, R., and Boden, J. A., 1995, Plasma clearance of an antibody-enzyme conjugate in ADEPT by monoclonal anti-enzyme: Its effect on prodrug activation *in vivo*, *Br. J. Cancer* **72:**1357–1363.

Sehon, A. H., 1991, Carl Prausnitz Memorial Lecture—Suppression of antibody responses by chemically modified antigens, *Int. Arch. Allergy Appl. Immunol.* **94:**11–20.

Sharma, S. K., Bagshawe, K. D., Burke, P. J., Boden, J. A., Rogers, G. T., Springer, C. J., Melton, R. G., and Sherwood, R. F., 1994, Galactosylated antibodies and antibody-enzyme conjugates in antibody-directed enzyme prodrug therapy, *Cancer* **73:**1114–1120.

Sharma, S. K., Bagshawe, K. D., Burke, P. J., Boden, J. A., and Rogers, G. T., 1990, Inactivation and clearance of an anti-CEA carboxypeptidase G_2 conjugate in blood after localisation in a xenograft model, *Br. J. Cancer* **61:**659–662.

Sharma, S. K., Bagshawe, K. D., Melton, R. G., and Begent, R. H. J., 1996, Effect of cyclosporine on immunogenicity of a bacterial enzyme carboxypeptidase G_2 in ADEPT, *Transplant. Proc.* **28:**3154–3155.

Sharma, S. K., Bagshawe, K. D., Melton, R. G., and Sherwood, R. F., 1992, Human immune response to monoclonal antibody-enzyme conjugates in ADEPT, *Cell Biophys.* **21:**109–120.

Shepherd, T. A., Jungheim, L. N., Meyer, D. L., and Starling, J. K., 1991, A novel targeted delivery system utilizing a cephalosporin-oncolytic prodrug activated by an antibody β-lactamase conjugate for the treatment of cancer, *Bioorg. Medicinal Chem. Let.* **1:**21–26.

Springer, C. J., Bagshawe, K. D., Sharma, S. K., Searle, F., Boden, J. A., Antoniw, P., Burke, P. J., Rogers, G. T., Sherwood, R. Y., and Melton, R. G., 1991, Ablation of human choriocarcinoma xenografts in nude mice by antibody-directed enzyme prodrug therapy (ADEPT) with three novel compounds, *Eur. J. Cancer* **27:**1361–1366.

Stevenson, G. T. and Stevenson, F. K., 1975, Antibody to a molecularly defined antigen confined to a tumour cell surface, *Nature* **254:**714–716.

Tagliabue, E., Porro, G., Barbanti, P., Torre, G. D., Menard, S., Rilke, F., Cerasoli, S., and Colnaghi, M. I., 1986, Improvement of tumor cell detection using a pool of monoclonal antibodies, *Hybridoma* **5:**107–115.

2

Prodrugs in Cancer Chemotherapy

Tom Connors

1. INTRODUCTION

The first chemical in a series that has interesting pharmacodynamic properties is called the "lead" chemical. In developing the series, the next stage is the derivatization of this lead chemical to enhance its properties, for example, by reducing side effects and improving selectivity of action. In some instances the derivative has no intrinsic activity but is converted to the active drug, *in vivo*, at the appropriate time or place. Such analogs are called prodrugs and are used extensively in medicine.

Conversion of the prodrug to the active form takes place by a number of mechanisms depending, for example, on changes of pH, oxygen tension, temperature, or salt concentrations, or by spontaneous decomposition of the drug, internal ring opening, or cyclization. A major approach in prodrug design is synthesizing inert analogs that are converted to the active drug by enzymatic action.

2. USES OF PRODRUGS IN CANCER CHEMOTHERAPY

2.1. Optimizing Pharmacokinetics and Tissue Distribution

The use of prodrugs in medicine has been extensively reviewed (Bundegaard, 1985; Connors and Knox, 1995) and their special use in cancer (Connors,

Tom Connors • Centre for Polymer Therapeutics, The School of Pharmacy, London WC1N 1AX United Kingdom.

Enzyme-Prodrug Strategies for Cancer Therapy, edited by Melton and Knox. Kluwer Academic/Plenum Publishers, New York, 1999.

Table I. The Uses of Prodrugs in Cancer Chemotherapy

To improve solubility
To assist formulation
To improve pharmacokinetics/tissue distribution
To prevent unfavorable metabolism
To overcome tissue barriers, *e.g.* blood–brain
To overcome resistance
To cause selective organ effects
To cause specific tumor toxicity

1986; Ross, 1962; Sinhababu and Thakker, 1996; Zee Cheng and Cheng, 1989). Prodrugs have been used clinically in cancer chemotherapy for various purposes ranging from analogs with better formulation properties to prodrugs designed to be selectively activated in the tumor environment (Table I).

When tumor-bearing animals are used in drug discovery programs, the test drugs are often administered in solvents inappropriate for humans or as suspensions, for example, in arachis oil. If active agents are discovered, a suitable human formulation must be devised, preferably for intravenous administration, because this is the usual route for the first clinical trials of anticancer agents. Pancratastatin, first isolated in 1984, had interesting activity in a number of tumor types and in particular melanoma. However, all attempts to provide a suitable formulation for intravenous injection, for example, by complexation with cyclodextrins, failed (Torres Labandeira *et al.*, 1991). A simple phosphate solved the problem because the prodrug is rapidly converted to the drug *in vivo*, is several thousand times more soluble than the free drug, and is readily formulated by conventional techniques (Fig. 1) (Pettit *et al.*, 1995). Taxol isolated from the bark of the yew tree is unique among clinically used tubulin binders. It acts by stabilizing rather than disassem-

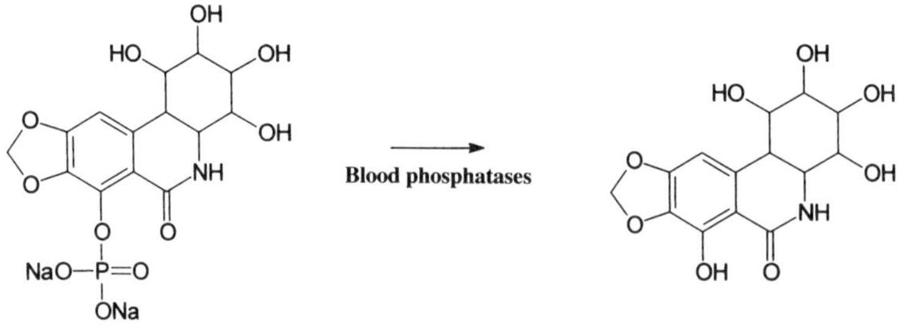

Figure 1. The formation of pancratastatin from a phosphate prodrug.

bling the microtubules. The promising clinical results of taxol and related taxanes has led to the synthesis of many analogs because of the insolubility of the parent compound and the probability that some of its toxicity may be result from using Chremophor EL in the formulation. A soluble prodrug form of taxol, in early clinical trial, comprises a polymer linked to molecules of taxol through amino acid side chains and which should act a slow release form of taxol after targeting to the tumor because of extended permeability and retention (Pesenti *et al.*, 1995). Other insoluble anticancer agents that have been successfully formulated after synthesizing more soluble prodrug derivatives include the phosphates of combretastatin and bisantrene (Murdock *et al.*, 1993), the glycinates of aphidhocholine (Beijnen *et al.*, 1995), rapamycin (Supko and Malspeis, 1994) and the acylated enol derivative of acronine (Bourne *et al.*, 1977). Adenosine arabinoside can be formulated but the need for high doses means that large volumes of solvent have to be given. The 5′-*O*-formate ester is some sixty times more soluble and is rapidly converted to the parent drug *in vivo* (Fig. 2) (Connors, 1986).

Some drugs may pose no formulation problem but are inherently unstable as the solid or when made up into solution for injection. This creates problems in storing the drug in the former case and in the latter case when the drug is administered as a prolonged infusion. DTIC (dimethylaminotriazene imidazole carboxamide) is a prodrug used extensively in treating melanoma and some sarcomas but it is unstable in light and requires special precautions for storage and administration because the breakdown products formed by photolysis have no antitumor action but are toxic. A further problem with DTIC is that it is activated by demethylation—primarily in the liver. This leads to large interpatient variation in blood levels of the active metabolite because patients have different drug metabolizing capacities (especially if liver metastases are present). Both problems have been overcome by introducing the prodrug temozolomide (Newlands *et al.*, 1992, 1997). Temozolomide is stable in light and releases the same active metabo-

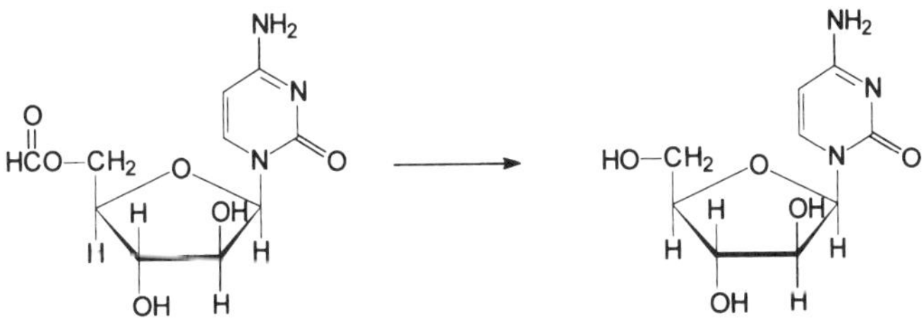

Cytosine arabinoside 5′ -*O*-Formate ester **Cytosine arabinoside**

Figure 2. The formation of an arabinoside from a formate ester.

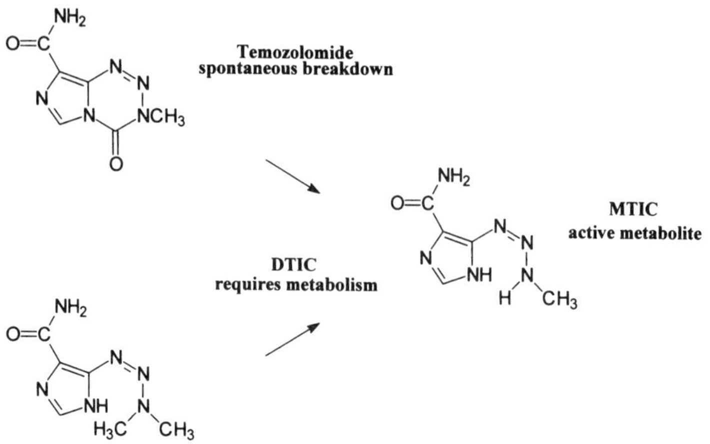

Figure 3. A comparison of the activation of DTIC and temozolomide to the active drug MTIC.

lite as DTIC. However, in this case it is by spontaneous ring opening thus overcoming the problems of variable metabolism. An additional advantage of temozolomide is that it is readily bioavailable orally, so that frequent doses can be given by mouth on an outpatient basis (Fig. 3).

Similar approaches have been used to improve bioavailability. In one clinical trial, it was shown that etoposide has good effects against lung cancer when administered orally. However, there was a large interpatient variation (ranging from 25% to 75%). Etopophos, a monophosphate of etoposide, is 100% bioavailable on oral administration but is so rapidly dephosphorylated *in vivo* by phosphatases in the plasma that its pharmacokinetic profile is identical to etoposide (Schilder *et al.*, 1994). Even when high doses of etopophos are given intravenously as a five-minute infusion, there is no saturation, and the pharmacokinetic profiles of etoposide administered as the parent drug or as the prodrug are similar (Budman *et al.*, 1994).

Prodrugs are also used to improve the intracellular concentration of therapeutic agents or to penetrate areas inaccessible to the parent drug. Purine and pyrimidine antimetabolites, for example, are administered as the free base or the nucleoside, although the active forms are the nucleotide mono-, di- or triphosphates because the ionized nucleotides are only poorly taken up by cells, whereas the bases and nucleosides are rapidly internalized. Methotrexate is also considered a prodrug because its intracellularly polyglutamylated forms are better inhibitors of dihydrofolate reductase and have a much longer intracellular residence time than methotrexate. Some chemicals have poor partition coefficients or are charged at physiological pH, which means that generally, unless they are taken up by a specific receptor mechanism, they are only poorly internalized. Esterification to reduce or abolish the charge and improve the partition coefficient often leads to big improvements in tissue distribution. A particular problem in cancer chemotherapy is to find drugs that penetrate the blood–brain barrier to treat primary brain cancers or intracranial metastases. Although esterification may assist passage through the blood brain barrier, it might also increase uptake by normal tissue and lead to increased toxicity. A novel approach developed by Bodor and colleagues (Bodor 1984) involves derivatizing drugs with a 1-substituted dihydronicotinic acid moiety. The increased lipophilicity achieved by the substitution allows penetrating the blood–brain barrier. Then the prodrug is acted upon by tissue oxidoreductases to form the methylpyridinium derivative. In extracerebral tissue, the charged metabolite is rapidly excreted. However, in the brain, the blood–brain barrier prevents rapid excretion, and the trapped prodrug is acted on by esterases to release active drug (Fig. 4). The technique has been used to concentrate chlorambucil in the brain (Bodor *et al.*, 1989). A similar approach

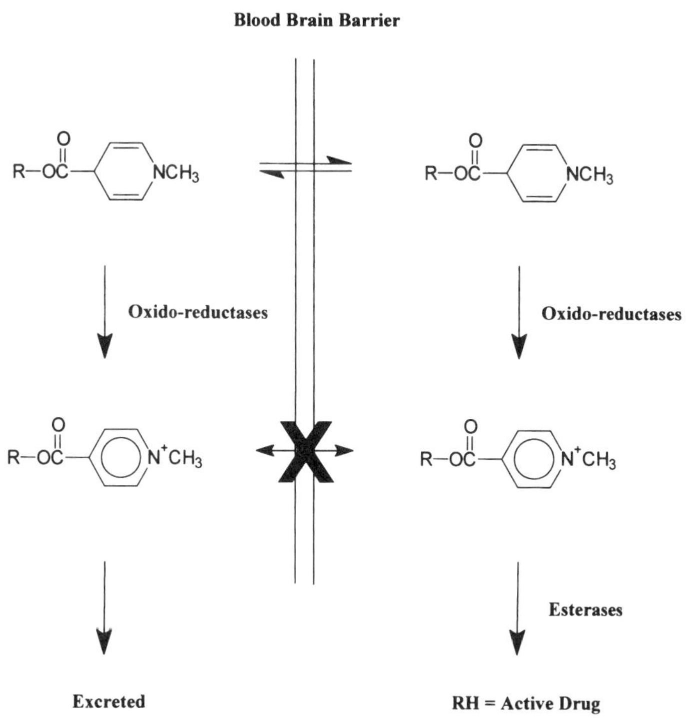

Figure 4. Penetration of the blood–brain barrier by prodrugs made by derivatization of drugs with a 1-substituted dihydronicotinic acid moiety.

relying on the impermeability of charged forms across a membrane barrier has been used to concentrate basic antitumor agents in tumors of low pH (Connors *et al.*, 1965).

Cytosine arabinoside is a useful drug for treating some leukemias, but it is phase-specific and acts only on cells in the S-phase of the cycle. This means that the drug must be at a therapeutic concentration in the blood for the period of time during which all cells in the tumor pass through the S-phase. Ara-C acts only after forming its triphosphate, but it is also rapidly excreted and metabolized by deoxycytidine deaminase in the liver (Fig. 5). Therefore, to obtain optimum antitumor action, the drug must be given in a high concentration and as a slow infusion over several hours to maintain therapeutic blood levels. Prodrugs have been designed that persist in the body and slowly release Ara-C. These are lipophilic 3′ and 5′ esters, N^4-substituted derivatives and 2,2-anhydro AraC (Hadfield and Sartorelli, 1984; Wip and Li, 1994). A prodrug that releases Ara-C at an appropriate rate and also prevents its rapid breakdown by hepatic deaminase would be of interest, although no prodrug form has yet taken the place of Ara-C in the clinic. Like Ara-C, adenosine arabinoside (Ara-A) is degraded rapidly by adenosine deaminase to an inactive metabolite. Using the same approach as for Ara-C, many analogs have been synthesized as potential slow release forms of Ara-A and include 2′, 3′, and 5′ mono or di-substituted O-acyl derivatives (Alexander and Holly, 1994). Of particular interest is the 5′-O-valeryl analog, which

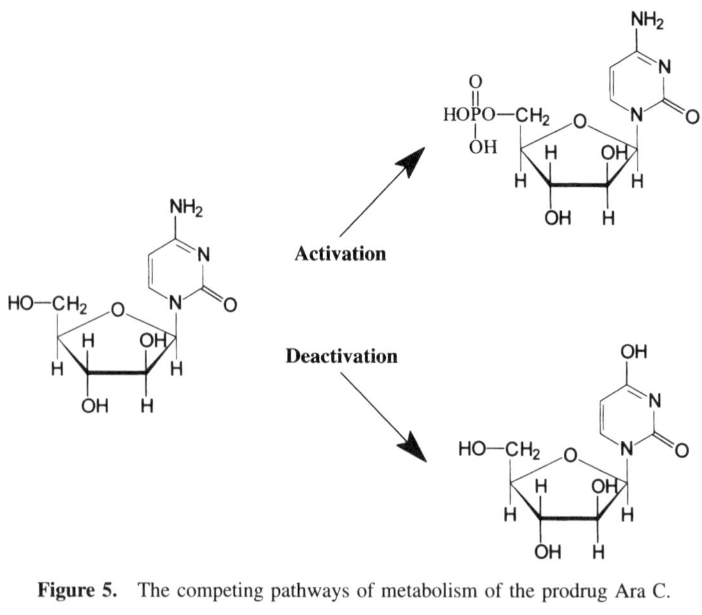

Figure 5. The competing pathways of metabolism of the prodrug Ara C.

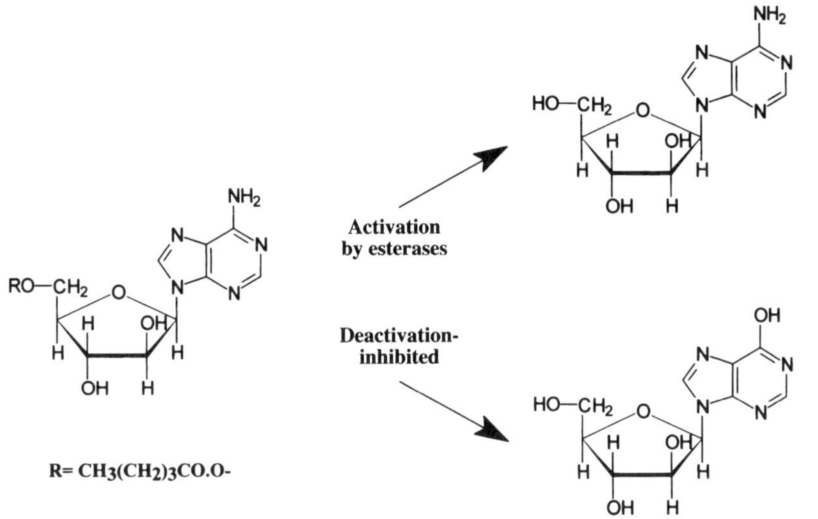

Figure 6. 5′-*O*-valeryl analogue of Ara-A. The drug acts as a slow release form of Ara-A, and it also inhibits the degradation of Ara-A to hypoxanthine arabinoside.

shows a marked increase in potency compared to Ara-A. The drug acts as a slow release form of Ara-A, and it also inhibits the degradation of Ara-A to hypoxanthine arabinoside (Fig. 6). Thus, as long as the prodrug is present there is no degradation of Ara-A and there is a corresponding increase in potency (Baker *et al.*, 1979). A recent patent (Chasalow, 1996) describes a general method for preventing the first-pass inactivation of drugs in the liver. This allows drugs to reach their sites of action and could conceivably be used to prevent the extensive hepatic metabolism of Ara-C. The method involves conjugating drugs to phosphocholine moieties that presumably restrict uptake by hepatocytes on the first pass but then are hydrolyzed to release the active drug (Fig. 7). 5-Fluorouracil is the only base analog that has activity in the clinic against solid tumors and has been used to treat colon cancer for many years. As with Ara C, there have been numerous attempts to reduce the toxicity of 5-FU, to prevent its catabolism and to optimize its pharmacokinetics by prodrug design (Anantham 1995; Jones 1985).

Newly formed tumor blood vessels are more permeable to macromolecules than normal blood vessels because their endothelial cells are not tightly continuous and have more gaps through which large molecules pass and enter tumor interstices. The macromolecules penetrate tumor tissues more readily and because of poor or nonexistent lymphatic drainage, have a longer than normal residence time (a phenomenon known as extended permeability and retention) (Maeda and Matsumura, 1989) (Fig. 8). This property has been exploited by using prodrugs

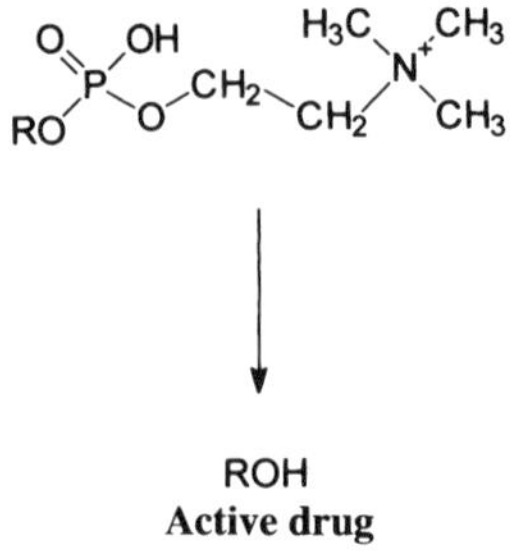

Figure 7. A general method for preventing the first-pass inactivation of drugs in the liver by conjugating them to phosphocholine moieties. These restrict uptake by the hepatocytes on the first pass but then are hydrolyzed to release the active drug.

based on a polymeric carrying structure. Anticancer drugs are linked to the polymer via peptide side chains. The linkage is such that the drug is slowly released in the extracellular spaces or, because the linkage is stable, only after uptake by endocytosis and hydrolysis in secondary lysozomes by a specific enzyme to release the drug intracellularly (Duncan *et al.*, 1996).

Cancer cells rapidly acquire resistance to many agents. Once the mechanism by which acquired resistance is understood it may be possible to design prodrugs to circumvent the problem. For example, 6-mercaptopurine is a prodrug that is taken up by cells and converted to the nucleotide by the enzyme hypoxanthine guanine phosphoribosyltransferase (HGPRT). Cells acquire resistance by losing the enzyme, maintaining their level of nucleotides by the alternative *de novo* biosynthetic pathway. Resistance cannot be overcome by administering 6-MP

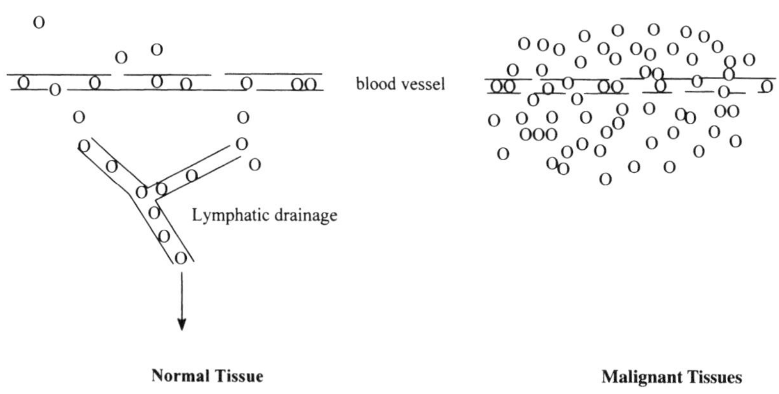

Figure 8. Extended permeability and retention (EPR) of tumor cells.

nucleotide because the phosphate is rapidly broken down extracellularly and as a charged molecule is only poorly taken up by cells. A prodrug, in which the charge is neutralized by cyclizing the phosphate groups and lipophilicity is increased by esterifying one of the sugar hydroxyls, will be readily taken up and will release the active nucleotide under the influence of intracellular phosphatases and esterases (Fig. 9) (Meyer *et al.*, 1979).

2.2. Prodrugs Activated by Tumor Enzymes

Results from animal experiments and dose-intensification studies in humans have indicated that some tumor types, for example, ovarian cancer, might be completely eradicated by chemotherapy if the dose of anticancer agent to which they respond could be increased by a hundredfold. Attempts to increase the dose administered using dose intensification by, autologous bone marrow transplantation after high doses of myelotoxic therapy, by rescue experiment (e.g., folinic acid after methotrexate or by isolated limb perfusions, allow a greater total dose to be given, but not of this order of magnitude. However, there are many examples where this level of dose intensity can theoretically be achieved by using prodrugs that are selectively activated by enzymes present in tumors. Experiments on tumor-bearing animals have shown that when a prodrug is activated uniquely in the tumor environment, cures are obtained in mice bearing large primary tumors and extensive metastases (Connors and Whisson, 1966; Double and Workman, 1977; Whisson and Connors, 1965). Given that the prodrug is a good substrate for the enzyme specifically expressed in the tumor and that the difference in toxicity between prodrug and drug is a hundredfold or more, then once a candidate enzyme

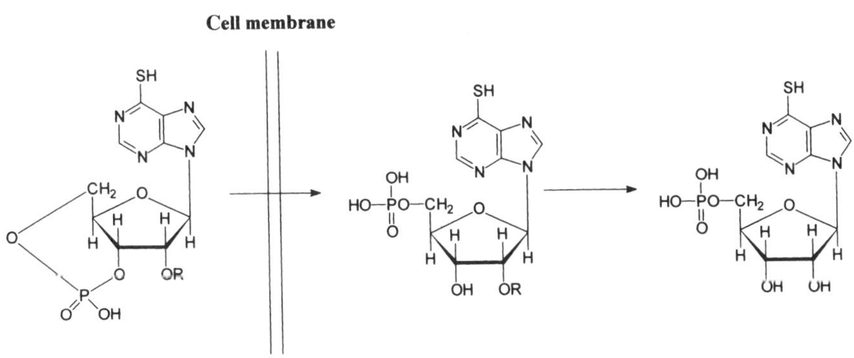

Figure 9. Release of 6-mercaptopurine from a prodrug in which the charge is neutralized by cyclizing the phosphate groups and lipophilicity is increased by esterifying one of the sugar hydroxyls.

is identified (especially if there is also a high concentration of the enzymes in the extracellular spaces of the tumor), in most cases many different classes of anticancer drugs can often be derivatized to form appropriate prodrugs. This can be demonstrated in approaches for designing prodrugs of cytotoxic alkylating agents. Because this class of anticancer agent acts predominantly by covalent alkylation of adjacent strands of DNA, then the basic requirements for cytotoxicity are first, that the agent has an optimal level of chemical reactivity that enables it to reach the tumor site after injection and is reactive enough to aklylate DNA. If the reagent is too reactive it may hydrolyze before reaching the tumor and if too unreactive may be excreted before sufficient DNA alkylation has taken place. Secondly it must pass through the endothelium and the cell and nuclear membrane to reach its target. Finally, because the predominant reaction that leads to cytotoxicity is a cross-linking reaction, the alkylating agent must have a minimum of two alkylating arms. Once a unique tumor enzyme has been identified, a prodrug can be synthesized which is lacking one or more of the features described but is acted upon by the enzyme to produce an appropriate drug. Thus, many alkylating prodrugs are chemically unreactive and nontoxic but are substrates for enzymes that metabolize them to highly reactive and toxic products. The ability of an alkylating agent to react with biological molecules depends on a minimal level of chemical reactivity, and this level of activity varies greatly depending on chemical structure (Connors, 1986; Ross, 1962). Small changes in electron or withdrawing properties greatly alter chemical reactivity. Thus, cyclophosphamide is a prodrug because the electron-withdrawing properties of the oxygen are such that it is chemically inert as an alkylating agent and nontoxic to cells in culture. It is converted *in vivo* via an enzymatically hydroxlylated intermediate to the highly cytotoxic phosphoramide mustard (Fig. 10). Cyclophosphamide is not an ideal

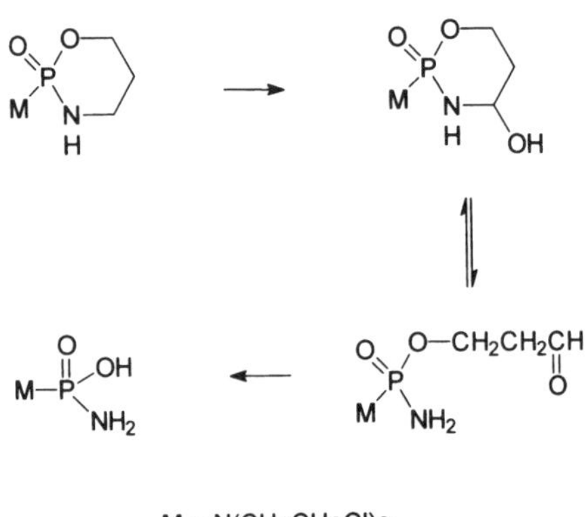

Figure 10. Activation of cyclophosphamide.

prodrug because the initial hydroxylation takes place in the liver rather than in the tumor and its selectivity results from more efficient detoxification of the 4-hydroxy derivative in normal sensitive tissues, compared to the tumor, where it breaks down to form a greater amount of phosphoramide mustard (Connors *et al.*, 1974). Gene transfer experiments in which the activating cytochrome P_{450} enzyme is expressed in tumor cells suggest that cyclophosphamide would be a much more effective agent if activation took place only in the tumor (Chen *et al.*, 1996; Wei *et al.*, 1994). The chemically inert azomustards were specifically designed as prodrugs that would be activated in tumors that express azoreductase activity (Bukhari *et al.*, 1973; Ross and Warwick, 1955). As in the case of cyclophosphamide, there is a difference of orders of magnitude in toxicity between the prodrug and the reduced drug (Fig. 11). An added advantage of this type of structure is that the drug formed from the prodrug is extremely reactive and might be expected to hydrolyze very quickly to the nontoxic hydroxyethyl derivative, so that any drug escaping from the tumor detoxified before reaching any sensitive host tissue. Preliminary clinical trials of an azomustard were carried out against primary hepatocellular carcinoma, but early reports that cancers of this type have high levels of azoreductase were never substantiated. The use of azo derivatives as prodrugs was also suggested some years previously by Tsou and Su based on the observation that Ehrlich ascites cells are very efficient in reducing tetrazolium

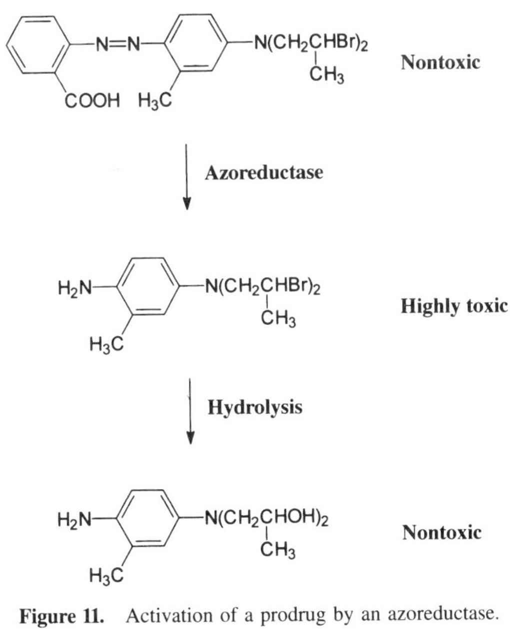

Figure 11. Activation of a prodrug by an azoreductase.

salts to formazans and beyond. A series of tetrazolium analogs was made which would be inactive as alkylating agents because of their impermeability and their low chemical reactivity due to the conjugated ring system. On reduction to the formazan the chemical reactivity increases and further reduction to the amine leads to the highly toxic mustard (Fig. 12) (Tsou *et al.*, 1966).

Generally, chemicals that are charged at physiological pH are only poorly taken up by cells which is why glucuronide and sulfate formation are major excretory pathways of foreign chemicals. Conjugation forms water soluble products that are not taken up by tissues and rapidly excreted. Therefore, impermeable prodrugs, that are substrates for enzymes and are converted to metabolites which are readily taken up by cells, are good candidates for prodrug design. However, in this situation, the activating enzyme must be in the extracellular spaces of the tumor or at least expressed on the outer cell membrane rather than intracellularly. Because tumor cells are frequently dying and releasing their contents into the

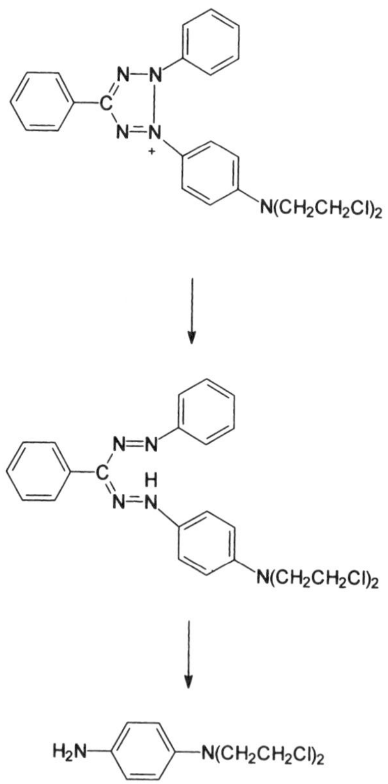

Figure 12. Release of an active mustard by reduction of its tetrazolium derivative.

interstitial spaces of the tumor, this may not be a problem and indeed one of the most selective prodrugs used experimentally is aniline mustard which after preliminary hydroxylation in the liver, is conjugated to glucuronic acid. In this form it does not enter cells and is nontoxic and rapidly excreted except when it enters a tumor environment high in β-glucuronidase. Here the para-hydroxy form is released which is highly toxic and rapidly internalized (Fig. 13). The low interstitial pH of solid cancers may be particularly important in this respect because β-glucuronidase, like many other lysosomal enzymes that are released extracellularly from dying cells, has an optimum pH of 5.0. It is claimed that the interstitial pH of tumors is as low as 6.5 and at this pH the enzyme would still function but might not at normal pH. Another example of using prodrugs that are not taken up well by cells involves derivatives that are substrates for plasmin. Plasmin is a key enzyme in a cascade of proteases that are involved in tumor spread and metastasis. Many tumors have high level of plasminogen activator that

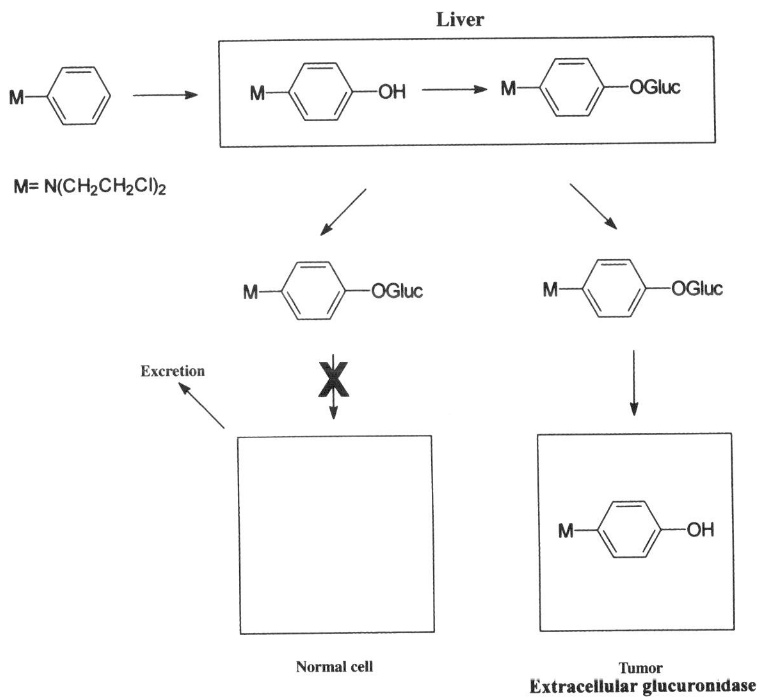

Figure 13. Activation of a prodrug by β-glucuronidase. Aniline mustard is conjugated to glucuronic acid after preliminary hydroxylation in the liver. In this form it does not enter cells and is nontoxic and rapidly excreted. However, when it enters a tumor environment which is high in β-glucuronidase, the para-hydroxy form is released which is highly toxic and rapidly internalized.

hydrolyzes plasminogen at a leu-lys specifier sequence to release plasmin. Prodrugs with similar specifier sequences may also be activated by plasminogen activator to release an active drug. In one example, *p*-phenylenediamine mustard is linked to D-val-leu-lys through the amine group (Chakravarty *et al.*, 1983a, b). The prodrug is made less chemically reactive by this substitution and therefore less toxic and, as a tri-peptide, is probably not taken up efficiently by tumors. Upon hydrolysis the highly reactive metabolite is formed which is rapidly taken up by cells close to the site of activation (Fig. 14). Another example of impermeable prodrugs is the family of glutamate derivatives of nitrogen mustards used in ADEPT (Antibody Directed Enzyme Prodrug Therapy). Glutamates are not taken up by cells and are nontoxic. The glutamate derivative of *p*-hydroxyaniline mustard (Fig. 15) is acted upon by bacterial carboxypeptidase G2 linked to antibodies on the cell surface to form *p*-hydroxy aniline mustard that is very permeable and orders of magnitude more toxic than the glutamate prodrug (Melton and Sherwood, 1996). Like the azomustards previously described, the released drug is very reactive chemically and breaks down very quickly to nontoxic products, so that there should be no systemic toxicity. Another prodrug based on glutamate, γ-glutamyl-*p*-phenylenediamine mustard is nontoxic to cells in culture but thirty times more potent if they express γ-glutamyl transferase. The glutamate is not normally taken up by cells but the transferase enzyme is expressed on the outer surface of the cell and thus hydrolyzes the prodrug outside the cell to a metabolite that is rapidly internalized (Smith *et al.*, 1984). Similarly, sulfates and phosphates that are charged at physiological pH are potential candidates for prodrug design. There are no good examples of sulfate prodrugs and although many phosphates, ranging from diethylstilbestrol to etoposide phosphate (Brock,

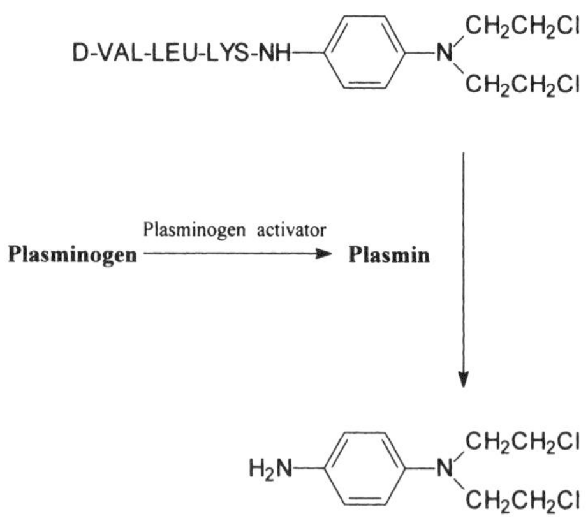

Figure 14. Activation of a prodrug by plasmin.

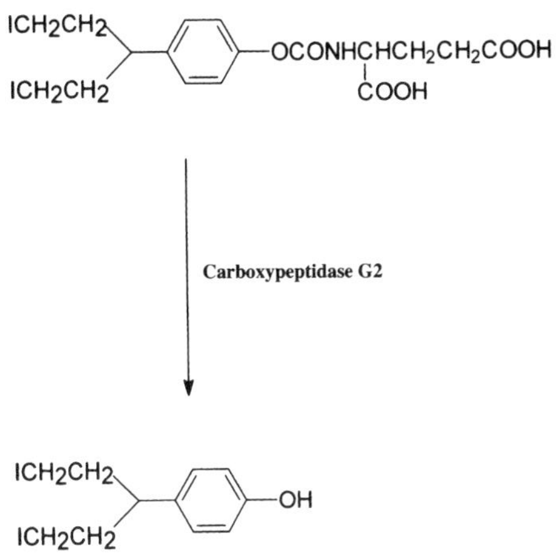

Figure 15. Activation of the glutamate derivative of *p*-hydroxyaniline mustard by carboxypeptidase G2.

1957; Senter *et al.*, 1988), have been proposed as prodrugs, the ubiquitous distribution of extracellular phosphatases in most cases ensures their rapid conversion to the drug, and no selective release of the active drug occurs at the tumor site. Of course, synthesizing impermeable prodrugs is not restricted to alkylating agents. Any anticancer agent or indeed any toxic agent that can be derivatized to an impermeable and nontoxic product, which is a substrate for an enzyme uniquely expressed in the extracellular environment of tumor, is a potentially useful prodrug. Thus, γ-glutamyl derivatives of AraC, doxorubicin, acivicin, AT125, and cycloleucine might all be potentially useful prodrugs if they were good substrates for γ-glutamyl transferase. In fact, glucuronides of 6-mercaptopurine, 5-fluorouridine, epirubicin and doxorubicin (Bosslet *et al.*, 1995; Haisma *et al.*, 1992; Parker and Fedor, 1982; Watanabe *et al.*, 1981) and the plasmin-activated prodrugs of acivicin and doxorubicin have all been examined as potential prodrugs (Chakravarty *et al.*, 1983a, b). Large numbers of antitumor alkylating agents have been tested experimentally, and almost without exception active derivatives must be at least difunctional, that is, have at least two alkylating arms. Although monofunctional agents may be carcinogens, they are usually much less toxic and if they can be converted enzymatically to difunctional agents, might be effective prodrugs. An example of this is CB 1954, a monofunctional aziridine that is highly effective against the rat Walker tumor normally sensitive only to difunctional alkylating agents (Cobb *et al.*, 1969). The tumor has a high concentration of the enzyme DT-diaphorase that reduces the 4-nitro group to a hydroxylamine that is then converted (probably by acetyl CoA) to a difunctional agent (Fig. 16). Knox

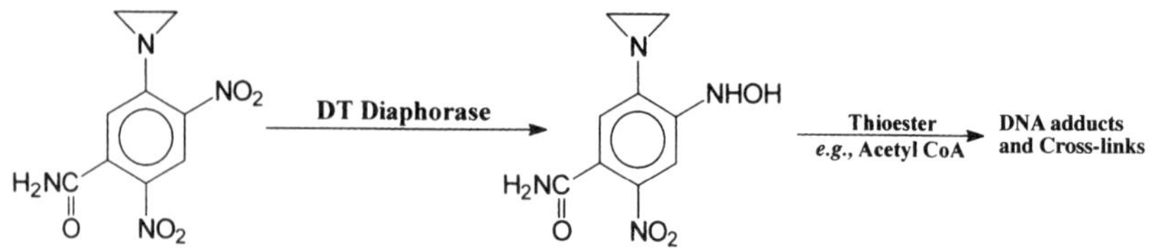

Figure 16. Bioactivation of the prodrug CB (1954).

et al., 1993). A nitroimidazole designed as a bioreductive agent is similarly converted from a monofunctional to a bifunctional agent after reduction of the nitro group (Binger and Workman, 1991). The reduction of nitro groups *in vivo* has led to the description of a general method for designing prodrugs, relying on the reduction of the nitroaromatic derivatives, and has been extended to substituted amines of related structures. This self-immolative approach has been used to design drugs active in tumor hypoxic cells and in ADEPT therapy (Fig. 17) (Mauger *et al.*, 1994).

Other strategies can be adopted for prodrug design for different classes of antitumor agents. If an anticancer agent acts by high affinity for an enzyme (e.g., an antimetabolite) or another receptor (e.g., DNA minor groove binder), prodrugs may be designed which no longer have affinity for the receptor. Thus alanyl methotrexate is a poor substrate for dihydrofolate reductase but can be converted

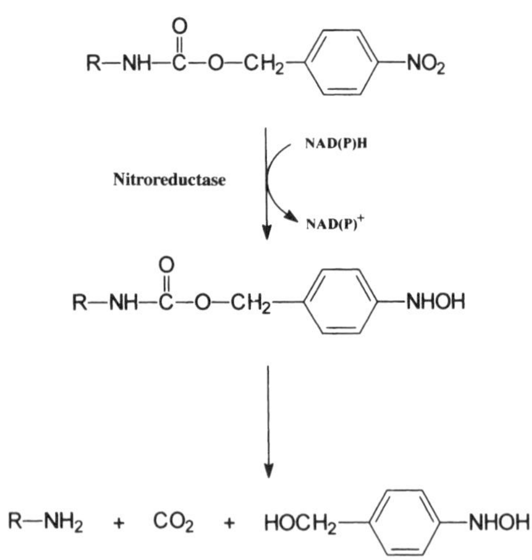

Figure 17. Activation of a prodrug by self-immolation following nitro reduction.

to methotrexate by the action of carboxypeptidase A (Kuefner *et al.*, 1989). Actinomycin D is a cytotoxic agent that acts by binding to the minor groove of DNA. Although only hydrogen bonding is involved, the binding is so tight that it is essentially irreversible. However, small changes in structure lead to a dramatic loss in binding activity with greatly reduced toxicity. Thus, actinomycin D acetylated on the aromatic amine does not bind to DNA and is nontoxic. An N-substituted actinomycin D might be an excellent prodrug if there were tumor enzymes that could remove the N-substituted group.

2.3. Bioreductive Prodrugs

Cancers grow peripherally and induce the formation of new blood vessels by secreting angiogenic factors. The invasive nature of malignant cells destroys surrounding tissue including blood vessels, so that that at the center of even small metastatic tumors there is necrosis as the cells are deprived of oxygen and nutrients. Between the actively growing peripheral zones that are well supplied by newly formed blood vessels and on the margin of the necrotic center are a few layers of hypoxic cells that are resistant to conventional therapy (Fig. 18). The existence of hypoxic cells in tumors and cell spheroids and their resistance to radio- and chemotherapy has been amply proven over the years by physical and chemical measurements and also by direct visualization. Chapman and colleagues injected tumor-bearing animals with a labeled nitroimidazole and showed covalent binding in autoradiographic sections exactly where they had been predicted,

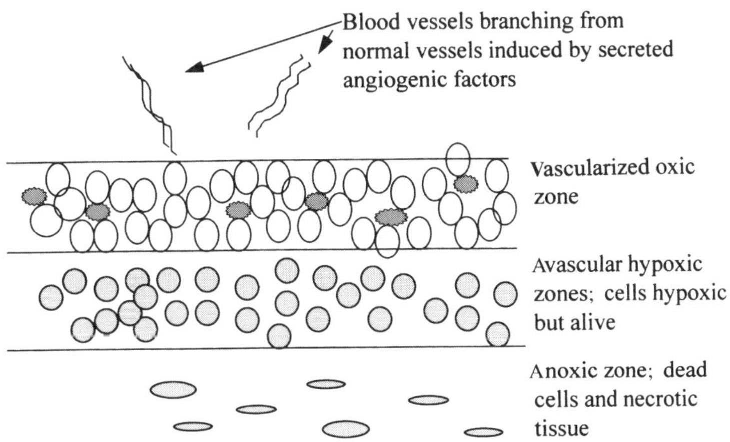

Figure 18. The zones of a solid tumor.

namely, in cells on the borders of necrotic zones (Chapman *et al.*, 1996). Because covalent binding only occurs after the nitro group is reduced and converted into an electrophilic reactant, the nitroimidazole acts as a prodrug that is selectively activated by enzymes in hypoxic cells. Theoretically bioreductive prodrugs have a crucial role in chemotherapy. By the very nature of cancer, all solid tumors have a hypoxic cell zone that is resistant to treatment and may regrow and cause failure of therapy. Combining conventional therapy with bioreductive agents that remove a resistant cell fraction might greatly improve the responses of solid tumors to therapy (Connors, 1996).

The earliest bioreductive agents examined in detail were the nitroimidazoles. These were originally investigated as electron-affinic radiation sensitizers. However, it was soon shown that, on reduction, metabolites are formed which are electrophilic reactants (Denny, 1995; Kedderis and Miwa, 1988). Selectivity for hypoxic cells occurs because the first reduction product, the nitro radical anion, reacts with oxygen in normally oxygenated tissue to reverse the process and regenerate the parent compound. The rate of reduction of a drug can depend on the reduction potential, with the result that although many nitroimidazoles are selectively activated in hypoxic tissue because their reduction potential is in the optimal range, other chemicals containing nitroaromatic groups are not reduced and may be excreted with the nitroaromatic ring intact although these may be substrates for bacterial nitroreductases used in ADEPT and GDEPT (Gene Directed Enzyme Prodrug Therapy). Also, some prodrugs that are activated by reduction might be equally well metabolized by cells of normal oxygen tension so no hypoxic cell selectivity results.

A number of chemical classes have been developed as bioreductive agents including nitroimidazoles, other classes of nitroaromatics, mono- and di-N-oxides and quinones. Misonidazole was the first bioreductive to have a clinical trial but, because of its poor potency and peripheral neurotoxicity, many more analogs were examined including etanidazole and RSU 1069. The latter compound shows excellent differential cytotoxicity to hypoxic cells *in vitro* and is a latent bifunctional alkylating agent. RSU 1069 is orally bioavailable but caused serious gastrointestinal tract toxicity. However, the problem may have been overcome by introducing the bromoethylamino analog of RSU 1069 to which it slowly cyclizes under physiological conditions (Fig. 19). Other nitroaromatics investigated as hypoxia selective prodrugs include acridines, such as nitacrine and pyrazolo-

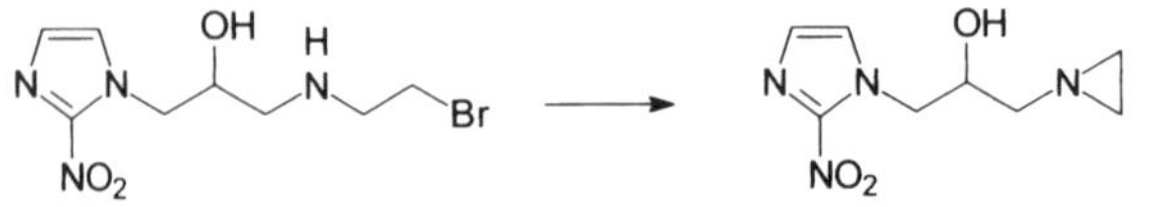

Figure 19. The formation of RSU 1069 by cyclizing a bromoethylamino derivative.

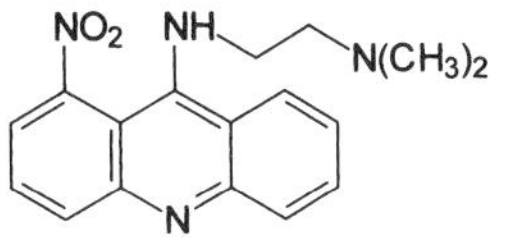

Nitacrine

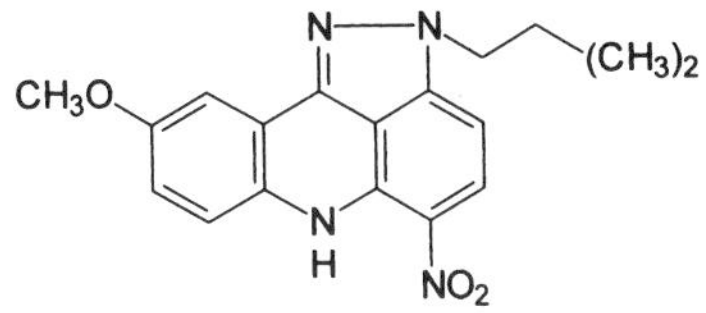

Pyrazoloacridine

Figure 20. Examples of hypoxia selective prodrugs.

acridine (Fig. 20). However, although they show some evidence of activity as bioreductive prodrugs *in vitro*, they have not been very effective as antitumor agents. Nitroaromatic nitrogen mustards have long been proposed as potentially useful prodrugs because the electron-withdrawing properties of the nitro group strongly deactivate the alkylating activity of the bischloroethylamine. The fully reduced reduction product, the primary amine, has the opposite effect and its electron-releasing properties make structures of this type among the most reactive and toxic of the nitrogen mustards. However, whether they are suitable for development as bioreductive agents depends on their rate of reduction *in vivo*. Thus, the simplest member of the series, *p*-nitrophenylamine mustard, is only poorly reduced under hypoxic conditions and is not very active, whereas an analog of CB 1954 referred to previously, where the aziridine group is replaced by a di-2-chloroethylamino group is very highly selective against hypoxic cells in culture (Palmer *et al.*, 1992). An interesting suggestion has been to design quaternary nitrogen mustards which would not act as alkylating agents, because the lone pair of electrons is no longer available for activating the chlorine atoms, but which might be reduced *in vivo* to release the highly reactive and toxic HN2 (methyl di-2-chloroethylamine), which would be toxic for the cells carrying out the reaction but which could also diffuse to oxygenated cells. (Fig. 21). Because the prodrug is charged, it is predicted that it would not enter cells and require activation extracellularly. Surprisingly, it is very highly selective *in vitro* for hypoxic cells (Denny *et al.*, 1994).

It has long been known that the N-oxide of HN2 (nitromin) is similarly a very unreactive alkylating agent because the lone pair of electrons on the nitrogen atom is similarly involved in bonding. It has low *in vivo* toxicity compared to HN2, is not very toxic to cells in culture, and has none of the vesicant action associated with the reactive aliphatic nitrogen mustards. However, over forty years ago, it was shown that it is highly active in tumor-bearing animals [references quoted in (Ross, 1962) p. 164]. If it indeed acts by conversion to HN2 in hypoxic cells and diffusion of the drug to normoxic cells, this was the first hypoxic selective prodrug to be synthesized. However, nitromin shows only a modest hypoxia selectivity to

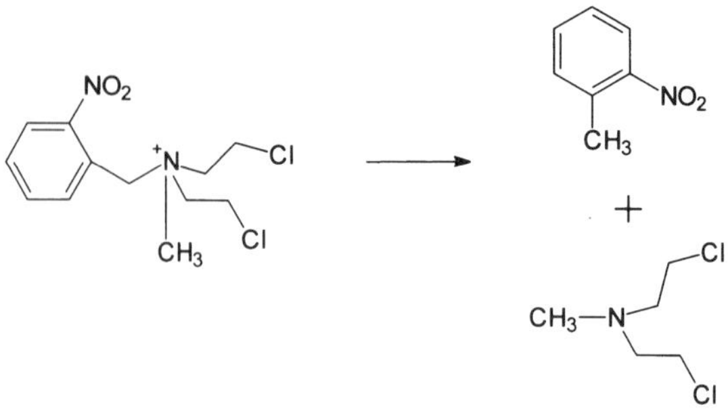

Figure 21. The formation of methyl di-2-chloroethylamine by reducing a quaternary nitrogen mustard prodrug. The prodrug is not an alkylating agent because the lone pair of electrons is no longer available for activating the chlorine atoms of the mustard ligand.

cells in culture (White *et al.*, 1989), and the N-oxide of chlorambucil (*p*-di-2-chloroethylaminophenylbutyric acid) shows none (Mann and Shervington, 1991). Thus, the reason for its high selectivity against animal tumors remains to be elucidated. More recent N-oxides with good bioreductive activity are the benzotriazine di-N-oxides, such as SR 4233 (Brown, 1993). Agents of this class are reduced by a number of enzymes through one- and two-electron reductions. The toxic species produced in hypoxic cells is likely to be the one-electron reduction product, an oxidizing radical that breaks DNA strands following abstraction of hydrogen. Although the predominant enzyme carrying out this reaction is microsomal cytochrome P450, the free radicals formed have a very short halflife and, assuming DNA is the target, then radicals generated in the vicinity of DNA are the most important in causing cytotoxicity. In this respect, it is of interest that cytochrome P450 reductase, which also metabolizes SR 4233, is present in the nucleus. Thus, it may be that, although this enzyme metabolizes only a fraction of the bioreductive agent entering the cells, the amount of this enzyme in the nucleus determines cytotoxicity (Cahill and White, 1990). The radical is converted back to the parent compound in cells at normal oxygen tension before DNA damage occurs. Many other N-oxides have been investigated for their bioreductive activity, including fused pyrazine mono-N-oxides, and quinoline 1,4-di-N-oxides. A chemical of particular interest, which unfortunately has not yet had a clinical trial, is a di-N-oxide of a DNA affinic anthraquinone, which is related to mitoxantrone and activated by cytochrome P450. The compound shows good hypoxia-selective toxicity, is an excellent radiation sensitizer, and releases a DNA intercalating agent on activation that might be active against solid cancers (McKeown *et al.*, 1995).

Quinones represent a third class of agents that may show selective toxicity to hypoxic tumor cells, although their activity may vary greatly with small changes in structure. Thus, the naturally occurring mitomycin C, which has been widely used clinically in treating various cancers, is active only after reduction of the quinone but shows only a small increase in toxicity in hypoxic cells. However, the N-methyl analog porfiromycin has much greater selectivity (Keyes *et al.*, 1985). A synthetic indolquinone, EO9, shows excellent selectivity against hypoxic cells although it had no activity in a number of clinical trials as a single agent (Dirix *et al.*, 1996; Smitskamp Wilms *et al.*, 1996). The mechanism of activation of quinones of this type may rely on several mechanisms dependent on the structure of the chemical. One-electron reduction forms reactive oxygen species that may damage DNA, but two-electron reduction by enzymes, such as DT-diaphorase, may also be necessary to convert such compounds to DNA cross-linking agents (Siegel *et al.*, 1992; Smitskamp Wilms *et al.*, 1996). On reduction quinones that have a side chain bearing a leaving group in the 2-position form quinone methides which alkylate DNA (Fig. 22) (Moore *et al.*, 1983).

"Self-immolative" chemicals that break down to release a cytotoxic drug on deacylation of an amino or reduction of a nitrobenzyl group can also be used as bioreductive agents if the preliminary reaction is specific for hypoxic cells. In one example, a nitroimidazole is combined through a self-immolative linkage with *p*-phenylenediamine mustard. On reduction of the nitro group to the amine, the highly reactive nitrogen mustard is formed (Prakash *et al.*, 1994). Other classes of cytotoxic chemicals containing primary amines could be made into prodrugs by the same procedure, providing substitution of the amine deactivates the drug. In a

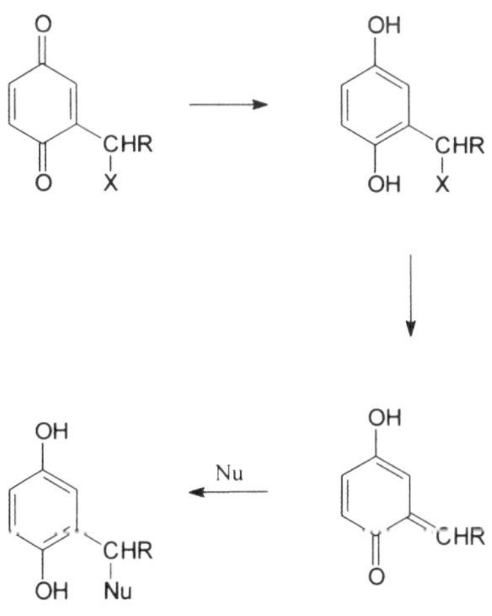

Nu = nucleophile

Figure 22. Bioreductive activation of quinone prodrugs.

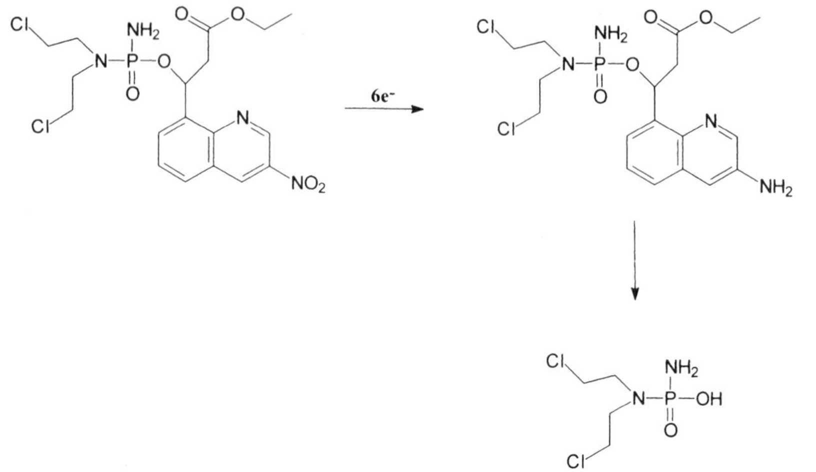

Figure 23. Release of an active mustard from a nitroaromatic carbamate following reduction and rapid cyclization.

related approach, a rapid cyclization of a nitroaromatic carbamate occurs on reduction followed by release of the drug (Fig. 23) (Atwell *et al.*, 1994). In another example, on reduction a nitroquinoline releases phosphoramide mustard the active metabolite of cyclophosphamide (Firestone *et al.*, 1991). Similarly, quinones have been used as carriers of cytotoxic agents in an inert form that releases the active agent following hydroquinone formation followed by cyclization (Fig. 24) (Carpino *et al.*, 1988). Quinone methides, referred to previously, might also be used as carriers of active drugs in an inactive form.

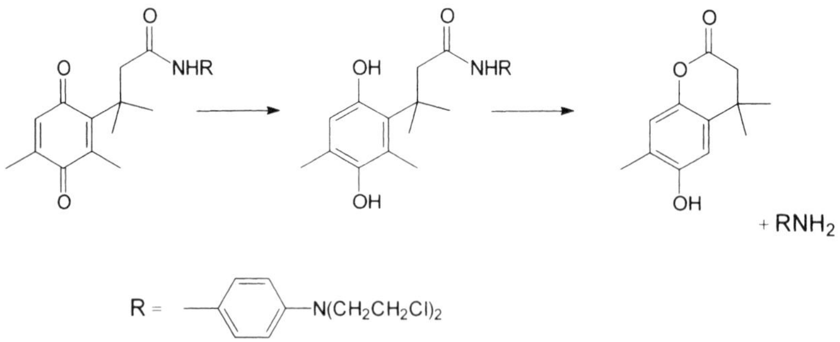

Figure 24. Formation of aniline mustard following reduction of a quinone prodrug.

3. CLINICAL USE OF PRODRUGS

Theoretically prodrugs could lead to effective treatments for solid cancers which at present respond poorly to conventional therapy. In tumor-bearing animals, where the right conditions are fulfilled, prodrugs are vastly superior to normal anticancer agents and cause complete regression of advanced primary tumors and metastases that are resistant to chemotherapy. However, prodrugs specifically designed to be activated in tumors that uniquely express an enzyme, have not been successful clinically for a variety of reasons. The main reason is that it is difficult to identify human tumors that express an enzyme not expressed at similar concentrations in normal tissue. In attempting to exploit aniline mustard glucuronide as a prodrug, it was found that human tumors with high glucuronidase levels occur only infrequently and are distributed randomly over different types of cancer (Young *et al.*, 1976). Because the enzyme activity in tumors can only be measured in biopsy specimens, it is not feasible to carry out large numbers of biopsies to identify a small number of tumors that may respond to treatment with the appropriate prodrug. There is certainly a need to carry out further trials of the bioreductive prodrug concept because all solid cancers contain hypoxic cells and combining conventional therapy with bioreductive drugs may well lead to cures of cancers which now respond only moderately to treatment. What is really required are comparative trials against tumors with high and low hypoxic fractions combined with appropriate therapy. One also needs to know the enzyme content of the tumor but it must be remembered that there are at least five enzymes that may be involved in bioreductive transformation and depending on the prodrug, some may be deactivating rather than activating (Ross *et al.*, 1996; Workman, 1994). The site of the enzyme may also be more important than the actual amount. The development of prodrugs would be considerably enhanced if noninvasive methods using, for example, PET scanning could be developed to estimate rapidly the content of relevant enzymes in human cancers.

The future for prodrugs clinically might well lie in the ADEPT or GDEPT approaches although in the latter case not until methods have been developed for delivering the gene to all tumor stem cells. Certainly, in the ADEPT approach, a unique enzyme can be delivered selectively to the tumor in appropriate amounts to overcome the major problem encountered with prodrug therapy in the clinic.

REFERENCES

Alexander, P., and Holly, A., 1994, Prodrugs of analogues of nucleic acid components, *Coll. Czech Commun.* **59:**2127–2165.

Anantham, S., 1995, Fluoropyrimidines, in: *Fluoropyrimidines* (W. O. Foye, ed.), American Chemical Society, Washington, DC, pp. 49–58.

Atwell, G. J., Sykes, B. M., O'Connor, C. J., and Denny, W. A., 1994, Relationships between structure and kinetics of cyclization of 2-aminoaryl amides: Potential prodrugs of cyclization-activated aromatic mustards. *J. Med. Chem.* **37:**371–380.

Baker, D. C., Haskell, T. H., Putt, S. R., and Sloan, B. J., 1979, Prodrugs of 9-(β-D-arabinofuranosyl)-adenine 2. Synthesis and evaluation of a number of 2′,3′- and 3′,5′-di-O-acyl derivatives, *J. Med. Chem.* **22:**273–279.

Beijnen, J. H., Flora, K. P., Halbert, G. W., Henrar, R. E., and Slack, J. A., 1995, CRC/EORTC/NCI Joint Formulation Working Party: Experiences in the formulation of investigational cytotoxic drugs, *Br. J. Cancer* **72:**210–218.

Binger, M., and Workman, P., 1991, Pharmacokinetic contribution to the improved therapeutic selectivity of a novel bromoethylamino prodrug (RB 6145) of the mixed-function hypoxic cell sensitizer/cytotoxin α-(1-aziridinomethyl)-2-nitro-1H-imidazole-1-ethanol (RSU 1069), *Cancer Chemother. Pharmacol.* **29:**37–47.

Bodor, N., 1984, Novel approaches to the design of safer drugs; soft drugs and site specific chemical delivery systems, *Adv. Drug Res.* **13:**225–331.

Bodor, N., Vasudevan, V. D. W., and Estes, K., 1989, Improved delivery through biological membranes. XLI: Brain enhanced delivery of chlorambucil, *Int. J. Pharmacol.* **53:**195–208.

Bosslet, K., Czech, J., and Hoffman, D., 1995, A novel one step tumor selective prodrug activation system, *Tumor Targeting* **1:**45–50.

Bourne, D. W., Higuchi, T., and Repta, A. J., 1977, Acetylacroninium salts as soluble prodrugs of the antineoplastic agent acronine, *J. Pharm. Sci.* **66:**628–631.

Brock, N., 1957, Transport und wirkform als chemotherapeutisches prinzip in der tumortherapie, *Zschr. Krebsforsch* **62:**9–24.

Brown, J. M., 1993, SR 4233 (tirapazamine): A new anticancer drug exploiting hypoxia in solid tumours, *Br. J. Cancer* **67:**1163–1170.

Budman, D. R., Igwemezie, L. N., Kaul, S., Behr, J., Lichtman, S., Schulman, P., Vinciguerra, V., Allen, S. L., Kolitz, J., and Hock, K., 1994, Phase I evaluation of a water-soluble etoposide prodrug, etoposide phosphate, given as a 5-minute infusion on days 1, 3, and 5 in patients with solid tumors, *J. Clin. Oncol.* **12:**1902–1909.

Bukhari, A., Connors, T. A., Gilsenan, A. M., Ross, W. C., Tisdale, M. J., Warwick, G. P., and Wilman, D. E., 1973, Cytotoxic agents designed to be selective for liver cancer, *J. Natl. Cancer Inst.* **50:** 243–247.

Bundegaard, H., 1985, *Design of Prodrugs*, Elsevier, Amsterdam, New York, Oxford.

Cahill, A., and White, I. N. H., 1990, Reductive metabolism of 3-amino-1,2,4-benzotriazine-1, 4-dioxide (SR 4233) and the induction of unscheduled DNA synthesis in rat and human derived cell lines, *Carcinogenesis* **11:**1407–1411.

Carpino, L. A., Triolo, S. A., and Berglund, R. A., 1988, Reductive lactonisation of strategically methylated quinone propionic acid esters and amides, *J. Org. Chem.* **54:**3303–3310.

Chakravarty, P. K., Carl, P. L., Weber, M. J., and Katzenellenbogen, J. A., 1983a, Plasmin-activated prodrugs for cancer chemotherapy. 2. Synthesis and biological activity of peptidyl derivatives of doxorubicin, *J. Med. Chem.* **26:**638–644.

Chakravarty, P. K., Carl, P. L., Weber, M. J., and Katzenellenbogen, J. A., 1983b, Plasmin-activated prodrugs for cancer chemotherapy. 1. Synthesis and biological activity of peptidylacivicin and peptidylphenylenediamine mustard, *J. Med. Chem.* **26:**633–638.

Chapman, J. D., Franco, A. J., and Sharplin, J., 1996, A marker in hypoxic cells for tumours with potential clinical applicability, *Br. J. Cancer* **45:**546–550.

Chasalow, F. I., 1996, Phosphocholine drug derivatives, Amur Research Corp, WO 9616680.

Chen, L., Waxman, D. J., Chen, D., and Kufe, D. W., 1996, Sensitization of human breast cancer cells to cyclophosphamide and ifosfamide by transfer of a liver cytochrome P450 gene, *Cancer Res.* **56:**1331–1340.

Cobb, L. M., Connors, T. A., Elson, L. A., Khan, A. H., Mitchley, B. C., Ross, W. C., and Whisson, M. E., 1969, 2,4-Dinitro-5-ethyleneiminobenzamide (CB 1954): A potent and selective inhibitor of the growth of the Walker carcinoma 256, *Biochem. Pharmacol.* **18:**1519–1527.

Connors, T. A., 1986, Prodrugs in cancer chemotherapy, *Xenobiotica* **16:**975–988.

Connors, T. A., 1996, Bioreductive agents, hypoxic cells and therapy, *Eur. J. Cancer* **32A:**1833–1834.

Connors, T. A., Cox, P. J., Farmer, P. B., Foster, A. B., and Jarman, M., 1974, Some studies of the active intermediates formed in the microsomal metabolism of cyclophosphamide and isophosphamide, *Biochem. Pharmacol.* **23:**115–129.

Connors, T. A., Jeney, A., Warwick, G. P., and Whisson, M. E., 1965, Some factors influencing the sensitivity of tumours to nitrogen mustards, in *Some Factors Influencing the Sensitivity of Tumours to Nitrogen Mustards* (L. J. Roth, ed.), University of Chicago Press, Chicago and London, pp. 433–438.

Connors, T. A. and Knox, R. J., 1995, Prodrugs in medicine, *Expert Opinion on Therapeutic Patents* **5:**873–885.

Connors, T. A. and Whisson, M. E., 1966, Cure of mice bearing advanced plasma cell tumours with aniline mustard: The relationship between glucuronidase activity and tumour sensitivity, *Nature* **210:**866–867.

Denny, W. A., 1995, Hypoxia selective cytotoxins, in *Hypoxia Selective Cytotoxins* (W. O. Foye, ed.), American Chemical Society, Washington DC, pp. 483–497.

Denny, W. A., Wilson, W. R., Tercel, M., Van Zijl, P., and Pullen, S. M., 1994, Nitrobenzyl mustard quaternary salts: A new class of hypoxia-selective cytotoxins capable of releasing diffusible cytotoxins on bioreduction, *Int. J. Radiat. Oncol. Biol. Phys.* **29:**317–321.

Dirix, L. Y., Tonnesen, F., Cassidy, J., Epelbaum, R., ten Bokkel Huinink, W. W., Pavlidis, N., Sorio, R., Gamucci, T., Wolff, I., Te Velde, A., Lan, J., and Verweij, J., 1996, EO9 phase II study in advanced breast, gastric, pancreatic and colorectal carcinoma by the EORTC Early Clinical Studies Group, *Eur. J. Cancer* **32a:**2019–2022.

Double, J. A. and Workman, P., 1977, A new high-glucuronidase mouse tumor curable by aniline mustard therapy, *Cancer Treat Rep.* **61:**909–911.

Duncan, R., Dimitrijevic, S., and Evagorou, E. G., 1996, The role of polymer conjugates in the diagnosis and treatment of cancer, *STP Pharm. Sci.* **6:**237–263.

Firestone, A., Mulcahy, R. T., and Borch, R. F., 1991, Nitroheterocycle reduction as a paradigm for intramolecular catalysis of drug delivery to hypoxic cells, *J. Med. Chem.* **34:**2933–2935.

Hadfield, A. F. and Sartorelli, A. C., 1984, The pharmacology of prodrugs of 5-fluorouracil and 1-beta-D-arabinofuranosylcytosine, *Adv. Pharmacol. Chemother.* **20:**21–67.

Haisma, H. J., Boven, E., van Muijen, M., de Jong, J., van der Vijgh, W. J., and Pinedo, H. M., 1992, A monoclonal antibody-β-glucuronidase conjugate as activator of the prodrug epirubicin-glucuronide for specific treatment of cancer, *Br. J. Cancer* **66:**474–478.

Jones, G., 1985, Decreased toxicity and adverse reactions via prodrugs, in *Decreased Toxicity and Adverse Reactions via Prodrugs* (H. Bundgaard, ed.), Elsevier, Amsterdam, p. 199.

Kedderis, G. L. and Miwa, G. T., 1988, The metabolic activation of nitroheterocyclic therapeutic agents, *Drug Metab. Rev.* **19:**33–62.

Keyes, S. R., Rockwell, S., and Sartorelli, A. C., 1985, Porfiromycin as a bioreductive alkylating agent with selective toxicity to hypoxic EMT6 tumor cells *in vivo* and *in vitro*, *Cancer Res.* **45:**3642–3645.

Knox, R. J., Friedlos, F., and Boland, M. P., 1993, The bioactivation of CB 1954 and its use as a prodrug in antibody-directed enzyme prodrug therapy (ADEPT), *Cancer Metastasis Rev.* **12:**195–212.

Kuefner, U., Lohrmann, U., Montejano, Y. D., Vitols, K. S., and Hunnekens, F. M., 1989, Carboxypeptidase-mediated release of methotrexate from methotrexate-α-peptides, *Biochemistry* **28:** 2288–2297.

Maeda, H. and Matsumura, Y., 1989, Tumoritropic and lymphotropic principles of macromolecular drugs, *Crit. Rev. Ther. Drug Carrier Syst.* **6:**193–210.

Mann, J. and Shervington, L. A., 1991, Synthesis of novel N- and S-mustards as potential prodrugs activated by bioreductive processes, *J. Chem. Soc. Perkin Trans* **1:**2961–2964.

Mauger, A. B., Burke, P. J., Somani, H. H., Friedlos, F., and Knox, R. J., 1994, Self-immolative prodrugs: Candidates for antibody-directed enzyme prodrug therapy in conjunction with a nitroreductase enzyme, *J. Med. Chem.* **37:**3452–3458.

McKeown, S. R., Hejmadi, M. V., McIntyre, I. A., McAleer, J. J., and Patterson, L. H., 1995, AQ4N: An alkylaminoanthraquinone N-oxide showing bioreductive potential and positive interaction with radiation *in vivo*, *Br. J. Cancer* **72:**76–81.

Melton, R. G. and Sherwood, R. F., 1996, Antibody-enzyme conjugates for cancer therapy, *J. Natl. Cancer Inst.* **88:**153–165.

Meyer, R. B., Stone, T. E., and Ullman, B., 1979, 2′-*O*-Acyl-6-thioinosine cyclic 3′,5′-phosphates as prodrugs of thioinosinic acid, *J. Med. Chem.* **22:**811–815.

Moore, H. W., West, K. F., and Srinivasacher, K., 1983, Bioreductive alkylation- naturally occurring quinones as potential candidates, in *Bioreductive Alkylation- Naturally Occurring Quinones as Potential Candidates* (D. N. Reinhoudt, T. A. Connors, H. M. Pinedo, and W. van de Poll, eds.), Martinus Nijhoff, The Hague, pp. 93–110.

Murdock, K. C., Lee, V. J., Citarella, R. V., Durr, F. E., Nicolau, G., and Kohlbrenner, M., 1993, N-phosphoryl derivatives of bisantrene. Antitumor prodrugs with enhanced solubility and reduced potential for toxicity, *J. Med. Chem.* **36:**2098–2101.

Newlands, E. S., Blackledge, G. R., Slack, J. A., Rustin, G. J., Smith, D. B., Stuart, N. S., Quarterman, C. P., Hoffman, R., Stevens, M. F., Brampton, M. H. *et al.*, 1992, Phase I trial of temozolomide (CCRG 81045: M&B 39831: NSC 362856), *Br. J. Cancer* **65:**287–291.

Newlands, E. S., Stevens, M. F. G., Wedge, S. R., Wheelhouse, R. T., and Brock, C., 1997, Temozolomide: A review of its discovery, chemical properties, pre-clinical development and clinical trials, *Cancer Treat. Rev.* **23:**35–61.

Palmer, B. D., Wilson, W. R., Cliffe, S., and Denny, W. A., 1992, Hypoxia-selective antitumor agents. 5. Synthesis of water-soluble nitroaniline mustards with selective cytotoxicity for hypoxic mammalian cells, *J. Med. Chem.* **35:**3214–3222.

Parker, A. and Fedor, L., 1982, Ammonium 7H-purin-6-yl 1-thio-β-D-glucopyranosiduronate, a latent, selective anticancer agent, *J. Med. Chem.* **25:**1505–1507.

Pesenti, E., Franzetti, C., and Biasoli, G., 1995, Synthesis and biological activity of water soluble polymer bound taxol derivatives, *Proc. Am. Assoc. Cancer Res.* **36:**1824.

Pettit, G. R., Freeman, S., Simpson, M. J., Thompson, M. A., Boyd, M. R., Williams, M. D., and Doubek, D. L., 1995, Antineoplastic agents 320: Synthesis of a practical pancratistatin prodrug, *Anticancer Drug Des.* **10:**243–250.

Prakash, S. R., Correa, I. D., and Gan, L. S., 1994, Tumor targeting by reductive activation of nitroimidazole carbamates, in *Tumor Targeting by Reductive Activation of Nitroimidazole Carbamates* (J. Allen, and R. Voges, eds.), Wiley, Chichester, pp. 573–576.

Ross, D., Beall, H. D., Siegel, D., Traver, R. D., and Gustafson, D. L., 1996, Enzymology of bioreductive drug activation, *Br. J. Cancer* **74(S XXVII):**S1–S8.

Ross, W. C. J., 1962, *Biological Alkylating Agents*, Butterworth, London.

Ross, W. C. J. and Warwick, G. P., 1955, Reduction of cytotoxic azo compounds by hydrazine and by the xanthine oxidase-xanthine system, *Nature* **176:**298–299.

Schilder, R. J., Fields, S. Z., McAleer, C., Himpler, B. S., Igwemezie, L. N., Schacter, L., and O'Dwyer, P. J., 1994, Phase I pharmacokinetic study of Etoposide phosphate (BMY-40481), *Proc. Am. Assoc. Cancer Res.* **35:**245.

Senter, P. D., Saulnier, M. G., Schreiber, G. J., Hirschberg, D. L., Brown, J. P., Hellstrom, I., and Hellstrom, K. E., 1988, Anti-tumor effects of antibody-alkaline phosphatase conjugates in combination with etoposide phosphate, *Proc. Natl. Acad. Sci. USA* **85:**4842–4846.

Siegel, D., Beall, H., Senekowitsch, C., Kasai, M., Arai, H., Gibson, N. W., and Ross, D., 1992, Bioreductive activation of mitomycin C by DT-diaphorase, *Biochemistry* **31:**7879–7885.

Sinhababu, A. K. and Thakker, D. R., 1996, Prodrugs of anticancer agents, *Adv. Drug Delivery Rev.* **19:**241–273.

Smith, G. D., Chakravarty, P. K., Connors, T. A., and Peters, T. J., 1984, Synthesis and preliminary characterization of a novel substrate for γ-glutamyl transferase. A potential anti-hepatoma drug, *Biochem. Pharmacol.* **33:**527–529.

Smitskamp Wilms, E., Hendriks, H. R., and Peters, G. J., 1996, Development, pharmacology, role of DT-diaphorase and prospects of the indoloquinone EO9, *Gen. Pharmacol.* **27:**421–429.

Supko, J. G. and Malspeis, L., 1994, Dose-dependent pharmacokinetics of rapamycin-28-N, N-dimethylglycinate in the mouse, *Cancer Chemother. Pharmacol.* **33:**325–330.

Torres Labandeira, J. J., Davignon, P., and Pitha, J., 1991, Oversaturated solutions of drug in hydroxypropylcyclodextrins: Parenteral preparation of pancratistatin, *J. Pharm. Sci.* **80:**384–386.

Tsou, K. C., Su, H. C., Turner, R. B., and Mirachi, U., 1966, Synthesis of β-(4-hydroxy-1-naphthyl)-DL-alanine, *J. Med. Chem.* **9:**57–61.

Watanabe, K. A., Matsuda, A., Halat, M. J., Hollenberg, D. H., Nisselbaum, J. S., and Fox, J. J., 1981, Nucleosides. 114. 5′-*O*-Glucuronides of 5-fluorouridine and 5-fluorocytidine. Masked precursors of anticancer nucleosides, *J. Med. Chem.* **24:**893–897.

Wei, M. X., Tamiya, T., Chase, M., Boviatsis, E. J., Chang, T. K., Kowall, N. W., Hochberg, F. H., Waxman, D. J., Breakefield, X. O., and Chiocca, E. A., 1994, Experimental tumor therapy in mice using the cyclophosphamide-activating cytochrome P450 2B1 gene, *Hum. Gene Ther.* **5:** 969–978.

Whisson, M. E. and Connors, T. A., 1965, Cure of mice bearing advanced plasma cell tumours with aniline mustard, *Nature* **206:**689–691.

White, I. N., Suzanger, M., Mattocks, A. R., Bailey, E., Farmer, P. B., and Connors, T. A., 1989, Reduction of nitromin to nitrogen mustard: Unscheduled DNA synthesis in aerobic or anaerobic rat hepatocytes, JB1, BL8 and Walker carcinoma cell lines, *Carcinogenesis* **10:**2113–2118.

Wip, F. and Li, W., 1994, Prodrugs of Ara-C, *Drugs Future* **19:**49–54.

Workman, P., 1994, Enzyme-directed bioreductive drug development revisited: A commentary on recent progress and future prospects with emphasis on quinone anticancer agents and quinone metabolizing enzymes, particularly DT-diaphorase, *Oncol. Res.* **6:**461–475.

Young, C. W., Yagoda, A., Bittar, E. S., Smith, S. W., Grabstald, H., and Whitmore, W., 1976, Therapeutic trial of aniline mustard in patients with advanced cancer. Comparison of therapeutic response with cytochemical assessment of tumor cell β-glucuronidase activity, *Cancer* **38:**1887–1895.

Zee Cheng, R. K. and Cheng, C. C., 1989, Delivery of anticancer drugs, *Methods Findings Exp. Clin. Pharmacol.* **11:**439–529.

3

Factors Influencing Tumor-Selective Localization of Antibody Conjugates

M. A. Sims and Roger G. Melton

"Someday perhaps it will turn out to be one of the ironies of nature that cancer, responsible for so many deaths, should be so indissolubly connected with life"
—Oberling, 1946 (*The Riddle of Cancer*, 1946).

1. INTRODUCTION

Almost a century ago the concept of targeting diseased cells using "magic bullets" was proposed by Paul Ehrlich. He envisioned that antibodies, newly described by Emil Behring, which could "in the manner of magic bullets, seek out the enemy," might be used to create highly specific magic bullet drugs. However, it was not until 1978 that a tumor was localized with radiolabeled antibodies in a patient (Goldenberg *et al.*, 1978). The search for suitable targets for those magic bullets, in the form of tumor-specific antigens, started in earnest in the late 1970s with the development of hybridoma technology by Köhler and Milstein (1975). This ability to produce large quantities of monoclonal antibodies (Mabs), reagents of a defined and monospecific nature, greatly facilitated the immunological mapping of the cell plasma membrane and provided a means for targeting cytotoxic drugs specifically at cancer cells.

M. A. Sims • Centre for Applied Microbiology & Research, Salisbury SP4 0JG, United Kingdom. *Roger G. Melton* • Enzacta Ltd., Salisbury SP4 0JQ, United Kingdom.

Enzyme-Prodrug Strategies for Cancer Therapy, edited by Melton and Knox. Kluwer Academic/Plenum Publishers, New York, 1999.

2. TUMOR PHYSIOLOGY

Antibody Directed Enzyme Prodrug Therapy (ADEPT), like other monoclonal antibody-targeted strategies for cancer therapy, depends for its success on specifically localizing the antibody and its conjugated payload at tumor sites. It is something of a misnomer to term this targeted therapy, if the term targeting is taken to imply an active process. The process of antibody uptake by tumor is, essentially the same as that for other macromolecules, including synthetic polymers and particles used for drug delivery: is passive and dependent primarily on the physiology of the tumor. Once an antibody conjugate arrives at its intended destination, however, the antigen-binding properties of the antibody come into play, helping to retain the conjugate at the tumor by binding to tumor-associated antigen, in contrast to most other delivery agents which do not bind to the target cells, except bioactive agents, such as growth factors and lectins. In this chapter, therefore, first we discuss the unique properties of tumors which cause them to accumulate macromolecules by passive uptake, we examine the physicochemical properties of antibodies and relate them to their biodistribution and, finally we consider the suitability of the many potential tumor-associated antigens for use in ADEPT-type applications.

The concept of using monoclonal antibodies (Mabs) to deliver prodrug-activating enzymes specifically to tumors and retain them there to achieve site-specific generation of active drugs is attractive, but deceptively simple. In practice there are many factors that influence the effectiveness of this type of approach. As yet, there are very limited clinical data available for the antibody-enzyme conjugates used in ADEPT studies. However, there is a wealth of experience derived from work with radiolabeled antibodies and antibodies conjugated to conventional cytotoxic drugs, protein toxins, or cytokines. The behavior of antibody-enzyme conjugates is not expected to be radically different. In experimental models, the ADEPT approach works in a range of systems using a variety of enzymes and a more limited range of antibodies and target antigens (Melton and Sherwood, 1996). These preliminary results encourage further exploration of the parameters which might influence targeted enzyme therapy. In dealing with this topic, it is assumed that administered antibody-enzyme conjugates are almost exclusively constructed with immunoglobulins of the IgG subclass, as are the vast majority of the radiolabeled antibodies and antibody-toxin conjugates. The prodrug component has not been taken into account in this chapter. As molecules of low molecular mass, their biodistribution is rather more dependent on other factors, such as chemical half-life and lipophilicity than on considerations of molecular mass and immunoreactivity which are important for antibody-enzyme conjugates.

2.1. Relevance of Murine Tumor Models

In patients, typical localization values of 0.0001%–0.01% injected dose per gram of tumor are the norm with Mabs (Begent, 1985), which very are low compared with the 10%–30+% injected dose g^{-1} achievable in mouse models. However, straight comparisons cannot be meaningfully made. For example, the tumor burdens commonly used in rodent models are relatively high compared with those encountered in patients. Tumors weighing 0.5–1 g and greater are common in mouse studies and represent 2.5–5% or more of the total body mass of the animal. In contrast, tumors approaching 1 kg (about 1.5% of the body mass of an average adult) are almost always fatal in humans. Where the tumor is constricted by space and/or proximity to critical organ, for example, in the brain or adjacent to major blood vessels, a tumor mass of a few grams may be fatal. Very large tumors do not represent a realistic target for approaches, such as ADEPT. With such tumors the primary form of treatment is usually surgical resection. For animal studies, therefore, barely palpable tumors represent more realistic models, but present difficulties in obtaining quantitative experimental data. The dose levels of agents under test are also far higher than those permissible in patients in both single- and multiple-dose regimens. Although body weight loss of 5–10% is common in experimental animals undergoing therapy, such conditions would not be tolerated in humans. In addition, because human tumor models are typically studied in immunocompromised host animals (nude or SCID mice), the adverse immunological reactions that may halt treatment in humans do not occur.

Finally, the usual subcutaneous site of transplanting model tumors is unrepresentative of human diseases (Jain, 1996a), and the mouse vasculature sustaining the implanted tumor may not be similar to that found in a patient. Typically, human tumor xenografts do not metastasize when implanted in nude mice, and this is a severe limitation of such models because most deaths from cancer are caused by metastases that are resistant to conventional therapies (Fidler and Balch, 1987). Some workers have used tumor models, where the site of implantation is more representative of human disease and the tumors are able to metastasize. For example, human ovarian cystadenocarcinoma cells implanted intraperitoneally in SCID mice form distant metastases in the lungs (Schumacher *et al.*, 1996). Orthotopic tumor models, where tumor is seeded in the original host organ, are becoming available and these also retain the ability to metastasize. Such models have been described for colon (Fu *et al.*, 1991; Togo *et al.*, 1995; Wang *et al.*, 1994), stomach (Furukawa *et al.*, 1993a,c), lung (Wang *et al.*, 1992a,b), breast (Fu *et al.*, 1993), pancreas (An *et al.*, 1996; Furukawa *et al.*, 1993b), ovarian (Astoul *et al.*, 1993; Schumacher *et al.*, 1996), bladder (Fu *et al.*, 1991), and bone (Crnalic *et al.*, 1997) tumors, although their use for testing of putative therapeutic agents has been very restricted as yet. They have the potential to mimic the surgical resection of a human primary tumor and subsequent treatment of metastases but require further development work before they become a routine tool for testing therapeutic agents.

It is now recognized that the physiology of tumors offers opportunities and obstacles to antibody-based therapy, quite apart from any considerations relating to the nature and distribution of the target antigen, which are discussed in the latter part of this chapter.

2.2. Route of Administration

The route by which an antibody-enzyme conjugate is administered has significant implications for its biodistribution. The most common method of administration is by intravenous injection as a bolus or as a prolonged infusion. When conjugates are administered by this route, they are greatly diluted in the plasma and almost immediately encounter the principal organs of metabolism (the liver and spleen) and excretion (the kidneys) before reaching the tumor site. On arrival at the tumor, they must escape from the circulation by crossing the vascular endothelium and diffuse into the interstitial space of the tumor. This process is influenced primarily by the size and, possibly, the charge of the protein. Therefore, it follows that simplifying the route taken by the conjugate may have significant impact on its biodistribution and selectivity of tumor uptake. In some cases there may be possibilities for administering an antibody-enzyme conjugate directly to a tumor. A classic example of this is ascitic tumors, which develop in the peritoneal cavity. Direct administration of a radiolabeled Mab to the peritoneal cavity has been used to treat ovarian ascitic carcinomas with some success (Epenetos *et al.*, 1987; Griffin *et al.*, 1990; Hird *et al.*, 1993). Intravesical administration of radiolabeled Mabs has also been used for bladder tumors (Bamias *et al.*, 1993; Malaitsi *et al.*, 1995) albeit with mixed success. Although the opportunities for direct regional administration are limited, some workers have injected Mabs directly into the lymphatic duct to target cancerous lymph nodes (Keenan *et al.*, 1987), and the possibility of targeting lung tumors by directly administering Mab aerosols to the luminal surface of the bronchial epithelium has been explored (Mulshine, 1990). There may also be opportunities with relatively large tumors to inject conjugate directly into tumors. Such an approach has been used with subcutaneous tumor implants in animal models and very high levels of localization are attained in such models (Rowlinson-Busza *et al.*, 1991). However where there are extensive metastatic lesions or tumors are less accessible, the technique is of limited value. Intratumoral injection has been described for patients with relapsed malignant gliomas (Hopkins *et al.*, 1995; Riva *et al.*, 1995), but this is an unusual example and normally intravenous injection or infusion is the predominant route of administration.

2.3. Vascularization of Tumors

As solid tumors grow, they recruit a blood supply in a process known as angiogenesis. It is a complex process, involving extracellular matrix remodeling, endothelial cell migration and proliferation, and capillary proliferation. These processes are controlled by a range of angiogenic factors which are normally tightly regulated, and occur only in certain transient circumstances, such as wound healing, reproduction, and development. In tumors, the normal regulatory processes break down, leading to disordered growth of blood vessels, the nature and distribution of which are important factors to be considered.

The distribution of blood vessels in a tumor is not uniform, and this has important implications that influence the accessibility of tumor cells to cytotoxic agents. The process by which tumors grow is illustrated diagrammatically in Figs. 1 and 2, which show that in the early stage of growth, to about 1 mm diameter, all cells remain healthy without the need for a separate blood supply. As the tumor grows to 1–3 mm in diameter, necrosis begins to develop at its center because the supply of oxygen and nutrients by diffusion alone is insufficient to sustain the cells in the middle which are furthest from blood vessels (Fig. 1). For growth to continue beyond 3 mm, a new vascular supply is necessary, and capillary buds are

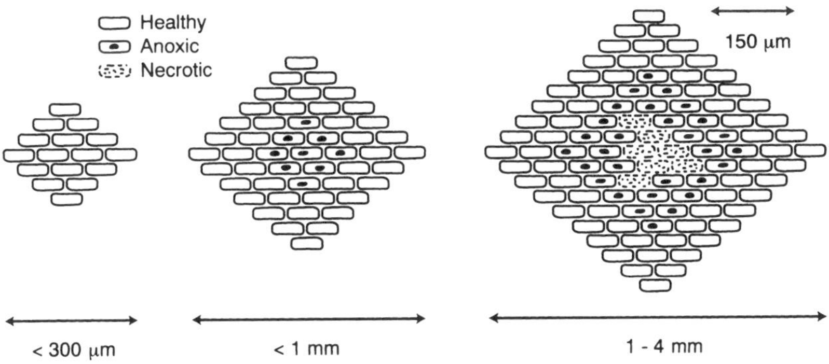

Figure 1. Schematic representation of the growth of a tumor nodule through prevascular and vascular phases. As the tumor enlarges, cells in the center become unhealthy and then necrotic as growth continues. To grow beyond about 1 mm in diameter, the tumor has to acquire a blood supply in the form of capillaries branching off from adjacent blood vessels. The well-vascularized layer is confined to the surface of the tumor and is about 150 μm thick, but the quiescent anoxic cells regrow if exposed to nutrients and oxygen by the death of the surface layer.

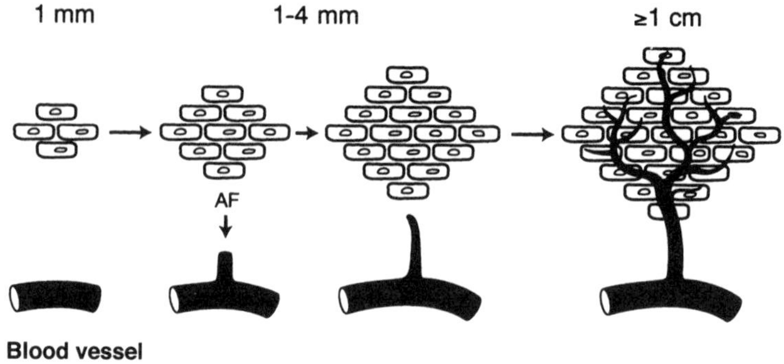

Figure 2. Schematic representation of the development of tumor angiogenesis. As tumor cells proliferate and the tumor grows, they need to recruit a blood supply. They do this by secreting angiogenic factor proteases, resulting in the induction of endothelial cell migration, digestion of basement membranes, and formation of new capillary tubes by vascular endothelial cells.

recruited from surrounding tissue (Fig. 2). The tumor cells produce both positive and negative angiogenic factors to initiate and control the process of neoangiogenesis by procuring an outgrowth from an existing capillary bed (Fidler and Ellis, 1994; Jain, 1988; Klagsbrun and D'Amore, 1991). The growth process is disordered and results in irregular capillaries which may be dilated, saccular, sinusoidal, and tortuous and may contain multiple bifurcations, loops, and blind ending sprouts. These differ from those found in normal capillary beds (Gazit *et al.*, 1995; Less *et al.*, 1991; Warren, 1979) and provide an obstacle to smooth blood flow. Despite the recruitment of these new vessels, the center of the tumor continues to be necrotic because lymphatic drainage is not recruited. The function of the lymphatic capillaries is to drain macromolecules from tissues, filter them through the regional lymph node, and return them to the circulation (Weinstein, 1990). In the absence of functional lymphatic drainage, osmotically active macromolecules accumulate and thereby generate high interstitial fluid pressures, although these fall to normal values at the periphery of the tumor (Jain, 1996a) and give rise to a pressure gradient across the tumor. The high pressures at the necrotic center impede nutrition and oxygenation in a number of ways: by decreasing flow to local blood vessels and ultimately by closing them off, and by opposing percolation of molecules out of the vessels (Jain and Baxter, 1988; Rippe and Haraldsson, 1987). Taken together, these factors help to explain the experimental observation that uptake of antibodies and conjugates is markedly better in terms of percentage injected dose per gram, in small than in large tumors. The majority of antibody or conjugate localized in the tumor accumulates in the relatively thin, well-vascularized, periphery which forms a smaller percentage of the total volume as the tumor grows.

2.4. Vascular Permeability of Tumors

To gain access to the target tumor cells, an antibody-enzyme conjugate has to traverse the vascular capillary endothelium and pass into the interstitial space. In particular, tumor capillaries may have gaps in the vascular endothelium, termed fenestrations, which form pores by which macromolecules escape from the capillary. Such fenestrations are not normally found in blood vessels, other than in the sinusoidal tissue of the liver, spleen and bone marrow, together with the gastrointestinal tract and the renal glomerulus. Most tissues have vascular endothelia which are lined by continuous endothelia, underlain by a fully formed basement membrane, and macromolecules cross them only by transcytosis or by passage through the cell–cell junction gaps. The size of the openings in both sinusoidal and fenestrated capillaries (100 nm across and 30–80 nm across, respectively) is such that antibodies and similar sized macromolecules should be able to pass freely through them, although tumor uptake of macromolecules is influenced by their molecular mass. Given that this type of fenestrated capillary is frequently found in tumors, it is readily appreciated why tumor capillaries are commonly described as "leaky" compared with those in normal tissue. In some cases, the endothelium may be discontinuous, so that tumor cells may even line the capillary (Dvorak *et al.*, 1988; Jain, 1988). Continuous, fenestrated, and discontinuous endothelia are represented in Fig. 3. It is important to realize, moreover, that vascular permeability varies considerably from one tumor to another and it can vary within the same tumor (Jain, 1988). As mentioned earlier, model tumors grown as subcutaneous implants in nude mice may exhibit different properties when grown in a location more representative of their natural origin, but the experimental evidence for this is limited (Jain, 1996a). For example, the mechanism by which fenestrations develop is not fully understood but may be related to interaction of the tumor cells and their products with the host defense mechanisms, especially the clotting and fibrinolytic systems (Nagy *et al.*, 1989), and such processes may be different in the mouse.

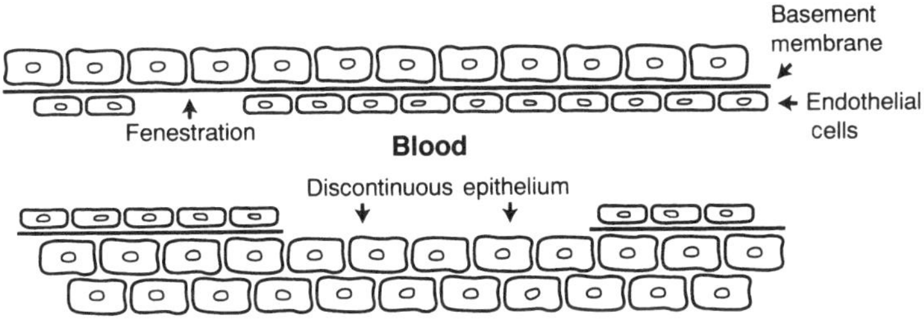

Figure 3. Types of openings found in capillary walls that give rise to "leakiness."

2.5. Movement of Macromolecules within Tumors

Once a macromolecule has escaped from the vascular system and migrated into the interstitial space of a tumor, its subsequent movements through the interstitial space occur by a combination of diffusion and convection (Jain, 1996b). Diffusion is proportional to the concentration gradient of osmotically active solutes in the interstitial fluid, whereas convection is proportional to the interstitial fluid flow velocity, which in turn is proportional to the fluid pressure gradient in the interstitium. In solid tumors, neither of these factors favor rapid and uniform distribution of administered macromolecules within the tumor because the underlying trend is for fluid to exude from the tumor. The accumulation of osmotically active solutes at the tumor center due to poor lymphatic drainage has already been discussed. For the same reason, the interstitial fluid flow is in an outward direction from the high pressure region at the necrotic center outward toward the periphery, which is at lower pressure. A molecule escaping from the tumor vasculature has to travel in the direction opposite to this flow to diffuse into the tumor. The problem is compounded by the fact that, although the interstitial volume of tumors is relatively large compared with normal tissue (Jain, 1996b), the structure of the interstitial space is predominantly composed of a collagen and elastin fiber network, which presumably imparts structural integrity. This is interspersed with macromolecular polysaccharides, such as glycosaminoglycan and proteoglycans, which, together with the interstitial fluid, form a hydrophilic gel that provides a further obstacle to free diffusion (Swabb *et al.*, 1974). The precise composition of this gel varies, affecting its resistance to fluid flow. Development of a "provisional stroma" arising from deposition of fibrinogen and its clotting to form a fibrin gel also helps to impose a structure which is eventually supplemented by the development of mature connective tissue (Nagy *et al.*, 1989). Just as retention times of macromolecules are determined by gel pore size in gel permeation chromatography, it is probable that smaller molecules pass through the interstitial gel more readily than larger ones. An important difference between ascitic and solid tumors is that fibrin gel is not normally deposited in the body cavities and consequently, does not form a provisional stroma to impose an initial structure, thereby improving access to the cells in such tumors (Nagy *et al.*, 1989).

2.6. Effect of Tumor Size and Necrosis

The presence of a necrotic center in a tumor does not markedly affect the pressure gradients within the tumor as a whole, because it is not vascularized. Therefore the main pressure gradient runs across the well-vascularized outer regions. The distances over which macromolecules must diffuse to attain even distribution are relatively large, and because the process is relatively slow, significant concentrations of targeted macromolecules do not accumulate in necrotic regions far from the blood vessels. Equally, however, material which does accumulate in necrotic regions may be retained for a long time (Baxter and Jain, 1990). Some success has been claimed with antihistone antibodies designed to target necrotic regions of tumors, but the degree of tumor uptake achieved with these antibodies is not comparable with that attainable with antibodies directed against antigens found in well-vascularized regions (Chen *et al.*, 1991; Gaffar *et al.*, 1991; Miller *et al.*, 1993).

The overall size of tumors has no significant effect on the concentration or pressure gradients once they grow to the point where a necrotic center develops because the pressure gradients are confined to the thin, well-vascularized layer at the tumor periphery. However, a number of studies have shown apparently higher uptake in smaller tumors than large ones (Boerman *et al.*, 1992; Chiou, 1991; Hagan *et al.*, 1986; Pedley *et al.*, 1987). This is predominantly caused by the fact that the well-vascularized periphery in which uptake occurs corresponds approximately to the surface area of the tumor, which increases proportionately less than the volume it encloses. At least one study has shown that total uptake is proportional to the surface area of the tumor (Pedley *et al.*, 1987).

3. PROPERTIES OF THE MONOCLONAL ANTIBODY VECTOR

The important attributes to be considered for the antibody vector used are summarized in Table I.

3.1. Affinity for Antigen

In ADEPT, one is using the antibody component of an antibody-enzyme conjugate to bind to cell surface antigens to retain the conjugate at the tumor, and therefore the migration of the antibody-enzyme conjugate is influenced by the

Table I. Properties of Antibodies that May Influence Tumor Localization

Immunochemical	Physicochemical	Physiological
Specificity	Size	Clearance rate
Immunoreactivity	Charge	Intracellular metabolism
Affinity constant	Isotype	
Association/dissociation rate constant	Carbohydrate content	
Immunogencity	Stability of immunoconjugate linkage chemistry	

amount of antigen it encounters as it diffuses through the tumor. Theoretical and practical studies by Weinstein and co-workers (Fujimori *et al.*, 1990; Weinstein and van Osdol, 1992) have demonstrated that antibody-antigen binding in tumors retards antibody percolation, leading to heterogeneous distribution, particularly for antibodies that have high affinity for antigen. In theory, therefore, high affinity for antigen is an undesirable property for achieving uniform distribution throughout a tumor, and a low-affinity antibody should show superior distribution. In practice there is evidence that high-affinity antibodies show distinct advantages in animal models (Schlom *et al.*, 1992; Velders *et al.*, 1996) although another report has suggested that a 10-fold difference in binding affinity makes little or no difference to tumor uptake on the basis of a % injected dose g^{-1} tumor (van Duk *et al.*, 1991). The problem can also be addressed by the superficially straightforward approach of giving an increased dose of antibody to saturate readily available sites at the tumor periphery (Fujimori *et al.*, 1990). This allows excess antibody to percolate further into the tumor, although this in turn decreases the specificity of uptake in terms of tumor to normal tissue and tumor to blood ratios and may also give rise to problems with toxicity or immunological responses.

3.2. Antibody Class and Isotype

The class of immunoglobulin used almost exclusively for antibody targeting applications is IgG, and the IgG_1 isotype predominates within this class. Some workers have used IgMs for radioisotope targeting but there appears to be no significant advantages over IgGs. The IgG_1 isotype predominates because of the ease of producing antigenically active but lower molecular mass fragments. The fragments produced by digestion of IgG_1 are also stable, whereas a $F(ab')_2$ fragment of IgG_4 was very unstable when compared to an IgG_1 and an IgG_2 raised against the same antigen. In the same experiment, the IgG_2 gave the best localization (Buchegger *et al.*, 1992). These results may be attributable in part to the differing numbers of disulfide bridges at the hinge region in the different isotypes. The isotypes that have the least number of disulfide bridges are the least stable *in vivo*.

3.3. Physical Properties of Immunoconjugates

An important factor in determining tumor penetration by an administered macromolecule is its size. Experimental evidence suggests that the rate of tumor penetration by antibody fragments increases with decreasing molecular mass in the sequence: intact Ig $<$ $F(ab')_2$ $<$ Fab $<$ single chain Fv (Fjeld *et al.*, 1991; Sutherland *et al.*, 1997; Yokota *et al.*, 1992). Against this, the rate of clearance of antibodies and fragments also increases as molecular mass decreases, so that absolute amounts of antibodies and conjugates actually localizing at tumors are lower (Adams *et al.*, 1993; Owens *et al.*, 1991; Rowlinson-Busza *et al.*, 1996; Wahl *et al.*, 1983). The rate of clearance from the serum is more rapid than the rate of loss of localized material from the tumor. Thus the specificity of localization is better, even though absolute levels are lower (Begent *et al.*, 1996b; Rowlinson-Busza *et al.*, 1996). Attachment of enzymes to Mabs or Mab fragments inevitably leads to a conjugate of increased molecular mass and might be expected to increase persistence times because of reduced susceptibility to glomerular filtration, but this is not always so. In part this discrepancy may be unrelated to molecular mass and more due to glycosylation of the antibody on the Fc domain, because clearance may result from interaction with Fc receptors which interact primarily with carbohydrate residues. This domain is absent from fragmented immunoglobulin, thus conjugate made with the smaller fragment may persist longer. Deglycosylated antibodies have clearance properties which differ from those of the native glycosylated form (Hand *et al.*, 1992). Glycosyl groups on enzymes may also influence the biodistribution of conjugates. A Mab-β-glucuronidase conjugate prepared with deglycoslated human enzyme enhanced plasma persistence of the conjugate around 20-fold (Houba *et al.*, 1996). There is also the possibility that attaching an enzyme to an antibody may distort the steric conformation of the antibody because the linkage occurs at random locations on both the antibody and enzyme using currently available chemical conjugation technology and such deformation may affect antigen-binding properties. Clearly the presence or absence of glycosylation on the enzyme component of a conjugate may also have implications in this respect. For ADEPT applications some loss of activity is acceptable for both components because the inherent amplification properties of the system offset the requirement for high levels of localization, and specificity of localization is probably the more important requisite if activation of prodrug in normal tissues is to be avoided.

3.4. Metabolism of Tumor-Bound Immunoconjugates

Metabolism of antibody conjugates which bind to target antigens is a phenomenon that is easy to overlook. It is well known that although some antigens are expressed at the cell surface, they are internalized upon becoming complexed with antibody. In fact this process forms an essential step in the mode of action of toxin-Mab conjugates (immunotoxins) because the commonly used ribosome inactivating toxins, such as ricin, exert their toxic effect within the cytosol. This is not the case for the ADEPT systems described to date and indeed most workers emphasize the desirability for the antibody-enzyme conjugate to remain attached to the cell surface. This avoids the need for the prodrug to cross an additional barrier in the form of the cell membrane before reaching the activating enzyme. Once the prodrug has been activated, then it needs to diffuse into adjacent cells to exert a bystander effect, which in turn makes an extracellular location for the antibody-enzyme conjugate desirable, although a bystander effect has been demonstrated for intracellularly activated prodrugs (Bridgewater *et al.*, 1997; Knox *et al.*, 1988).

Antibody binding is not irreversible, and metabolically active cells release bound antibodies at a slow rate. This release occurs irrespective of whether the antibody is bound to the cell surface or transported into the cell (internalized). The rate of release is related to the competition for antigen binding because it is accelerated by unbound antibody (Matzku *et al.*, 1988). The gradual loss of antibody or conjugate from a tumor, which is commonly observed in biodistribution experiments, is empirical evidence that binding of the antibody to antigen does not mark the onset of an irreversible event, although it is not axiomatic that the loss of antibody from a tumor denotes dissociation of the antibody-antigen complex. In the majority of experiments, radioiodinated materials are used to follow the fate of antibodies and conjugates, but the phenomenon of dehalogenation is well established and makes this an unreliable method of measurement unless one of the more recent iodination methods specifically designed to minimize dehalogenation is used (Reist *et al.*, 1996). A superior approach for ADEPT applications is to measure enzyme activity directly in the tissue of interest, assuming that the enzyme being used does not have an endogenous counterpart. When such an approach was used to measure radiolabeled carboxypeptidase G_2 conjugates, the enzyme activity was consistently in excess of that predicted from radiolabel measurements (Antoniw *et al.*, 1991).

Evidence suggests that the majority of antibodies undergo degradation before being released, and that study of the process necessitates observation over a prolonged period (Kyriakos *et al.*, 1992). Certain antibodies, for example, those directed against the transferrin receptor, are internalized and degraded very rapidly because they enter the coated pit/coated vesicle pathway, a route which is not typical of the majority of tumor-associated antigens. The process occurred rapidly in the earlier systems studied, for example, antitransferrin receptor. Short study periods were initially used, and it was concluded that essentially all antibody remains bound over a one day period if it does not undergo rapid internalization (Giacomini *et al.*, 1985; Masuho *et al.*, 1985). This erroneous conclusion may have arisen because loss of antibody from cells surfaces may, in part, result from shed antigen, but the assay used may not have distinguished between antibody bound to shed antigen and that bound to cells.

The work of Kyriakos *et al.* (1992) suggests that the majority of antibody binds essentially irreversibly and that antibody which appears to binds reversibly may possibly be binding monovalently. This provides an important indicator to suggest that Fab or single chain Fv-based conjugates may be of limited value unless constructed as bivalent pseudo-F(ab′)$_2$ forms. These results also suggest that antigen density on the target cell may be an important factor because if bivalent binding is important for retention of conjugate, it is necessary for the individual antigen molecules to be sufficiently close to one another to permit it, unless the antigen contains repeated epitopes.

Kyriakos *et al.* (1992) also showed that internalization and catabolism of antibodies bound to cell surface antigens occurs over a number of days, although there are considerable variations in the rate of degradation. *In vitro* experiments showed that degradation of bound antibody is prevented by two inhibitors of lysosomal enzymes, pointing to this as the likely degradative pathway. But one might expect some degradation *in vivo* to arise from attack by proteolytic enzymes in the interstitial fluid on antibody exposed at the cell surface, quite apart from any degradation which would be expected to occur on internalization.

Despite the evidence of these observations, it is dangerous to make generalizations. In an ADEPT experiment using an antibody conjugate directed at the c-erbB-2 receptor, Eccles *et al.* (1994) found that it was necessary to allow a period of 14 days between administering antibody-enzyme conjugate and prodrug because the clearance of conjugate was very slow. Despite this prolonged period, there was still sufficient conjugate available to turn over the prodrug and an impressive antitumor effect was achieved. It is possible that, in part, this may result from accumulation of shed antigen in the interstitial space of the tumor, to which conjugate binds and generates active drug from prodrug without having to bind to the tumor. Such accretions of antigen are found in carcinoembryonic antigen (CEA)-expressing tumors (Boxer *et al.*, 1992), and their presence may be very beneficial for ADEPT applications.

3.5. Clinical Findings with ADEPT Systems

The information presented in the first part of this chapter, on the face of it, is rather discouraging, but in reality one can get tumor-specific uptake of Mabs and antibody enzyme conjugates with sufficient differentials between tumor and normal tissue to achieve effective selective therapy (Bagshawe *et al.*, 1995).

Theoretical studies by Jain and co-workers (Baxter and Jain, 1996; Yuan *et al.*, 1991) suggest that the absolute level of localization of antibody-enzyme conjugate in a tumor is not the limiting factor in achieving successful ADEPT systems, but that the *differential* in uptake between tumor and normal tissue is crucial. Typical tumor to normal tissue ratios achieved in preclinical studies ranged from 1.5:1 to 80:1 six days after conjugate administration in two systems where second antibody clearance was not used (Antoniw *et al.*, 1991). Jain and co-workers concluded that an enzyme of low V_{max} provides optimum specificity because turnover rates are low in all tissues, but relatively higher levels of drug would be produced in a tumor containing appreciably higher concentrations of enzyme. Thus, exposure of normal tissues to high levels of active drug would be minimized because the overall rate of activity would be low. However, the pool of unbound antibody-enzyme conjugate in the blood and other tissues greatly outweighs that in the tumor because of the much greater mass of these tissues. In practice toxicity was observed in early clinical trials, albeit with an enzyme having a high V_{max} (Bagshawe *et al.*, 1995). Some of this toxicity may also be attributable to the use of a prodrug which gave rise to an active drug of long chemical half-life, allowing active drug to diffuse away from the tumor and attack normal tissues. The use of an enzyme with low Km would be desirable, irrespective of V_{max} optima, to achieve the maximum possible amplification effect because such an enzyme would function more efficiently at low prodrug concentrations. Therefore, it is encouraging to note that in experimental systems where second antibody clearance was not used, the ratio of the area under the curve for active drug generated compared with the prodrug source had a value of 29, compared with values of 0.9–5.5 for blood and normal tissues (Antoniw *et al.*, 1991). In other experiments where second antibody clearance was used, generation of active drug remained high in tumor, but levels in normal tissues were reduced compared with controls (Rogers *et al.*, 1995), indicating reduced turnover by antibody-enzyme conjugate not specifically bound. Although no other pharmacological studies have been reported, antitumor studies by other groups have confirmed the usefulness of second antibody clearing systems in enhancing therapeutic efficiency (Haisma *et al.*, 1995; Kerr *et al.*, 1993). Nevertheless, it is possible that residual antibody-enzyme conjugate in normal tissues could diffuse back into the circulation, particularly where short-lived clearing agents, such as galactosylated second antibodies, which are designed to be confined to the circulation, are used (Baxter and Jain, 1996). This problem can be addressed by administering the clearing agent by infusion rather than by bolus injection (Bagshawe *et al.*, 1995).

3.6. Manipulation of Tumor Physiology

To date we have dealt with the inherent physiological properties of tumors, but opportunities exist to modify these properties favorably for antibody-based targeting systems. A considerable amount of work has been carried out on these lines.

The origins of this approach are found in the field of radiation sensitizers. These are drugs which were originally designed to mimic the chemical properties of oxygen and thus render cells in hypoxic regions of tumors sensitive to ionizing radiation. It was discovered that one of the earlier agents tested as a radiosensitizer, misonidazole, also potentiates the action of chemotherapeutic agents, such as melphalan (Siemann, 1984). This property was subsequently attributed to misonidazole, which acts as a vasoactive agent, causing a rapid and prolonged drop in tumor blood flow and thereby altering the distribution and retention of melphalan (Murray *et al.*, 1987). A number of agents, both vasoconstrictors and vasodilators, have been tested as potential modifiers of tumor blood flow. Strictly speaking, however, the response caused by most of these agents is a relative shift between the normal tissues and the tumor. The tumor blood vessels lack both innervation and smooth muscle, rendering them incapable of response. Thus the tumor remains unaffected and the surrounding normal tissue responds (Jirtle, 1988). Some groups have shown that the uptake of monoclonal antibody-drug conjugates by murine E3 thymomas increases by administering the β-adrenergic agent propanolol (Smyth *et al.*, 1987) although other workers have found that such agents have little or no effect in human colon carcinoma or osteosarcoma models (Pimm, 1990). Co-administration of the cytokine tumor necrosis factor-alpha (TNF-α) enhances tumor uptake of Mabs (Rowlinson-Busza *et al.*, 1995), antibody-drug conjugates (Smyth *et al.*, 1988) and antibody-enzyme conjugates (Melton *et al.*, 1993) *in vivo* in the E3 murine thymoma model. Twofold increases or greater in tumor localization were typically achieved without concomitant increases in other tissues, as illustrated for antibody-carboxypeptidase G_2 conjugate in Fig. 4. The timing of administration of TNF-α was critical, as shown in other studies using antibody-aminopterin conjugates (Russell *et al.*, 1990) where administering the TNF-α 4 hr after conjugate resulted in significantly lower tumor uptake than in animals which received TNF and conjugate simultaneously. TNF-α may differ somewhat from the vasoactive agents described earlier in its mode of action. It has been postulated that TNF-α induces tumor-specific vascular damage (Russell *et al.*, 1990). Evidence of increased vascular permeability to IgG following TNF-α administration was noted in mice bearing HT29 human colon adenocarcinoma xenografts, but the vascular volume of the tumor was unaffected (Rowlinson-Busza *et al.*, 1995). There is some evidence that antibody uptake by a tumor is proportional to its vascular permeability (Sands *et al.*, 1988). In contrast, however, other workers have also found some evidence of reduced tumor blood flow in human gastric carcinoma and osteosarcoma models following TNF-α administration. In these experiments TNF-α did not enhance tumor uptake of a radiolabeled antibody, although it was not co-administered with the TNF-α (Pimm *et al.*, 1991).

Some synthetic vasoactive agents, such as flavone acetic acid (FAA) and 5,6-dimethylxanthenone 4-acetic acid (DMXAA), induce massive hemorrhagic necrosis of up to 95% of a tumor analogously to TNF-α (Zwi *et al.*, 1994). The observed necrosis is mediated by induction of TNF-α production by infiltrating lymphocytes because the vascular effects are blocked by antibodies to TNF-α (Ching *et al.*, 1994). With agents, such as FAA and DMXAA, as with TNF-α itself, scheduling is important because the antibody accumulation must be allowed to

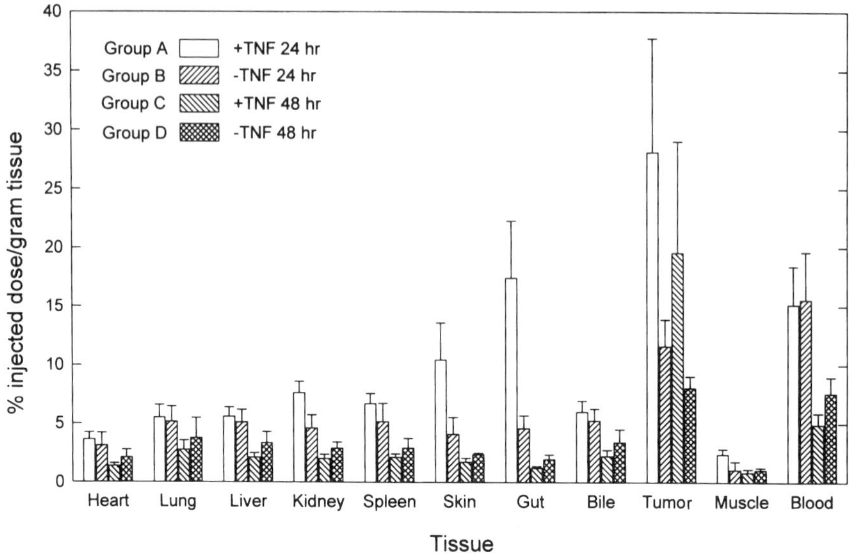

Figure 4. Tumor localization of anti-Ly2.1-CPG_2 conjugate in mice bearing Ly 2.1^+ E3 thymoma xenografts. Mice (four per group) in Groups A and C received 3.5 μg of ^{125}I-labelled conjugate mixed with 2 μg of recombinant TNF-α. Mice in groups B and D received the same conjugate dose without TNF. Groups of animals were sacrificed at 24 hours (Groups A and B) or 48 hours to determine tissue uptake, which was calculated as the mean percentage injected dose per gram of tissue.

reach its maximum before initiating drug-induced blood flow inhibition. Such considerations also suggest, however, that this type of approach may be of limited value for ADEPT systems because achieving high levels of tumor localization, although important, is not the sole criterion for success. Reduced tumor blood flow subsequently may restrict access of prodrug to the tumor, negating the value of high levels of tumor uptake of conjugate.

Other cytokines also affect tumor localization of antibody conjugates by inducing alterations in tumor vascular permeability. It has been shown that antibody-interleukin 2 immunoconjugates, directed against a lymphoma or a colon carcinoma model by an antibody of appropriate specificity, increase tumor uptake of a subsequently administered radiolabeled tracer antibody in a time- and dose-dependent manner. Tumor blood flow and vascular permeability studies carried out with this material demonstrated that there was induction of specific vascular leakage at the tumor site, but tumor blood flow remained unchanged (Leberthon *et al.*, 1991). More recently, a conjugate constructed from an antibody against a tumor blood vessel-associated antigen showed similar properties (Epstein *et al.*, 1995) and similar effects are also obtained using interleukin 2 by itself (Schultz *et al.*, 1992).

An interesting alternative approach has been to target cobra venom factor (CVF) to tumors as an antibody-toxin conjugate (Juhl *et al.*, 1995). CVF induces local inflammation within tissues by initiating activation of the complement cascade, including the release of both C3a and C5a, both of which mediate an inflammatory reaction and attract immunocytes, such as NK cells and macrophages, to the tumor, and induce hypervascularization, increase of blood flow, and vascular leaks (Erdei *et al.*, 1991). A secondary effect may be the formation of a cell membrane attack complex that causes cell lysis (Vogel, 1991). When animals bearing human pancreatic PancTu1 xenografts were pretreated with the CVF conjugate followed by a radiolabeled anti-CEA antibody, tumor localization of the anti-CEA increased by of 80–90% compared with controls (Juhl *et al.*, 1995). This mechanism of action may form the basis of a useful system for use with ADEPT, where retention or enhancement of tumor blood flow may be valuable for increasing doses of prodrug to enzyme localized at a tumor.

3.7. Vascular Targeting

A novel approach to cancer therapy is to target the endothelial cells of the tumor vasculature rather than the tumor cells themselves (Burrows and Thorpe, 1994). This approach has a number of advantages, one of which is the inherent amplification effect. Like normal tissue, tumors depend on their vasculature for continued survival. Therefore destroying even a small part of the tumor vasculature could lead to destroying thousands of tumor cells in the periphery. Secondly, vascular targeting should be applicable to many different solid tumors, and the nature of the target, its location on the endothelium itself, means that it is easily accessible to administered Mab-conjugate.

The validity of this approach has been demonstrated in an *in vivo* model, where experimentally derived antigen was used as the target for two different but complementary therapies using Mab-toxin constructs. The antivasculature approach produced dramatic tumor regressions, whereas the more conventional immunotoxin approach produced only a minor, transient response. When the two modalities were combined, complete remissions were achieved in 50% of animals (Burrows and Thorpe, 1993). If vasculature targeting is going to progress out of the laboratory and into the clinic, the key requirement is for an absolutely specific tumor vasculature antigen/Mab combination. A number of potential targets have been identified and are currently being evaluated. It is anticipated that in an ADEPT situation the accessibility of the target antigen on the tumor endothelium would overcome the problem of poor tumor penetration. The inherent amplification effect would flood drug into any avascular areas.

4. TUMOR ANTIGENS

Advances in recombinant technology have allowed us to identify and quantify the genetic material responsible for the expression of tumor antigens. Despite this, few truly "tumor-specific" antigens have been identified. Most of the tumor cell surface antigens recognized are more correctly defined as "tumor-associated antigens" because they are weakly expressed on normal tissue or have a restricted normal tissue distribution.

Although the term "antigen" has been used throughout this chapter, it should be appreciated that most of these "antigens" are normal cellular constituents, many of which were identified using xenogenic polyclonal antibodies, murine Mabs, and human Mabs generated *in vitro*.

The ideal ADEPT antigen would have the properties listed in Table II. To date the ideal ADEPT antigen has not been identified! Thus we are left to use a range of antigens that are less than ideal.

4.1. Heterogeneity of Antigenic Expression

One of the most important factors to be considered when using Maabs to target tumor-associated antigens is the extreme immunophenotypic heterogeneity of human cancer. Normal cell populations are not completely uniform, but tumors display extreme heterogeneity in antigenic expression and also in morphology, biochemistry, karyotype, oncogene products, drug resistance, and ability to invade and produce metastases. This heterogeneity results from a combination of factors. It reflects the increasing genetic instability that develops as the tumor evolves and epigenetic influences, such as the result of intercellular communication between

Table II. Properties of the Ideal ADEPT Antigen

Tumor specific
Uniform distribution throughout the tumor
Reasonably high concentration on the plasma membrane
Homogenous expression in all tumors and their metastases
Permanent constituent of the cell membrane

cells within the tumor, interactions between the tumor and the extracellular matrix, and reaction to the host's immune surveillance.

Tumors should not be considered static entities, but subpopulations of cells with different growth rates and phenotypes. Antigenic heterogeneity is observed intra- and intertumorally, and it is well established that multiple lesions and metastatic populations may have a distinct antigenic phenotype compared to the primary tumor (Enblad *et al.*, 1987; Liao *et al.*, 1989; Natali *et al.*, 1983). A patchwork pattern of antigenic heterogeneity is commonly observed with clusters of antigen-positive cells immediately adjacent to groups of antigen-negative cells (Hand *et al.*, 1983; Mattes *et al.*, 1990). The differences in antigen expression between patients bearing the same tumor are equally well documented. For example, when samples of infiltrating ductal carcinoma from 39 patients were analyzed using four Mabs, 10 different antigenic profiles were obtained, ranging from those that expressed all four antigens to those that expressed none (Hand *et al.*, 1983). Another factor which might contribute to antigen heterogeneity is the nature of the tumor-associated antigen observed. Many are normal constituents of the cell membrane and reflect the normal dynamic mechanism of antigenic expression seen during cell development and differentiation. Kufe and co-workers studied the cell-cycle expression of a membrane-bound antigen in MCF-7 breast cancer cells using flow cytometry. The expression of one antigen was explored through the cell cycle and was at its highest during the S-phase (Kufe *et al.*, 1983). This phenomenon may explain some of the patchwork of antigenic expression observed within tumors when they are mapped using Mabs.

Many tumor-associated antigens are differentiation antigens that are expressed on the cell membrane during cell development and differentiation. Many of the epitopes are derived from the carbohydrate components of membrane-bound glycoproteins and glycolipids, and their aberrant expression in cancer is well established (Feizi, 1985; Hakomori and Kannagi, 1983; King *et al.*, 1989). Nakasaki and co-workers (1989) described as "mosaicism" the expression of tumor-associated carbohydrate antigens in human colonic and gastric tumor samples. Using immunohistochemistry they showed a correlation between the pattern of antigenic expression and the degree of differentiation of the tumor section.

Those antigens associated with maintaining the transformed state are more likely to be homogeneously expressed and retained in any subsequent metastases. Conversely, those antigens expressed on the cell membrane that are related to differentiation may be lost during tumor progression. Another factor to be considered is the selection of antigen-negative populations in response to antigen-targeted therapy (Epenetos *et al.*, 1986b; Esteban *et al.*, 1990). The problem of antigenic heterogeneity in primary tumors between patients and in metastases is one of the limitations encountered in Mab-targeted therapies. It has been proposed that cocktails of Mabs may overcome this problem and increase the sensitivity of this approach (Durrant *et al.*, 1989; Liao *et al.*, 1989; Mangili *et al.*, 1996).

4.2. Antigenic Modulation

The fate of the Mab after it binds to its target is critically important in ADEPT and therefore the appropriate selection of antigenic target and antibody is vital. Antigenic modulation is defined as the redistribution of plasma membrane antigens induced by the binding of antibody. This redistribution results in the loss of antigenic expression on the cell surface either through internalization or shedding of the resultant immune complex. Antigenic modulation is a transient phenomenon because removing the antibody results in the reexpression of the antigen on the cell surface. The susceptibility to modulation depends on the molecule and the epitopes recognized by the antibody (Campbell *et al.*, 1986; Pesando *et al.*, 1986). Therefore, the factors to be considered are the nature of the antigen, its valency, density and interaction with the cytoskeleton, the nature and amount of the ligand, and the type of cell. Antigenic modulation is frequently encountered in studies using hematological and endothelial cells, and is one of the causes of failure of Mab-targeted therapy of hematological malignancies (Schroff *et al.*, 1984). As previously discussed, the rat Mab ICR12 has been successfully used to target the c-erbB-2 receptor in an *in vitro* ADEPT setting (Eccles *et al.*, 1994), whereas other Mabs can induce receptor-mediated endocytosis (De Santes *et al.*, 1992; Maier *et al.*, 1991).

4.3. Membrane Expression and Circulating Antigen

The ideal tumor-associated antigen has been identified as one which is homogeneously expressed at a reasonably high cell-surface density and is stably bound to the plasma membrane. It is unlikely that any of the tumor-associated antigens identified to date fulfill this requirement. Most are expressed heterogeneously, and many are released from the membrane into the circulation, sometimes in the form of membrane vesicles. The effect of circulating antigen on the pharmacokinetics of radiolabeled Mab localization from animal and clinical studies has been reviewed by Pimm (1995). The effects of secreted antigen on radiolabeled Mab localization are usually investigated before clinical evaluation in animal models by using human xenografts in immunodeficient mice. In these studies the formation and clearance of immune complexes significantly reduces tumor uptake of murine Mabs (Pimm, 1995). Clinical trials using radiolabeled Mabs have demonstrated that although some secreted antigen in the serum forms immune complexes with administered murine Mabs, this does not inhibit Mab localization. One possible explanation for the different behavior in animal models and clinical studies is that most clinical data were obtained using xenogeneic Mabs, which is not the case in the murine xenograft models. It has been suggested that the animal model better imitates the next stage in the evolution of Mab therapy, that is, the administration of human, or humanized Mabs to patients to circumvent the generation of human antimouse antibodies (HAMA).

Some studies report reduced Mab localization in patients with excessively high serum antigen concentrations. Boxer *et al.* (1992) studied the factors influencing the localization of an anticarcinoembryonic antigen (CEA) Mab in patients and determined the percentage of administered Mab localized in resected tumor samples. There was a positive relationship between serum CEA concentration and antibody localization in the tumor. Localization was not inhibited at moderate CEA serum levels ([CEA]:[Mab] > 1), but in two patients with very high CEA concentrations both the percentage localized and tumor to blood ratios were reduced. There was no correlation between CEA concentration in the tumor and localization of the Mab, which highlights the multifactorial nature of the localization process. This study also illustrates the problems encountered in targeting with Mabs, such as the heterogeneity of antigenic expression and their inefficient penetration into the tumor, as discussed earlier.

Release of target antigens by antigen-expressing cells can result in the accumulation of antigen in the periphery of the tumor where it is more accessible to circulating conjugate and thus can increase the concentration of conjugate localized to the tumor. In some instances, tumor-associated antigens are used as serum tumor markers to evaluate and monitor the response to therapy. It has been demonstrated that there is a direct correlation between serum levels of tumor-associated glycoprotein 72 and colorectal tumor levels. This important finding allows patients suitable for Mab-targeted therapies, such as ADEPT, to be identified (Guadagni *et al.*, 1996). The reverse is true for CEA, however, where there is no correlation between serum levels and colorectal tumor concentration, as illustrated by the radioimmunodetection of tumors in (CEA) serum-negative patients (Roselli *et al.*, 1996).

The behavior of immune complexes is also influenced by the nature of the antigen. For example the clearance of anti-CEA immune complexes is not straightforward and depends on the location of the tumor. It has been demonstrated that CEA from patients with colorectal tumors clears at a much faster rate than CEA secreted from tumors in other organs (Behr *et al.*, 1996). In addition, some Mabs do not recognize tumor-associated antigen in the serum: this may result from the low affinity of the Mab or structural changes in the released antigen resulting in the loss of an epitope. For example, the anti-CEA Mab PR1A3 does not bind significant amounts of circulating CEA. The conformational epitope recognized by this Mab includes part of the glycosyl–phosphatidyl–inosital linkage and the B3 domain of CEA which, it is suggested, is absent or sterically hindered on secreted CEA (Durbin *et al.*, 1994).

Clinical experience with ADEPT is limited. To date only one target antigen has been evaluated. The pharmacokinetics of conjugate localization and the effect of antigen in the circulation need to be evaluated in Phase I clinical trials for each individual target antigen. The problems posed by the development of human anti-mouse antibodies (HAMA) following the administration of xenogeneic Mab-conjugates are discussed in Chapter 8.

This is not intended as an exhaustive review of all potential tumor-associated antigens but rather an examination of some of the more commonly used antigens, outlining what they are and why/how tumor-specific epitopes can be derived. Some of the antigens described in this chapter are considered "pancarcinoma," that is, the same antigen is expressed by a number of carcinomas originating from a variety of sites. Examples of this category include CEA and the mucin antigens. On the other hand, the more specific tumor markers are related to the differentiation of a particular cell type and are expressed during malignant transformation, for example, prostate-specific membrane antigen (PSM).

Tumor-associated antigens are frequently normal cellular constituents, inappropriately expressed on the cell surface, overexpressed, expressed in a mutated form, or not expressed at all. A critical consideration is the accessibility of the tumor-associated antigen to the Mab (Kennel, 1992; Pervez *et al.*, 1989). For example, an antigen normally expressed in the lumen of the gastrointestinal tract is less likely to be accessible to administered Mabs (Boxer *et al.*, 1994). During malignant transformation, both normal tissue architecture and antigen expression are frequently disrupted and altered. To reject a tumor-associated antigen on the basis of its gross normal expression is unwise and underestimates the power of the rodent immune system or more frequently nowadays, the scientist's ability to generate antibodies against tumor-specific epitopes expressed on the tumor-associated antigen. A good demonstration of this is illustrated by CEA.

4.4. Antigens Potentially Useful for ADEPT Approaches

4.4.1. CARCINOEMBRYONIC ANTIGEN

The presence of CEA in carcinoma of the colon was first described by Gold and Freedman in 1965. They identified it as an oncofoetal antigen on the basis of its expression in the foetal digestive tract and its absence from the normal colon (Gold and Freedman, 1965a,b). Today it is recognized as a tumor-associated antigen that is expressed in normal adult tissue and in a range of human epithelial malignancies. Although CEA is not a tumor-specific antigen, it is a valuable tool in localization and postoperative management, especially for cancers of the gastrointestinal tract, as reflected in the number of commercial assays available. It has been suggested that most tumors of epithelial origin overexpress CEA, and it is used in radioimmunodetection to identify many cancers, including those of the gastrointestinal tract, lung, breast, ovary, prostate, pancreas, and thyroid (medullary type) (Goldenberg *et al.*, 1978, 1980, and 1990). Phase I clinical trials, often using humanized or chimeric Mabs, are being undertaken to evaluate the potential of anti-CEA Mabs for radioimmunotherapy (Sharkey *et al.*, 1995; Wong *et al.*, 1995).

Following the realization that CEA is a tumor-associated antigen, many laboratories began the search for specific anti-CEA Mabs, a search that led to the realization that CEA is the archetype of a large heterogeneous family of structurally related antigens. In 1987 the gene coding for CEA was cloned, which greatly facilitated our understanding of the structure and relationship between the various members of the CEA family and identified them as members of the immunoglobulin superfamily (Oikawa *et al.*, 1987; Paxton *et al.*, 1987). A total of 29 genes have been identified and are located in a densely packed cluster on the long arm of chromosome 19.

At the International Society for Oncodevelopmental Biology and Medicine workshop in 1989, this family of closely related cross-reactive glycoproteins was officially divided into two groups on the basis of their sequence homology: (1) the CEA subgroup (12 members) and (2) the PSA pregnancy-specific subgroup (11 members). Members of the CEA subgroup include multiple species of **N**on-specific **C**ross-reacting **A**ntigen (NCA50/90 NCA 95) and the biliary glycoprotein antigens (BGP). A third subgroup of six CEA-related genes has since been identified, CGM13–18 (**C**EA **G**ene-family **M**ember), but these are probably only pseudogenes (Teglund *et al.*, 1994). This review focuses on CEA and its role as a tumor-associated antigen.

A detailed ultrastructural study of the expression of CEA in normal and cancerous colon mucosa was carried out by Baranov *et al.* (1994). This study showed that in the normal adult colon a high density of CEA is in the form of microvesicles and filaments that make up the "fuzzy coat" of the apical surface of the epithelial cells at the free luminal surface. CEA is released from mature columnar cells as vesicles derived from the microvillous membrane. The expression of CEA is related to cellular differentiation. An increasing thickening of the surface layer of CEA is produced as the immature cells migrate up the crypts to the free luminal surface. CEA was not detected at the basolateral membranes of mature columnar cells or crypt cells. Thus, in normal colon mucosa CEA is present in considerable concentrations, but its expression is highly polarized toward the free luminal surface. In the moderately differentiated adenocarcinomas used in this study, the normal cellular architecture was typically disrupted by multilayered glandular structures and "intracellular lumina" in some of the neoplastic cells. CEA expression was heterogeneous and dependent on the location of the cell within the gland. Cells in a more normal environment displayed primarily apical membrane expression with some abnormal expression. On isolated cells with few or no intercellular junctions located in the center of the glandular structure, the antigen is expressed on most of the cell-surface membrane. Isolated pockets of CEA were observed in the intracellular spaces and in the lumen of tumor glands. Other immunohistochemical studies support these observations. Hamada *et al.* (1985) observed the loss of CEA polarization in moderately or poorly differentiated tumors and found that increased stromal distribution of CEA is accompanied by increased plasma CEA levels.

The overexpression of CEA on neoplastic transformation and in a number of benign conditions, such as inflammatory disease of the bowel and cirrhosis of the liver, is well documented. Extraction of CEA from colonic mucosa with perchloric acid showed a 35-fold increase in CEA expression in primary large bowel carcinomas and 230-fold increase in metastatic colon or rectum carcinomas, compared with normal colon mucosa concentrations (Fritsche and Mach, 1977). Quantitative analysis of CEA expression in colonic adenocarcinomas using Northern blot mRNA analysis and a commercial immunoassay for the protein product failed to demonstrate a correlation between mRNA levels and CEA content (Boucher *et al.*, 1989). In some cases the mRNA levels were equivalent to samples from adjacent normal tissue and in others sixfold to tenfold higher in the tumor samples. On average, the tumor to normal ratios of CEA were 65 times higher, but ranged between twofold and 113-fold higher. In addition, CEA expression was inversely correlated to the level of methylation in the 5′ region of the gene. These results confirm that the factors governing the regulation of CEA are complex and are probably controlled both transcriptionally and posttranscriptionally. Undoubtedly some of the differences in CEA concentration are related to the aberrant anatomical expression of CEA between normal and cancerous mucosa.

CEA is a cell-membrane glycoprotein which is generally described as having molecular weight of 180 kDa. The apoprotein is synthesized as a 60–80 kDa precursor which is then extensively glycosylated, so that it contains approximately 50–60% carbohydrate by weight. Twenty-eight putative N-glycosylation sites have been identified on CEA (Oikawa *et al.*, 1987). With 50–60% of the molecular mass composed of carbohydrate and only 28 N-glycosylation sites, the carbohydrate side chains must consist, on average, of 20 monosaccharide residues, resulting in an complex, extensively branched structure (Chandrasekaran *et al.*, 1983; Yamashita *et al.*, 1987). The other members of the subfamily contain 20–50% carbohydrate by weight. CEA and its related antigens belong to the immunoglobulin superfamily and consist of one immunoglobulin variable-like domain (IgV) and a varying number (zero, one, two, three or six) of immunoglobulin constant-like domains (IgC) of subtype A or B.

Structurally, CEA consists of a processed leader sequence of 34 amino acids followed by an N-terminal domain of 108 amino acids, which lacks the disulfide bonds characteristic of the immunoglobulin IgV domain but nevertheless resembles it. This is followed by six Ig C2-type constant region domain motifs, which in CEA consist of three highly conserved (67% homology) repeat domains of 178 amino acids (of which 1–3 are divided into A and B subdomains) each containing four conserved cysteine residues. The hydrophobic 26 amino acid C-terminal domain is processed to give a glycosyl phosphatidylinositol (GPI) type of linkage to the cell membrane. A schematic representation of CEA and the closely related NCA is depicted in Figure 5.

CEA has been subject to intensive immunological mapping with Mabs which have revealed the extensive heterogeneity of this glycoprotein (Hammarstrom *et al.*, 1989; Harwood *et al.*, 1986; Kuroki *et al.*, 1992; Muraro *et al.*, 1985). Although the epitopes recognized were derived from both polypeptide and carbohydrate moieties, most antibodies are directed against protein epitopes. The repeat domain structure of CEA may help to increase the amount of Mab localized by providing more than one epitope per CEA molecule (Hammarstrom *et al.*, 1989). The observed heterogeneity presumably results from failures or differences in protein and carbohydrate processing, which allows the generation of tumor-selective anti-CEA antibodies (Rogers, 1976; Seko *et al.*, 1996). Heterogeneity in the structure of CEA also identified depends on the location of the tumor (Behr *et al.*, 1996; Hernando *et al.*, 1994; Kuroki *et al.*, 1984).

In cancer the polarized expression of CEA is lost and there are both quantitative and qualitative differences in the expression of this tumor-associated antigen. The high concentrations and nature of CEA in normal colonic mucosa, coupled with its ability to bind *Escherichia coli* and *Salmonella* type I fimbriae, suggest

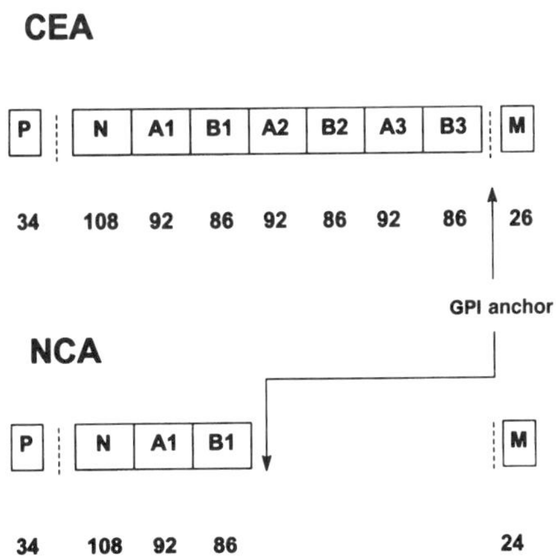

Figure 5. Structure of CEA and NCA based on deduced amino acid sequences. P = processed leader peptide, N = N-terminal (immunoglobulin IgV-like) domain, A/B = immunoglobulin Ig C2 subdomains (1–3), M = C-terminal hydrophobic domain replaced by glycosyl phosphatidylinositol anchor. The number of amino acids in each is shown below.

that it may have a protective role in recognizing and eliminating microorganisms. CEA mediates the Ca^{2+} independent aggregation of human adenocarcinoma cells and transfected rodent cells in culture, suggesting that it has a role as an intercellular adhesion molecule (Benchimol *et al.*, 1989; Oikawa *et al.*, 1989). The presence of CEA has been identified on cultured human umbilical vein endothelial cells, on the endothelial cell line Ea.hy926, and on tissue sections of human lymph nodes. It was demonstrated that CEA-positive, but not negative, breast carcinoma cells bind purified CEA. On the basis of this evidence it was proposed that CEA may play an important role in tumor cell extravasation and metastasis (Majuri *et al.*, 1994). There is no evidence from either immunohistochemical or radioimmunodetection studies of anti-CEA Mabs binding to normal or tumor endothelium. It is proposed that other members of the CEA family mediate intercellular adhesion. For example, NCA is up-regulated at the transcriptional level in a range of tumors, whereas expression of the *BGP* gene is known to be down-regulated in colorectal carcinomas.

CEA is the archetypal tumor-associated antigen and, despite its normal counterpart and the presence of so many closely related antigens, it has proved possible to generate tumor-specific Mabs. Polymorphic epithelial mucin is another highly glycosylated protein which shares many characteristics with CEA.

4.4.2. POLYMORPHIC EPITHELIAL MUCIN

Polymorphic epithelial mucin (PEM) is also known as episialin or MAM6, PUM, EMA-epithelial membrane antigen, NPG, and DF-3 antigen. Like CEA, PEM has a domain structure, is heavily glycosylated, and has a polarized luminal expression. The gene MUC-1 encodes the highly polymorphic mucin glycoprotein from breast (Gendler *et al.*, 1990; Ligtenberg *et al.*, 1990) and pancreas (Lan *et al.*, 1990). PEM is normally present on the apical surface of ductal epithelial cells during embryonic development of coelomic epithelium, and is present in adult structures derived from it. It is also present on other luminal surfaces, such as the mesothelium and in the lymph nodes on the high endothelial cells of the post-capillary venules (Zotter *et al.*, 1988). The extended extracellular domain creates a highly hydrated filamentous structure which extends into the luminal cavity. The physical properties of the molecule are dictated by the carbohydrate components. The precise biological function of the membrane-bound mucin glycoprotein (M_r > 400 kDa) encoded by MUC-1 is unclear. In adult tissues the genuine secreted mucins are the major component of mucus produced by specialized epithelial cells in the respiratory, gastrointestinal, and genitourinary tracts.

The mucins have a common structural motif consisting of a central region in which approximately two thirds of the protein is made up of tandem repeats of amino acids, particularly serine and threonine and, to a lesser extent, alanine, glycine, and proline (these five constitute 50% of the total amino acid content). This central domain is flanked by nonrepetitive domains with a more normal amino acid content, of which one constitutes a transmembrane domain and the other a 69 amino acid cytoplasmic amino terminus. Neither contains O-glycosylation sites, although there are some N-linked sites. The repeat units act as a scaffold, allowing the attachment of extensive O-glycosidic-linked oligosaccharides, resulting in the formation of an extended molecule. The apomucin core peptide serine and threonine residues are O-linked via N-acetylgalactosamine to extended polylactosamine side chains which terminate in sialic acid or fucose residues. The carbohydrate content is generally more than 50%. Differences in glycosylation are seen between histologic types because of differences in glycosyltransferase expression.

The biosynthesis of PEM has been studied in great detail in breast cancer cells (Hilkens and Buijs, 1988; Ligtenberg *et al.*, 1992). The apoprotein core is synthesized, N-glycosylated, and then cleaved just inside the extracellular domain. The two moieties form a stable complex. The mature product is not synthesized directly, but as a partially sialyated, O-glycosylated precursor which undergoes constitutive internalization and recycling to maintain the fully sialyated structure. The mature molecule is released from the cells into the culture medium. Human epithelia possess two MUC-1 genes which are codominantly expressed. Protein polymorphism arises from allelic variation in the tandem repeat area of the coding sequence (Swallow *et al.*, 1987). In breast tissue each repeat consists of 60 bp leading to a 20 amino acid sequence which is highly conserved. The number of repeats within the tandem repeat area varies between 21 and 125 in the Northern European population (Gendler *et al.*, 1990).

As described above for CEA, during the transformation to a malignant phenotype, the normal architecture of the tissue is lost, mucin accumulates, and the cell looses its polarity of expression. PEM is expressed in tumors of epithelial origin, in breast, pancreas, colon, ovarian, and lung adenocarcinomas. In breast cancer the increase in mucin relative to adjacent normal tissue was estimated to be 10-fold by RNA level on dot blots, *in situ* hybridization experiments and by immunohistochemistry (Zaretsky *et al.*, 1990). It may be that mucin is incompletely glycosylated and allows the recognition of otherwise cryptic epitopes that serve as specific tumor antigens because of the increased levels of mucin produced (Burchell *et al.*, 1987). During lactation the human milk fat globule (HMFG) membrane is produced in the breast at the apical surface of the breast epithelium. HMFG has been used as a source of breast epithelial mucin for generating polyclonal antibodies and Mabs. Chemically deglycosylated mucin has been used to mimic the aberrant glycosylation seen in the tumor. This technique was used to generate the Mab SM-3, which reacts with primary carcinomas of the breast, lung, colon and ovary, but with very little response in normal tissue, as shown by immunohistochemical analysis (Burchell *et al.*, 1987). The epitope was mapped to part of the polypeptide tandem repeat region, thus providing the opportunity for more than one Mab to bind per molecule of antigen.

4.4.3. TUMOR-ASSOCIATED GLYCOPROTEIN 72 (TAG-72) ANTIGEN

TAG-72 is a mucin-specific tumor antigen (CA 72) first identified using the Mab B72.3, which has been widely used or evaluated in the management of a range of human carcinomas. Using immunohistochemical, immunocytochemical, and radioimmunoassay techniques, it has been shown that Mab 72.3 reacts with more than 90% of colorectal, gastric, and ovarian carcinomas and 70% of breast carcinomas (Nuti *et al.*, 1982; Stramignoni *et al.*, 1983; Thor *et al.*, 1986). The epitope recognized by B72.3 has been identified as sialosyl-Tn (NeuAcα2 → 6GalNAcα1 → O-Ser/Thr) (where NeuAc = N-acetyl neuraminic acid, GalNAc = N-acetyl galactosamine, and Ser/Thr = serine or threonine residue on the polypeptide chain) (Kjeldsen *et al.*, 1988). TAG-72 has been immunologically mapped with a range of Mabs (Kuroki *et al.*, 1990). The core structure of normal simple O-linked (mucin type) carbohydrate chains α-glycosidically linked to either serine or threonine in the polypeptide chain are primarily those of the T (Galβ1 → 3GalNAcα → Ser/Thr) structure and Tn (GalNAcα1 → Ser/Thr) type. It has been shown in normal human colonic mucosa that the sialosyl Tn antigen is constitutively expressed and O-acetylated, whereas the O-acetylated sialosyl Tn structure was not detected in gastric and pancreatic tissues (Ogata *et al.*, 1995). Both Tn and sialosyl Tn are expressed uniquely at the cell membrane in carcinoma cells, compared to their intracellular localization in normal cells.

In those cancers where mucin is aberrantly expressed the T,Tn antigens are widely recognized as tumor-associated antigens, exposed because of incomplete synthesis leading to the accumulation of precursors rarely expressed in normal adult cells (Hirohashi *et al.*, 1985; Kjeldsen *et al.*, 1988; Springer, 1984). Why expression of the sialosyl Tn antigen increases in many carcinomas is unclear. It may result from this incomplete glycosylation, or it may be caused by premature α2 → 6 sialyation of the Tn structure, or in colon cancer, because of decreased O-acetylation.

The original B72.3 Mab was not reactive with normal adult tissues, inflammatory, or benign lesions. Overexpression of TAG-72 has been identified only in transitional colonic mucosa (normal-appearing mucosa adjacent to carcinoma) (Xu *et al.*, 1989) and secretory phase endometrium (Thor *et al.*, 1987). The sialosyl-Tn antigen has been detected only as a weak immunohistochemical reaction in Leydig cells of the testis, goblet cells of the colon, and parietal cells of the stomach (Kjeldsen *et al.*, 1988). A positive correlation has been observed between the expression of sialosyl Tn and the progression in colorectal carcinoma and gastric cancer, but not in cervical cancer.

4.4.4. EPIDERMAL GROWTH FACTOR RECEPTOR FAMILY

The two antigens discussed previously are well established, tumor-associated antigens. The next group of potential antigens to be discussed, the members of the epidermal growth factor receptor family, were more recently described and on initial evaluation appeared to be overexpressed in a much lower percentage of tumors. The study of oncogenic viruses led to the discovery that transforming retroviruses contain oncogenes homologous with, and/or derived from, cellular genes (proto-oncogenes). It is proposed that a major factor in the development of human cancers is the result of the increased expression of certain proto-oncogenes and the loss of function of tumor-suppressor genes and their protein products. Proto-oncogenes are activated to transforming oncogenes by a variety of mechanisms including point mutations, deletion mutations, amplification, chromosomal recombination, and aberrant gene expression. Proto-oncogene products include growth factors and growth inhibitory factors which control a wide range of key physiological processes, including cellular proliferation in normal cells. The binding of growth factors to specific receptors on the cell surface initiates an intracellular biochemical cascade leading eventually to DNA synthesis and cell division. It is proposed that the increased expression of cellular growth factors by the tumor itself (autocrine), cells at a distance (paracrine) and adjacent cells (juxtacrine), and their receptors results in feedback loops further stimulating tumor growth. The regulation of cellular proliferation can be subverted at several distinct levels. However, it is not known exactly how the aberrant expression of growth factor receptors give the tumor a growth advantage.

The Class 1 receptor tyrosine kinase subfamily, as proposed by Ullrich and Schlessinger (1990), consists of the epidermal growth factor receptor (EGFR), erbB-2, a receptor whose ligand is unknown, and erbB-3 (low-affinity) and erbB-4 (high-affinity) receptors which bind the heregulin family of growth factors. The EGFR family members show significant sequence homology and identical gross structural organization and there is evidence for heterodimer formation among the family members. The terminology for many oncogenes is complicated because of the history of their initial identification and recognition. Members of the epidermal growth factor receptor family are known as c-erbB (or ERBB) which reflects their homology to the product of v-*erb*B gene, the acutely transforming avian erythroblastosis viral gene, which encodes a truncated form of epidermal growth factor receptor. The term neu was adopted when they were recognized as the human counterparts of the rat *neu* oncogene first identified in ethylnitrosourea-induced adrenal **neu**roglioblastomas in neonatal rats. The third terminology HER derives from their homology with the **H**uman **E**GF **R**eceptor. The terms erbB, ERBB, and HER/neu are used interchangeably, but throughout this chapter we use the term EGFR for the product of the epidermal growth factor receptor gene and c-erbB terminology for the protein products of the other members of the family.

The expression of these receptors in human cancers has been studied extensively by a variety of techniques, and a wide range of values are reported in the literature. The nature of the target makes quantitation difficult. For example, the use of a radiolabeled ligand may underestimate the amount of EGF receptor present if the receptors in the tumor specimen already contain endogenous ligand and are internalized. Decreases in levels of detectable c-erbB-2 expression in formalin-fixed, paraffin-embedded sections compared to frozen tissue sections have been well documented (Press *et al.*, 1990; Slamon *et al.*, 1989). The specificity and sensitivity of some of the reagents used for immunohistochemical analysis of c-erbB-2 expression were studied by Press *et al.* (1994). Comparing 21 Mabs and seven polyclonal antisera using a multitumor paraffin-embedded tissue block, they showed that the relative sensitivity of these reagents varies between 6% and 82%. Some false positive responses were seen (relative specificity of antibodies were 92–99%). The lack of standardization in methodology and the subjective nature of immunohistochemical interpretation undoubtedly contributes to the wide range of target antigen values reported in the literature. Quantitative assays have been developed for detecting c-erbB-2 and EGFR in tissue samples (Press *et al.*, 1993; Robertson *et al.*, 1996).

4.4.4a. erbB-2

The *neu* oncogene was first identified during transfection experiments which showed that DNA derived from ethylnitrosourea-induced neuroglioblastomas induces malignant transformation of NIH3T3 cells (Shih *et al.*, 1981). The product of this gene was subsequently identified as a 185 kDa transmembrane phosphoglycoprotein which is homologous to the epidermal growth factor receptor (EGFR) (Schechter *et al.*, 1984). The human homologue of the rat *neu* gene was independently identified by three groups (Coussens *et al.*, 1985; King *et al.*, 1985; Semba *et al.*, 1985). A number of different mechanisms are responsible for the overexpression of c-erbB-2 in human cancers, including transcriptional deregulation, gene amplification, and possibly decreased receptor turnover, rather than a point mutation as seen in the rat (Kameda *et al.*, 1990; Press *et al.*, 1993; Robertson *et al.*, 1996).

The extracellular domain of c-erbB-2 contains two cysteine-rich areas and is 44% homologous to EGFR, and the intracellular domain contains a tyrosine kinase which is 82% homologous to EGFR (Yamamoto *et al.*, 1986). The native ligand for this putative receptor has not been identified. In normal adult tissues, c-erbB-2 is expressed on the membranes of epithelial cells throughout the gastrointestinal, respiratory, urinary, and reproductive tracts, and in the skin, breast, and placenta. The levels of expression were identified immunohistochemically and in all cases ranged from barely detectable to, in a few samples, moderate intensity (Press *et al.*, 1990). Expression of c-erbB-2 is typically observed on the basal and lateral membranes of epithelial cells, rather than the luminal membrane, like CEA and PEM. There was no evidence of gene amplification or rearrangement in normal tissue.

The widespread, albeit at low level, expression in normal tissue dictates that tumor specificity depends on the overexpression of the antigen. Many of the earlier studies reported in the literature are flawed by poor methodology. One difficulty encountered when using direct blotting techniques (i.e., Western, Northern, Southern blotting) to detect protein, RNA, and DNA is contamination of tumor specimens with normal cells. This is a particular problem in breast cancer where normal stromal, vascular cells, etc. can make up more than 50% of the tissue. Western blotting in particular suffers from the "diluting out effect" of normal tissue and, as previously discussed, tissue preparation for immunohistochemistry has a profound impact on antigen detection. The relationship between gene and protein expression in breast and ovarian tumors was analyzed using Western, Northern, Southern blotting and immunohistochemistry (Slamon *et al.*, 1989). Of a total of 187 breast samples analyzed, 27% showed amplification of the gene and correspondingly higher protein levels. In 10% of the samples there was increased expression of RNA and protein levels, but no evidence of gene amplification. A similar result was obtained from the primary ovarian tumor samples.

The expression of c-erbB-2 in breast cancer tissues has been studied using a rabbit polyclonal antiserum and quantified by using computerized image analysis (Press *et al.*, 1993). In breast cancer cells where there was neither amplification of c-*erb*B-2 nor overexpression, the concentration of c-erbB-2 was 0.18 $\pm$ 0.2 pg/cell (37 = number of samples); where it was overexpressed, but not amplified, 2.01 $\pm$ 0.73 pg/cell (8); where there was twofold to fivefold amplification, 2.60 $\pm$ 0.47 pg/cell (10 samples) and where there was >fivefold amplification, 3.38 $\pm$ 0.29 pg/cell (20 samples). Thus overexpression and amplification of the c-*erb*B-2 oncogene results in increased tumor expression of between 11- and 19-fold. More recently the expression of c-erbB-2 in breast cancer was studied using a radiolabeled rat anti-erbB-2 MAb (ICR12), and the concentration of receptor was quantified by radioimmunohistochemistry (Robertson *et al.*, 1996). Agreeing with earlier studies, the existence of two subpopulations were identified, but a much higher total number of tumors (91%; 81 samples) overexpressed c-erbB-2 compared to the normal breast tissue sample used. The majority of the tumors (79%) expressed c-erbB-2 between 0.33 and 19-fold (4.1-fold average), with no evidence of gene amplification. In 21% of the tumor samples the protein was overexpressed between 45- to 479-fold (187-fold average) and this correlated with gene amplification, which was detected by fluorescent *in situ* hybridization.

Thus, despite its limitations, immunohistochemistry could be used to determine which patients would be suitable for ADEPT, whereas quantitating the expression of the c-*erb*B-2 gene *per se* would underestimate the number of potential patients. One of the advantages of using c-erbB-2 as a target is its generally homogenous distribution throughout the tumor and its continued existence in metastases and recurrent lesions (Berchuck *et al.*, 1990; Niehans *et al.*, 1993).

4.4.4b. Use of c-erbB-2 Antigen in ADEPT

Although overexpression of c-erbB-2 has not been used in a clinical ADEPT setting, Eccles and co-workers (1994) used a human breast carcinoma cell line, in which c-erbB-2 is expressed at levels 17-fold higher than normal, to demonstrate the feasibility of using this antigen as the target for an ADEPT system. The bacterial enzyme CPG_2 was chemically conjugated to the rat Mab ICR12. Because of the stable binding of this Mab, the antibody-enzyme conjugate did not need to be cleared by the use of a second Mab before administering the prodrug, which was administered 14 days after the antibody-enzyme conjugate. A sustained dose-dependent tumor response was seen against large (6–7 mm) tumors, and at the highest prodrug concentration there was no evidence of regrowth at 90 days postprodrug administration. The Mab used in this model system displays many of the essential characteristics required for ADEPT. It recognizes a protein epitope on the external domain of c-erbB-2, to which it binds with high affinity (0.2 nm) with no significant cross-reactivity to normal tissues or sera. Binding of the Mab does not induce antigenic modulation *in vitro* or *in vivo*, either by shedding of complex or by internalization (Allan *et al.*, 1993; Dean *et al.*, 1993).

Overexpression of this oncogene product has also been identified in a number of other cancers, including ovarian cancer (Berchuck *et al.*, 1990; Slamon *et al.*, 1989), gastric carcinomas (Houldsworth *et al.*, 1990; Kameda *et al.*, 1990), lung adenocarcinomas (Rachwal *et al.*, 1995), bladder cancer (Rajkumar *et al.*, 1996), and thyroid tumors (Aasland *et al.*, 1988).

4.4.4c. Epithelial Growth Factor Receptor

The receptor for epidermal growth factor (EGFR) is a 170 kDa transmembrane protein that is expressed on a wide range of tissue, including normal epithelial cells in the digestive, urogenital, and respiratory tracts. The EGF receptor is overexpressed in tumors arising from these tissues and include bladder

(Rajkumar *et al.*, 1996), lung (Tateishi *et al.*, 1990), breast (Ro *et al.*, 1988), pancreas (Korc *et al.*, 1992), and brain tumors (Libermann *et al.*, 1984; Wong *et al.*, 1987).

Amplification of the EGFR gene is frequently observed in glioblastoma multiforme, the common and most malignant of the primary brain tumors. In one study there was evidence of gene amplification in six out of eight glioma biopsy samples, and five out of six samples showed evidence of gene rearrangement resulting in the expression of lower molecular weight mutant EGF receptors (120–145 kDa) (Humphrey *et al.*, 1988). A number of different truncation mutants have been isolated that differ in terms of the size and location of the deletion.

A mutated form of EGFR produced by a 267 amino acid deletion in the extracellular domain created a specific truncation mutant designated EGFRvIII (EGF receptor variant III). This 145 kDa receptor contains an additional glycine residue at the fusion point of the deleted region. A 14 amino acid synthetic peptide corresponding to this fusion region was conjugated to keyhole limpet hemocyanin and used to generate highly specific polyvalent sera (Humphrey *et al.*, 1990): the production of Mabs specific for EGFRvIII was technically more difficult (Wikstrand *et al.*, 1995). Using these Mabs, the tumor-specific nature of the mutant receptor was confirmed. EGFRvIII was not identified in a range of normal tissue. There was an increased incidence of EGFRvIII immunohistochemically positive glioma biopsies 52% (16/31), compared to the earlier evaluation 22% (6/27), and it was identified in 27–73% (3/11 by immunohistochemistry, 8/11 by mRNA analysis) breast carcinoma specimens. This particular mutant has also been identified in 16% (5/32) of non-small cell lung cancer samples examined in one study (Garcia de Palazzo *et al.*, 1993).

These anti-EGFRvIII Mabs are currently being evaluated for their potential use in Mab-targeted therapy. Binding of the Mab to EGFRvIII-expressing cells *in vitro* and *in vivo* leads to internalization of the complex (Reist *et al.*, 1995). Thus mutated proteins, such as EGFRvIII, provide a truly tumor-specific antigen that is most likely to be found in those tumors in which EGFR is overexpressed. The limiting factor may be its relatively low incidence and, for ADEPT, the necessity of producing a Mab that is not internalized.

4.4.4d. c-erbB-3

The third member of the type 1 receptor family, namely, c-*erb*B-3 was cloned by two groups (Kraus *et al.*, 1989; Plowman *et al.*, 1990). The gene encodes a 160 kDa transmembrane protein which is very highly homologous to EGFR and c-erbB-2, particularly in the tyrosine kinase domain. The expression of c-erbB-3 is distinct from that described for EGFR or c-erbB-2 , and it is expressed in a wide variety of normal tissues, including the digestive, urogenital, and respiratory tracts. It was also observed in endocrine tissues, particularly the adrenal cortex and the central nervous system (Prigent *et al.*, 1992). Both c-erbB-3 and c-erbB-4 have been localized to neuromuscular synapses in skeletal muscle (Zhu *et al.*, 1995). In these normal tissues the expression of c-erbB-3 was predominantly cytoplasmic, with no membrane accentuation.

In breast carcinoma specimens c-erbB-3 was heterogeneously expressed in 15 and 35% of the samples evaluated, the staining was predominantly cytoplasmic, and there was no evidence of gene amplification (Lemoine *et al.*, 1992a; Travis *et al.*, 1996). A similar pattern of expression was seen in ovarian tumors (Simpson *et al.*, 1995). Immunohistochemical staining of other tumors, such as thyroid, pancreatic, and gastrointestinal tumors revealed a higher level of membrane accentuation (Haugen *et al.*, 1996; Lemoine *et al.*, 1992b; Rajkumar *et al.*, 1993; Sanidas *et al.*, 1993).

The potential of c-erbB-3 to act as a target antigen in Mab therapies has yet to be addressed. The lack of membrane expression may reduce the concentration of available antigen, although in SKBR-3 breast cancer cells only 19% of the total c-erbB-2 was located at the cell surface (Kumar *et al.*, 1991). It is anticipated that c-erbB-2 will be the most appropriate target for ADEPT. However both the EGF receptor and c-erbB-3 may also be utilized. Tumor-specific type I growth factor receptors, such as EGFRvIII, would make excellent targets but only for a limited patient population.

4.4.5. HUMAN CHORIONIC GONADOTROPHIN

Human chorionic gonadotrophin (hCG) is a member of the glycoprotein hormone family that includes the pituitary hormones luteinizing hormone (hLH), follicle-stimulating hormone, and thyroid-stimulating hormone. It is a placental hormone normally produced in the placenta during early pregnancy. It consists of two noncovalently linked subunits (α and β). The α-subunit is common to the other hormones and is encoded by a single gene on chromosome 18. The β-subunit is unique and confers biological specificity on each of the hormones. In the placenta at least five, and possibly all six, of the genes located on chromosome 19 are active, although three are expressed preferentially *in vivo* (Bo and Boime, 1992). The single gene that encodes the β-subunit of hLH is linked to the hCGβ gene cluster.

This glycoprotein has a molecular mass of 38 ± 7 kDa, and it is heavily glycosylated (30% by weight). Both subunits contain N-linked oligosaccharides. The carboxyl terminal of the β-subunit is rich in proline and serine residues which are O-glycosylated with a minimum of 12 molecules of sialic acid (N-acetylneuraminic acid) per molecule of protein. The subunits are highly cross-linked by internal disulfide bridges, of which five are in the α-subunit and six are in the β-subunit (Pierce and Parsons, 1981).

In addition to its normal role during pregnancy, hCG and the individual subunits have been detected in normal healthy individuals in serum and other body fluids, and the β-subunits of hCG and hLH are produced in normal testis (Alfthan *et al.*, 1992; Berger *et al.*, 1994; Marcillac *et al.*, 1992). Early studies to evaluate the distribution of hCG, especially those using polyclonal antibodies, could not always distinguish between the closely related hormones and intact hormone and their subunits. In addition, it has been demonstrated that small amounts of α-hCG and intact hCG are synthesized by the pituitary gland. The expression of hCG and its subunits upon malignant transformation may be caused by increased expression in tissue, where it is normally present (eutopic expression), or aberrant expression in tissues where it is not usually expressed (ectopic expression). This glycoprotein is eutopically overexpressed in tumors arising from gestational trophoblastic cells in the placenta, namely hydatidiform mole, invasive mole, and choriocarcinoma. It is also an important tumor marker for diagnosing and managing patients with testicular cancer (Madersbacher *et al.*, 1994). Ectopic overexpression of the α- and β-subunit and its breakdown product have been detected in the serum and urine of patients with a wide variety of nontrophoblastic tumors, including bladder (47%), pancreatic (32%), and cervical (30%) tumors (Marcillac *et al.*, 1992).

Acevedo and colleagues have reported the presence of membrane-associated hCG, its subunits, and its fragments on cultured cancer cells and biopsy samples detected by quantitative analytical flow cytometry and mRNA analysis (Acevedo *et al.*, 1992, 1993, 1995; Krichevsky *et al.*, 1995). A total of 85 human cancer lines of different types and origins, including carcimomas, sarcomas, and lymphomas and 19 biopsy samples, were analyzed. The expression of these glycoproteins was also detected in cultured embryonic and foetal cells but not in normal cell populations. The β-subunit of hLH was identified together with the β-subunit of hCG which suggests that expression results from nonspecific activation of the β-subunit gene locus (Acevedo *et al.*, 1992). It was proposed that the synthesis and expression of hCG, its subunits, and its fragments is a common biochemical denominator of cancer. Other groups have identified the presence of the β-subunit in the culture media of a variety of cell types, usually of mucosal origin, and in cultured normal urothelium (Iles *et al.*, 1990). One of the original *in vitro* models for ADEPT was based on the CC3 human choriocarcinoma xenograft implanted in nude mice that actively secretes hCG (Bagshawe *et al.*, 1988). Despite high levels of circulating antigen, administration of the anti-hCG-carboxypeptidase G_2 conjugate resulted in significant differential retention at the tumor site and eradication of the established xenograft after one cycle of therapy (Melton *et al.*, 1990; Springer et al., 1991). In the future it is possible that hCG and the β-subunit in particular may provide an excellent target for ADEPT.

4.4.6. ALKALINE PHOSPHATASE

Human placental alkaline phosphatase (E.C. 3.1.3.1) (PLAP), like hCG, is produced in the placenta during pregnancy and at low levels in parts of the reproductive tract, thymus, and lung (McLaughlin *et al.*, 1984). PLAP is a dimer consisting of two identical subunits (65 kDa) bound to the plasma membrane by a glycosyl–phosphatidylinositol linkage, where its proposed primary role is the transportation of maternal IgG through the placenta. Three closely related tissue-specific alkaline phosphatases have 90% sequence identity, although their expression is not limited to these tissues. There is also a tissue nonspecific alkaline phosphatase which is ubiquitously expressed and has only 50% similarity to the tissue-specific enzymes. Germ cell alkaline phosphatase (GLAP) which is found at low levels in healthy testis and cervix has 98% sequence homology to PLAP. The genes encoding the three tissue-specific enzymes located on chromosome 2 and both PLAP and GLAP display a high degree of allelic polymorphism.

The terminology and abbreviations used to describe the various alkaline phosphatases is somewhat confusing. For example, some authors refer to placental alkaline phosphatase as PLAP and the germ cell alkaline phosphatase as PLAP-like, whereas others use PAP for placental alkaline phosphatase and PLAP for placental-like alkaline phosphatase, that is, GLAP! Initial studies with polyclonal antibodies could not distinguish between the closely related isoforms, and the sensitivity of the assays was quite variable. PLAP/GLAP-specific Mabs have been generated which do not cross-react with tissue nonspecific alkaline phosphatase, but because of the high degree of sequence homology, many Mabs cannot distinguish between PLAP and GLAP.

The ectopic production of PLAP was first observed in the serum of a patient with squamous cell carcinoma of the lung and has been used for radioimmunodetection, particularly for malignancies of the reproductive tract (Epenetos *et al.*, 1986a; Kalofonos *et al.*, 1994). Testicular germ cell tumors are subdivided clinically into two major groups, seminomas and nonseminomatous germ cell tumors. GLAP is eutopically expressed at high levels in classic testicular seminomas (25/25) (Dekker *et al.*, 1992) and (7/7) (with an antibody which does not distinguish PLAP and GLAP) and in malignant teratomas (9/9) (Epenetos *et al.*, 1984). The eutopic overexpression of GLAP in testicular seminomas is on average 80-fold higher than normal. PLAP/GLAP is detected in ovarian malignant tumors (10/11) (Nouwen *et al.*, 1985) and in primary breast cancers (7/7) (McDicken *et al.*, 1983), although not always at such a high incidence. It has also been observed in endometrial (1/1), colorectal (4/35), and gastric (25/107) carcinomas, (Epenetos *et al.*, 1984; Watanabe *et al.*, 1990). Tumors often coexpress more than one of the isoenzymes, such as PLAP and GLAP, and heterodimers of these two enzymes have been detected.

In addition PLAP/GLAP has been detected in the culture medium of a wide range of cell lines, such as bladder (8/17), cervical carcinoma (5/11), choriocarcinoma (3/3), colorectal (2/2), testicular germ cell (5/8), epithelial ovarian (1/5), oral and respiratory carcinomas, (7/8), and vulval carcinoma (1/1) (Iles *et al.*, 1994). The distribution of PLAP on cultured cancer cells and normal placental syncytial cells has been investigated with polyclonal antibodies and gold labeling by immunoelectron microscopy. PLAP was expressed in clusters throughout the membrane surface with a similar distribution pattern on both normal and cancer cells (Jemmerson *et al.*, 1985).

The ectopic and eutopic expression of PLAP/GLAP suggests that these oncofetal antigens may be suitable target antigens for ADEPT. As suggested by the cell culture studies, it is possible that, like hCG and lCG, PLAP/GLAP may be aberrantly expressed in many cancers, but either the reagents used to detect it do not detect all the polymorphic forms, or it is expressed in a mutated form.

4.4.7. PROSTATE-SPECIFIC MEMBRANE ANTIGEN

Prostate-specific membrane antigen (PSM) is one of the newer tumor-associated antigens. A membrane preparation of the human prostate cancer cell line LNCaP, derived from metastatic lymph node tissue, was used to immunize mice for Mab production (Horoszewicz *et al.*, 1987). The antigen has been characterized as a 100 kDa type II membrane protein, with a predicted 84 kDa apoprotein backbone. Two forms of prostate-specific membrane antigen have been identified, which arise from differential mRNA splicing, and the gene is located on chromosome 11 (Leek *et al.*, 1995; Su *et al.*, 1995). The alternatively spliced variant encodes a predicted 78 kDa protein, designated PSM′, that lacks 57 amino acids in the N-terminal domain. The current Mabs raised against PSM do not recognize PSM′, and so RNase protection assays were used to study the expression of these two variants *in vitro*. These experiments suggest that PSM′ is a cytosolic protein in normal prostate tissue, whereas PSM is expressed in a transmembrane form upon malignant transformation. The precise function of PSM is unknown, although *in vitro* it acts as a neuropeptidase, and there is 54% partial similarity to the human transferrin receptor (Carter *et al.*, 1996; Israeli *et al.*, 1993). Radiolabeled anti-PSM Mabs have been used in imaging prostatic cancer (Kahn *et al.*, 1994; Mardirossian *et al.*, 1996).

Using immunohistochemistry, the expression of PSM has been studied in a range of normal tissues and a limited number of carcinomas (Silver *et al.*, 1997). The anti-PSM Mab localized to normal prostate epithelial cells, the duodenal mucosal brush border, isolated cells in the colonic crypts, and a subset of proximal renal tubules, but in each case the staining was restricted to the luminal site. None of the other 28 normal tissue samples or the carcinoma samples, 17 renal cell, 13 bladder transitional cell and 19 colon, were immunoreactive. PSM expression in tumor tissues was restricted to primary prostrate carcinomas (33/35) and both bone (8/18) and lymph node (7/8) metastases. The staining pattern was typically heterogeneous and varied with the degree of differentiation. Other studies have identified PSM expression in primary prostate tumors (157/165) and bone (7/7) and lymph node metastases (72/79) (Wright *et al.*, 1995). PSM has not been detected in other carcinomas, such as the breast or lung (three samples of each), or sarcomas but is specific for prostate carcinoma (Lopes *et al.*, 1990). Thus, the restricted normal tissue distribution of this antigen and the high percentage of expression in both prostatic primary and metastatic deposits suggest that PSM could act as a target antigen in ADEPT.

An interesting observation in the immunohistochemical distribution of the anti-PSM Mab was its localization to the endothelial cells of peritumoral and endotumoral capillaries in the renal cell (8/17), bladder transitional cell (7/13), and colon (3/19) carcinomas but not adjacent normal vasculature. It was proposed that this neoexpression in the tumor vasculature may be related to tumor angiogenesis (Silver *et al.*, 1997). However, the identity of the antigen recognized by the anti-PSM Mab has not been established, so this response could be a cross-reaction.

5. MANIPULATION OF TUMOR ANTIGENIC EXPRESSION

The levels of tumor antigen expression discussed in this chapter to date relate to those measured in surgical specimens, etc., and typify those encountered in the clinic. However, it is possible to up-regulate the expression of tumor antigens on some target cells, which should lead to improved localization of monoclonal antibodies directed against these antigens. A diverse range of agents potentiate such an effect (Friedman *et al.*, 1987; Guadagni *et al.*, 1991; Hwang *et al.*, 1986; Ohannesian *et al.*, 1994), but among the most commonly studied are the recombinant human interferons (rHu-IFNs), which amplify tumor-associated antigens on a wide variety of human cell lines both *in vitro* and *in vivo* in nude mice bearing human tumor xenografts. They also bring about similar changes in the expression of class I and class II major histocompatibility complex (MHC) antigens and Fc receptors (Fellous *et al.*, 1979; Rowlinson *et al.*, 1986).

The degree of differentiation of the cell does not influence whether exposure to rHu-IFNs affects antigen expression and, just as there is heterogeneity of antigen expression, there are varying degrees of response to rHu-IFNs (Guadagni *et al.*, 1990). Where cells do respond *in vitro*, antigen expression increased by upward of 50%, compared with untreated controls. This led to concomitant increases in binding antibodies to specific antigens (Guadagni *et al.*, 1989) but cell lines which already express high levels of antigen do not always increase on treatment, suggesting an upper limit to the amount of antigen that is expressed by some types of cell (Sivinski *et al.*, 1995). In other experiments, the inability of rHu-IFNs to induce *de novo* expression of TAG-72 antigen has been confirmed, but cells which express TAG-72 and/or CEA constitutively have antigen expression enhanced by rHu-IFN treatment. Cell membrane expression of CEA could be increased threefold to fivefold in a dose- and duration-dependent manner, although the effects were transient and CEA levels returned to baseline levels over a four day period following discontinuation of rHu-IFN treatment. Measurement of the total CEA present in the cells indicated much higher increases, of up to 96-fold, depending on the cell line used. This discrepancy may have been caused by the ELISA assay used detecting shed antigen bound to the plastic of the culture vessel (Greiner *et al.*, 1992; Guadagni *et al.*, 1989).

The effects obtained *in vitro* are achievable *in vivo* in an animal model. Significantly enhanced levels of tumor uptake were observed in human colon carcinoma xenografts following administration of rHu-IFN, followed by an antibody to nonspecific cross-reacting antigen (NCA), an antigen found on >90% of colon carcinomas. Administration of 250,000 units of rHu-IFN-αA brought about a 2.7-fold increase in tumor uptake without a concomitant increase in other major organs, such as blood, liver, and spleen (Guadagni *et al.*, 1988).

Some clinical trials have also been carried out to determine whether rHu-IFNs up-regulate tumor antigen expression (in this instance TAG-72) in a group of eight patients with malignant secondary ascites of colon or ovarian carcinoma origin. In all but one patient there was a pronounced increase in antigen expression, both in terms of intensity of expression on cells and in the percentage of cells which were antigen-positive, as measured by binding of the anti-TAG-72 antibody B72.3 using flow cytometric and immunohistochemical methods. The cells from the patient who did not respond were TAG-72-negative before the therapy began. All of the patients who responded did so at relatively low entry level doses in this dose escalation study, where doses increased from 0.1–100 MU over an 8-week period, given as intraperitoneal injections (Greiner *et al.*, 1992). Other studies have shown similar effects, as measured by tumor uptake of radiolabeled Mabs (Guadagni *et al.*, 1988). Patients suffering from metastatic breast cancer have significantly elevated levels of TAG-72 expression following treatment with interferon-α. A mean increase of about 45% was recorded. In this group of patients there was no significant increase in tumor localization of a radiolabeled antibody, but this may possibly have been caused by other factors, such as limited tumor access (Murray *et al.*, 1995). In addition to enhanced cell surface expression, there is an increase in antigen shedding brought about by rHu-IFN treatments because serum levels of the marker antigens CEA, TAG-72, and CA19-9 increase in a number of gastrointestinal adenocarcinoma patients following administration of rHu-IFNs (Guadagni *et al.*, 1996). Although this observation is of limited value for therapy, it may be a useful approach in helping to develop sensitive assays for detecting minimal residual disease.

6. EVALUATION OF ANTIGENIC TARGETS FOR ADEPT

Some of the factors and properties that need to be considered during the evaluation of potential Mab/tumor-associated antigen combinations for ADEPT are listed in Table III.

One of the initial steps in the evaluation process is to undertake immunochemical methods of epitope detection. During the preparation of tissue samples for immunohistochemistry, a balance must be achieved between preserving antigenicity and the need to maintain the tissue morphology. In most histopathology laboratories formalin-fixed, paraffin-wax-embedded sections are the standard because this method of fixation maintains morphology, but it also results in the loss of many antigenic determinants. In research laboratories frozen tissue sections are often used because they preserve more antigenic features, but conversely, this approach results in some loss of morphology. Other variables include differences in technique, reagent specificity, and interobserver variation in the scoring and evaluation of stained sections. Although stained sections allow visualization of antigens and cellular structures, they allow only a semiquantitative assessment of antigen concentration. The use of well-defined reagents and computerized image analysis can provide a more quantitative approach, in addition to more quantitative techniques, such as autoradiography.

Immunohistochemistry is a valuable tool in evaluating target antigens. However it is well recognized that antibody interaction with an antigen *in vitro* does not necessarily reflect the situation *in vivo*. In some cases Mab binding to normal tissue *in vitro* is not observed when the Mab is administered to patients, and Mab interactions not identified *in vitro* can cause serious toxicity, which will be discussed in more detail later. In addition to immunohistochemistry, a range of solid matrix blotting techniques for DNA, RNA, and protein are frequently undertaken to evaluate antigen expression and genetic alteration.

Table III. Properties of Antigen and Monoclonal Antibody

Antigen	Monoclonal antibody
Normal tissue distribution/accessibility	Size
Tumor distribution/accessibility	Epitope specificity
Antigen density	Interaction with free antigen
Cellular location	Clearance of antigen-Mab complexes
Heterogeneity	
Biochemical structure	
Stability	
Number of available epitopes	
Biological function	

7. *IN VITRO* CELL CULTURE PROBLEMS

Evaluation of potential tumor antigens and their corresponding antibodies is initially performed using *in vitro* cell culture assays and *in vivo* models. Consideration must be given to the effect of tissue disruption techniques on antigen expression both for cell culture and quantitative analysis techniques, such as flow cytometry. An important factor to be borne in mind is that antigen expression in biopsy samples is frequently altered, sometimes by a complete loss of antigen expression during the transition to cell culture and subsequent passage. This phenomenon has been observed in a wide range of tumors and is caused in part by the loss of the spatial configuration of the cell population, that is, the normal tissue architecture. For example, TAG-72 expression is generally not observed on breast (1/25) or colon carcinoma cell lines (1/18) but is present in a high percentage of breast (25/56) and colon carcinoma biopsies (16/20). When the antigen-positive colon carcinoma line LS174T was grown in conditions that promoted three-dimensional growth, however, a pronounced increase in antigen expression was observed (Hand *et al.*, 1985). The decline in expression of cell surface antigens during the transition to tissue culture has been observed for other tissue types and antigens. Growing the cells under conditions which more closely mimic the normal environment overcomes this problem (Guadagni *et al.*, 1991; Tsuchida *et al.*, 1987). Contamination of cell cultures with mycoplasma also leads to altered antigen expression and its interaction with corresponding antibody (Oredipe *et al.*, 1992).

8. UNFORESEEN PROBLEMS OF SPECIFICITY

It cannot be emphasized strongly enough that many of the antigens described in this chapter are tumor-associated rather than tumor-specific antigens. Intensive investigation of the specificity of the Mab raised against a tumor-associated antigen is required to ensure that vital tissues are not targeted. A Phase I clinical trial of an antibreast cancer immunotoxin was stopped early because severe neurotoxicity developed (Gould *et al.*, 1989). Similarly, intraperitoneal administration of an immunotoxin in a Phase I clinical trial for ovarian cancer led to fatal neurotoxicity (Pai *et al.*, 1991). Both of these immunotoxins that had been extensively tested, including primate studies, hereby highlighting the need for closely monitored dose escalation studies, as seen in Phase I clinical trials. In contrast, many of the Mabs used in radioimmunotherapy bind to normal tissue *in vitro*. Despite this, high-dose radiation therapy has been administered with these Mabs without significant side effects. However, in radioimmunotherapy the limiting toxicity is the radiation dose to vital organs and tissues during the initial biodistribution phase. Multistep procedures and improved isotope selection should increase the efficacy of radioimmunotherapy. The difference between ADEPT and radioimmunotherapy is highlighted by comparing the tumor to blood ratios, which were estimated to be 16:1 and 1.5:1, respectively (Begent and Bagshawe, 1996a). There is obviously a wide gulf between the absolute specificity required for immunotoxin-targeted therapy and the lower requirements for radioimBmunotherapy. It is likely that the degree of Mab specificity for ADEPT falls somewhere between these two.

Another advantage that ADEPT has over other Mab-targeted therapies is the potential for generating a wide range of cytotoxic agents from a number of different enzyme/prodrug combinations or from one enzyme by using the self-immolative chemistry developed by Mauger and others (Azoulay *et al.*, 1995; Dowell *et al.*, 1996; Mauger *et al.*, 1994; Schmidt *et al.*, 1997). Coupled with the wide range of potential tumor-associated antigens, this means that achieving a high degree of selectivity is possible. Thus it should be possible to avoid generating large concentrations of active drug at inappropriate sites by judiciously choosing target antigen and enzyme/drug combinations. For example, the c-erbB-2 receptor is not expressed in bone marrow or heart tissue, so the combination of a β-lactamase-anti-c-erbB-2 fusion protein and a doxorubicin prodrug (Rodrigues *et al.*, 1995) is not expected to generate high concentrations of active drug at these doxorubicin-sensitive sites. The results from the first clinical trials of an ADEPT system are discussed in more detail in Chapter 8, but it is pertinent to note here that no toxicity has been associated with inappropriate targeting of the Mab-CPG_2 conjugate. Most of the toxicities in the trial were related to the long circulatory half-life of the tumor-generated drug.

REFERENCES

Aasland, R., Lillehaug, J. R., Male, R., Josendal, O., Varhaug, J. E., and Kleppe, K., 1988, Expression of oncogenes in thyroid tumours: Coexpression of c-erbB2/neu and c-erbB, *Br. J. Cancer* **57:**358–363.

Acevedo, H. F., Campbell Acevedo, E. A., Buffo, M. J., and Krichevsky, A., 1993, *In vivo* expression of membrane-associated human chorionic gonadotropin by cells isolated from cancerous human tissues, *Proc. Annu. Meet. Am. Assoc. Cancer Res.* **34:**27(Abstract).

Acevedo, H. F., Krichevsky, A., Campbell-Acevedo, E. A., Galyon, J. C., Buffo, M. J., and Hartsock, R. J., 1992, Expression of membrane-associated human chorionic gonadotropin, its subunits, and fragments by cultured human cancer cells, *Cancer* **69:**1829–1842.

Acevedo, H. F., Tong, J. Y., and Hartsock, R. J., 1995, Human chorionic gonadotropin-beta subunit gene expression in cultured human fetal and cancer cells of different types and origins, *Cancer* **76:**1467–1475.

Adams, G. P., McCartney, J. E., Tai, M. S., Oppermann, H., Huston, J. S., Stafford, W. F., Bookman, M. A., Fand, I., Houston, L. L., and Weiner, L. M., 1993, Highly specific *in vivo* tumor targeting by monovalent and divalent forms of 741F8 anti-c-erbB-2 single-chain Fv, *Cancer Res.* **53:**4026–4034.

Alfthan, H., Haglund, C., Roberts, P., and Stenman, U. H., 1992, Elevation of free beta subunit of human choriogonadotropin and core beta fragment of human choriogonadotropin in the serum and urine of patients with malignant pancreatic and biliary disease, *Cancer Res.* **52:**4628–4633.

Allan, S. M., Dean, C., Fernando, I., Eccles, S., Styles, J., McCready, V. R., Baum, M., and Sacks, N., 1993, Radioimmunolocalisation in breast cancer using the gene product of c-erb-B2 as the target antigen, *Br. J. Cancer* **67:**706–712.

An, Z., Wang, X., Kubota, T., Moossa, A. R., and Hoffman, R. M., 1996, A clinical nude mouse metastatic model for highly malignant human pancreatic cancer, *Anticancer Res.* **16:**627–631.

Antoniw, P., Springer, C. J., Bagshawe, K. D., Searle, F., Melton, R. G., Rogers, G. T., Burke, P. J., and Sherwood, R. F., 1991, Disposition of the prodrug 4-(bis[2-chloroethyl]amino) benzoyl L-glutamic acid and its active parent drug in mice, *Br. J. Cancer* **62:**909–914.

Astoul, P., Colt, H. G., Wang, X., and Hoffman, R. M., 1993, Metastatic human pleural ovarian cancer model constructed by orthotopic implantation of fresh histologically intact patient carcinoma in nude mice, *Anticancer Res.* **13:**1999–2002.

Azoulay, M., Florent, J. C., Monneret, C., Gesson, J. P., Jacquesy, J. C., Tillequin, F., Koch, M., Bosslet, K., Czech, J., and Hoffman, D., 1995, Prodrugs of anthracycline antibiotics suited for tumor-specific activation, *Anti-Cancer Drug Des.* **10:**441–450.

Bagshawe, K. D., Sharma, S. K., Springer, C. J., and Antoniw, P., 1995, Antibody directed enzyme prodrug therapy: A pilot-scale clinical trial, *Tumor Targeting* **1:**17–29.

Bagshawe, K. D., Springer, C. J., Searle, F., Antoniw, P., Sharma, S. K., Melton, R. G., and Sherwood, R. F., 1988, A cytotoxic agent can be generated selectively at cancer sites, *Br. J. Cancer* **58:**700–703.

Bamias, A., Bowles, M. J., Krausz, T., Williams, G., and Epenetos, A. A., 1993, Intravesical administration of indium-111-labelled HMFG2 monoclonal antibody in superficial bladder carcinomas, *Int. J. Cancer* **54:**899–903.

Baranov, V., Yeung, M. M., and Hammarstrom, S., 1994, Expression of carcinoembryonic antigen and nonspecific cross-reacting 50-kDa antigen in human normal and cancerous colon mucosa: Comparative ultrastructural study with monoclonal antibodies, *Cancer Res.* **54:**3305–3314.

Baxter, L. T., and Jain, R. K., 1990, Transport of fluid and macromolecules in tumors. II. Role of heterogeneous perfusion and lymphatics, *Microvascular Res.* **40:**246–263.

Baxter, L. T., and Jain, R. K., 1996, Pharmacokinetic analysis of the microscopic distribution of enzyme-conjugated antibodies and prodrugs: Comparison with experimental data, *Br. J. Cancer* **73:**447–456.

Begent, R. H. J., 1985, Recent advances in tumour imaging: Use of radiolabelled monoclonal antibodies, *Biochim. Biophys. Acta* **780:**151–166.

Begent, R. H. J., and Bagshawe, K. D., 1996a, Biodistribution studies, *Adv. Drug Delivery Rev.* **22:**325–329.

Begent, R. H. J., Verhaar, M. J., Chester, K. A., Casey, J. L., Green, A. J., Napier, M. P., Hopestone, L. D., Cushen, N., Keep, P. A., Johnson, C. J., Hawkins, R. E., Hilson, A. J. W., and Robson, L., 1996b, Clinical evidence of efficient tumor targeting based on single-chain Fv antibody selected from a combinatorial library, *Nature Med.* **2:**979–984.

Behr, T. M., Sharkey, R. M., Juweid, M. I., Dunn, R. M., Ying, Z., Zhang, C. H., Siegel, J. A., Gold, D. V., and Goldenberg, D. M., 1996, Factors influencing the pharmacokinetics, dosimetry, and diagnostic accuracy of radioimmunodetection and radioimmunotherapy of carcinoembryonic antigen-expressing tumors, *Cancer Res.* **56:**1805–1816.

Benchimol, S., Fuks, A., Jothy, S., Beauchemin, N., Shirota, K., and Stanners, C. P., 1989, Carcinoembryonic antigen, a human tumor marker, functions as an intercellular adhesion molecule, *Cell* **57:**327–334.

Berchuck, A., Kamel, A., Whitaker, R., Kerns, B., Olt, G., Kinney, R., Soper, J. T., Dodge, R., Clarke Pearson, D. L., Marks, P., McKenzie, S., Yin, S., and Bast, R., 1990, Overexpression of HER-2/neu is associated with poor survival in advanced epithelial ovarian cancer, *Cancer Res.* **50:**4087–4091.

Berger, P., Kranewitter, W., Madersbacher, S., Gerth, R., Geley, S., and Dirnhofer, S., 1994, Eutopic production of human chorionic gonadotropin beta (hCG beta) and luteinizing hormone beta (hLH beta) in the human testis, *FEBS Lett.* **343:**229–233.

Bo, M., and Boime, I., 1992, Identification of the transcriptionally active genes of the chorionic gonadotropin beta gene cluster *in vivo*, *J. Biol. Chem.* **267:**3179–3184.

Boerman, O. C., Sharkey, R. M., Blumenthal, R. D., Aninipot, R. L., and Goldenberg, D. M., 1992, The presence of a concomitant bulky tumor can decrease the uptake and therapeutic efficacy of radiolabeled antibodies in small tumors, *Int. J. Cancer* **51:**470–475.

Boucher, D., Cournoyer, D., Stanners, C. P., and Fuks, A., 1989, Studies on the control of gene expression of the carcinoembryonic antigen family in human tissue, *Cancer Res.* **49:**847–852.

Boxer, G. M., Abassi, A. M., Pedley, R. B., and Begent, R. H. J., 1994, Localisation of monoclonal antibodies reacting with different epitopes on CEA—implications for targeted therapy, *Br. J. Cancer* **69:**307–314.

Boxer, G. M., Begent, R. H. J., Kelly, A. M. B., Southall, P. J., Blair, S. B., Theodorou, N. A., Dawson, P. M., and Ledermann, J. A., 1992, Factors influencing variability of localisation of antibodies to carcinoembryonic antigen (CEA) in patients with colorectal carcinoma—implications for radioimmunotherapy, *Br. J. Cancer* **65:**825–831.

Bridgewater, J. A., Knox, R. J., Pitts, J. D., Collins, M. K., and Springer, C. J., 1997, The bystander effect of the nitroreductase CB 1954 enzyme prodrug system is due to a cell-permeable metabolite, *Hum. Gene Ther.* **8:**709–717.

Buchegger, F., Pelegrin, A., Hardman, N., Heusser, C., Lukas, J., Dolci, W., and Mach, J. P., 1992, Different behaviour of mouse-human chimeric antibody $F(ab')_2$ fragments of IgG_1, IgG_2 and IgG_4 sub-class *in vivo*, *Int. J. Cancer* **50:**416–422.

Burchell, J., Gendler, S., Taylor Papadimitriou, J., Girling, A., Lewis, A., Millis, R., and Lamport, D., 1987, Development and characterization of breast cancer reactive monoclonal antibodies directed to the core protein of the human milk mucin, *Cancer Res.* **47:**5476–5482.

Burrows, F. J., and Thorpe, P. E., 1993, Eradication of large solid tumors in mice with an immunotoxin directed against tumor vasculature, *Proc. Natl. Acad. Sci. USA* **90:**8996–9000.

Burrows, F. J., and Thorpe, P. E., 1994, Vascular targeting—A new approach to the therapy of solid tumors, *Pharmacol. Ther.* **64:**155–174.

Campbell, M., Bieber, M., Levy, R., and Teng, N. N., 1986, Influence of avidity and idiotope recognition on the modulation of surface immunoglobulin on malignant human B cells by rat monoclonal anti-idiotype antibodies, *J. Immunol.* **136:**2983–2988.

Carter, R. E., Feldman, A. R., and Coyle, J. T., 1996, Prostate-specific membrane antigen is a hydrolase with substrate and pharmacologic characteristics of a neuropeptidase, *Proc. Natl. Acad. Sci. USA* **93:**749–753.

Chandrasekaran, E. V., Davila, M., Nixon, D. W., Goldfarb, M., and Mendicino, J., 1983, Isolation and structures of the oligosaccharide units of carcinoembryonic antigen, *J. Biol. Chem.* **258:**7213–7222.

Chen, F. M., Hansen, E. B., Taylor, C. R., and Epstein, A. L., 1991, Binding of monoclonal antibody TNT-1 in multicellular tumor spheroids, *J. Natl. Cancer Inst.* **83:**200–204.

Ching, L.-M., Joseph, W. R., Zhuang, L., and Baguley, B. C., 1994, Interaction between endotoxin and the antitumour agent 5,6-dimethylxanthenone acetic acid in the induction of tumour necrosis factor and haemorrhagic necrosis of colon 38 tumors, *Cancer Chemother. Pharmacol.* **35:**153–160.

Chiou, R. K., 1991, The impact of tumor size on the efficacy of monoclonal antibody-targeted radiotherapy—Studies using a nude mouse model with human renal cell carcinoma xenografts, *J. Urol.* **146:**232–237.

Coussens, L., Yang Feng, T. L., Liao, Y. C., Chen, E., Gray, A., McGrath, J., Seeburg, P. H., Libermann, T. A., Schlessinger, J., Francke, U., Levinson, A., and Ullrich, A., 1985, Tyrosine kinase receptor with extensive homology to EGF receptor shares chromosomal location with neu oncogene, *Science* **230:**1132–1139.

Crnalic, S., Hakansson, I., Boquist, L., Lofvenberg, R., and Brostrom, L. A., 1997, A novel spontaneous metastasis model of human osteosarcoma developed using orthotopic transplantation of intact tumor tissue into tibia of nude mice, *Clin. Exp. Metastasis* **15:**164–172.

De Santes, K., Slamon, D., Anderson, S. K., Shepard, M., Fendly, B., Maneval, D., and Press, O., 1992, Radiolabeled antibody targeting of the HER-2/neu oncoprotein, *Cancer Res.* **52:**1916–1923.

Dean, C. J., Eccles, S. A., Valeri, M., Box, G., Allan, S., McFarlane, C., Sandle, J., and Styles, J., 1993, Rat Mabs to the product of the c-*erb*B-2 proto-oncogene for diagnosis and therapy in breast cancer, *Cell Biophys.* **22:**111–127.

Dekker, I., Rozeboom, T., Delemarre, J., Dam, A., and Oosterhuis, J. W., 1992, Placental-like alkaline phosphatase and DNA flow cytometry in spermatocytic seminoma, *Cancer* **69:**993–996.

Dowell, R. I., Springer, C. J., Davies, D. H., Hadley, E. M., Burke, P. J., Boyle, F. T., Melton, R. G., Connors, T. A., Blakey, D. C., and Mauger, A. B., 1996, New mustard prodrugs for antibody-directed enzyme prodrug therapy: Alternatives to the amide link, *J. Med. Chem.* **39:**1100–1105.

Durbin, H., Young, S., Stewart, L. M., Wrba, F., Rowan, A. J., Snary, D., and Bodmer, W. F., 1994, An epitope on carcinoembryonic antigen defined by the clinically relevant antibody PR1A3, *Proc. Natl. Acad. Sci. USA* **91:**4313–4317.

Durrant, L. G., Robins, R. A., and Baldwin, R. W., 1989, Flow cytometric screening of monoclonal antibodies for drug or toxin targeting to human cancer, *J. Natl. Cancer Inst.* **81:**688–696.

Dvorak, H. F., Nagy, J. A., Dvorak, J. T., and Dvorak, A. M., 1988, Identification and characterisation of the blood vessels of solid tumours that are leaky to circulating macromolecules, *Am. J. Pathol.* **133:**95–109.

Eccles, S. A., Court, W. J., Box, G. A., Dean, C. J., Melton, R. G., and Springer, C. J., 1994, Regression of established breast carcinoma xenografts with antibody-directed enzyme prodrug therapy against c-erbB2 p185, *Cancer Res.* **54:**5171–5177.

Enblad, P., Glimelius, B., Busch, C., Ponten, J., and Pahlman, L., 1987, Antigenic heterogeneity and individuality in adenocarcinomas of the rectum and their secondaries, *Br. J. Cancer* **55:**503–508.

Epenetos, A. A., Carr, D., Johnson, P. M., Bodmer, W. F., and Lavender, J. P., 1986a, Antibody-guided radiolocalisation of tumours in patients with testicular or ovarian cancer using two radioiodinated monoclonal antibodies to placental alkaline phosphatase, *Br. J. Radiol.* **59:**117–125.

Epenetos, A. A., Hooker, G., Krausz, T., Snook, D., Bodmer, W. F., and Taylor Papadimitriou, J., 1986b, Antibody-guided irradiation of malignant ascites in ovarian cancer: A new therapeutic method possessing specificity against cancer cells, *Obstet. Gynecol.* **68:**71S–74S.

Epenetos, A. A., Munro, A. J., Stewart, S., Rampling, R., Lambert, H. E., McKenzie, C. G., Soutter, P., Rahemtulla, A., Hooker, G., Sivolapenko, G. B., Snook, D., Courtenay-Luck, N. S., Dhokia, B., Krausz, T., Taylor-Papadimitriou, J., Durbin, H., and Bodmer, W. F., 1987, Antibody-guided irradiation of advanced ovarian cancer with intraperitoneally administered radiolabeled monoclonal antibodies, *J. Clin. Oncol.* **5:**1890–1899.

Epenetos, A. A., Travers, P., Gatter, K. C., Oliver, R. D., Mason, D. Y., and Bodmer, W. F., 1984, An immunohistological study of testicular germ cell tumours using two different monoclonal antibodies against placental alkaline phosphatase, *Br. J. Cancer* **49:**11–15.

Epstein, A. L., Khawli, L. A., Hornick, J. L., and Taylor, C. R., 1995, Identification of a monoclonal antibody, TV-1, directed against the basement membrane of tumor vessels, and its use to enhance the delivery of macromolecules to tumors after conjugation with interleukin 2, *Cancer Res.* **55:**2673–2680.

Erdei, A., Fust, G., and Gergely, J., 1991, The role of C3 in the immune response, *Immunol. Today* **12:** 332–337.

Esteban, J. M., Hyams, D. M., Beatty, B. G., Merchant, B., and Beatty, J. D., 1990, Radioimmunotherapy of human colon carcinomatosis xenograft with 90Y-ZCE025 monoclonal antibody: Toxicity and tumor phenotype studies, *Cancer Res.* **50:**989s–992s.

Feizi, T., 1985, Demonstration by monoclonal antibodies that carbohydrate structures of glycoproteins and glycolipids are onco-developmental antigens, *Nature* **314:**53–57.

Fellous, M., Kamoun, M., Gresser, I., and Bono, R., 1979, Enhanced expression of HLA antigen and β2-microglobin on interferon-treated human lymphoid cells, *Eur. J. Immunol.* **9:**446–449.

Fidler, I. J. and Balch, C. M., 1987, The biology of cancer metastasis and implications for surgery, *Curr. Prob. Surg.* **24:**137–209.

Fidler, I. J. and Ellis, L. M., 1994, The implications of angiogenesis for the biology and therapy of cancer metastasis, *Cell* **79:**185–188.

Fjeld, J. G., Michaelsen, T. E., Benestad, H. B., and Nustad, K., 1991, The effect of the biodistribution differences between IgG, $F(ab')_2$ and Fab′ on their immunotargeting potential for human tumor cells in immunocompetent mice, *Antibody Immunoconj. Radiopharm.* **4:**443–451.

Friedman, J., Seger, M., Levinsky, H., and Allolouf, D., 1987, Modulation of carcinoembryonic antigen release by HT29 colon carcinoma line in the presence of different agents, *Experimentia* **43:**1121–1129.

Fritsche, R. and Mach, J. P., 1977, Isolation and characterization of carcinoembryonic antigen (CEA) extracted from normal human colon mucosa, *Immunochemistry* **14:**119–127.

Fu, X., Besterman, J. M., Monosov, A., and Hoffman, R. M., 1991, Models of human metastatic colon cancer in nude mice orthotopically constructed by using histologically intact patient specimens, *Proc. Natl. Acad. Sci. USA* **88:**9345–9349.

Fu, X., Le, P., and Hoffman, R. M., 1993, A metastatic orthotopic-transplant nude-mouse model of human patient breast cancer, *Anticancer Res.* **13:**901–904.

Fu, X., Theodorescu, D., Kerbel, R. S., and Hoffman, R. M., 1991, Extensive multi-organ metastasis following orthotopic implantation of histologically intact human bladder carcinoma tissue in nude mice, *Int. J. Cancer* **49:**938–939.

Fujimori, K., Covell, D. G., Fletcher, J. E., and Weinstein, J. N., 1990, A modelling analysis of monoclonal antibody percolation through tumours: A binding site barrier, *J. Nucl. Med.* **32:**1191–1198.

Furukawa, T., Fu, X., Kubota, T., Watanabe, M., Kitajima, M., and Hoffman, R. M., 1993a, Nude mouse metastatic models of human stomach cancer constructed using orthotopic transplantation of histologically intact tissue, *Cancer Res.* **53:**1204–1208.

Furukawa, T., Kubota, T., Watanabe, M., Kitajima, M., and Hoffman, R. M., 1993b, A novel 'patient-like' treatment model of human pancreatic cancer constructed using orthotopic transplantation of histologically intact human tissue in nude mice, *Cancer Res.* **53:**3070–3072.

Furukawa, T., Kubota, T., Watanabe, M., Kitajima, M., and Hoffman, R. M., 1993c, Orthotopic transplantation of histologically intact clinical specimens of stomach cancer to nude mice: Correlation of metastatic sites in mouse and individual patient donors, *Int. J. Cancer* **53:**608–612.

Gaffar, S. A., Chen, F. M., Liu, C., and Epstein, A. L., 1991, Cell based radioimmunoassays to quantitate the immunoreactivity of TNT monoclonal antibodies directed against intracellular antigens, *J. Immunoassay* **12:**1–14.

Garcia de Palazzo, I. E., Adams, G. P., Sundareshan, P., Wong, A. J., Testa, J. R., Bigner, D. D., and Weiner, L. M., 1993, Expression of mutated epidermal growth factor receptor by non-small cell lung carcinomas, *Cancer Res.* **53:**3217–3220.

Gazit, Y., Berk, D. A., Leunig, M., Baxter, L. T., and Jain, R. K., 1995, Scale-invariant behaviour and vascular network formation in normal and tumor tissue, *Phys. Rev. Lett.* **75:**2428–2431.

Gendler, S. J., Lancaster, C. A., Taylor Papadimitriou, J., Duhig, T., Peat, N., Burchell, J., Pemberton,

L., Lalani, E. N., and Wilson, D., 1990, Molecular cloning and expression of human tumor-associated polymorphic epithelial mucin, *J. Biol. Chem.* **265:**15286–15293.

Giacomini, P., Natali, P., and Ferrone, S., 1985, Analysis of the interaction between a human high molecular weight melanoma-associated antigen and the monoclonal antibodies to three distinct antigenic determinants, *J. Immunol.* **135:**696–702.

Gold, P. and Freedman, S. O., 1965a, Demonstration of tumour-specific antigens in human colonic carcinomata by immunological tolerance and absorption techniques, *J. Exp. Med.* **121:**439–462.

Gold, P. and Freedman, S. O., 1965b, Specific carcinoembryonic antigens of the human digestive system, *J. Exp. Med.* **122:**467–481.

Goldenberg, D. M., DeLand, F., Kim, E., Bennett, S., Primus, F. J., van Nagell, J. R., Estes, N., DeSimone, P., and Rayburn, P., 1978, Use of radiolabeled antibodies to carcinoembryonic antigen for the detection and localisation of diverse cancers by external photoscanning, *N. Engl. J. Med.* **298:**1384–1388.

Goldenberg, D. M., Goldenberg, H., Sharkey, R. M., Higginbotham Ford, E., Lee, R. E., Swayne, L. C., Burger, K. A., Tsai, D., Horowitz, J. A., Hall, T. C., Pinsky, C. M., and Hansen, H. J., 1990, Clinical studies of cancer radioimmunodetection with carcinoembryonic antigen monoclonal antibody fragments labeled with ^{123}I or ^{99}mTc, *Cancer Res.* **50:**909s–921s.

Goldenberg, D. M., Kim, E. E., DeLand, F. H., Bennett, S., and Primus, F. J., 1980, Radioimmunodetection of cancer with radioactive antibodies to carcinoembryonic antigen, *Cancer Res.* **40:**2984–2992.

Gould, B. J., Borowitz, M. J., Groves, E. S., Carter, P. W., Anthony, D., Weiner, L. M., and Frankel, A. E., 1989, Phase I study of an anti-breast cancer immunotoxin by continuous infusion: Report of a targeted toxic effect not predicted by animal studies, *J. Natl. Cancer Inst.* **81:**775–781.

Greiner, J. W., Guadagni, F., Goldstein, D., Borden, E. C., and Schlom, J., 1992, Intraperitoneal administration of interferon-gamma to carcinoma patients enhances expression of tumor-associated glycoprotein-72 (TAG-72) and carcinoembryonic antigen (CEA) on malignant ascites cells, *J. Clin. Oncol.* **10:**735–746.

Griffin, T. W., Collins, J., Bokhari, F., Stochl, M., Brill, A. B., Ito, T., Emond, G., and Sands, H., 1990, Intraperitoneal immunoconjugates, *Cancer Res.* **50:**1031s–1038s.

Guadagni, F., Roselli, M., Cosimelli, M., Spila, A., Cavaliere, F., Tedesco, M., Arcuri, R., Abbolito, M. R., Casale, V., Pericoli, M. N., Vecchione, A., Casciani, C. U., Greiner, J. W., and Schlom, J., 1996, Correlation between tumor-associated glycoprotein 72 mucin levels in tumor and serum of colorectal patients as measured by the quantitative CA 72-4 immunoassay, *Cancer Res.* **56:**5293–5298.

Guadagni, F., Roselli, M., and Hoffman, R. M., 1991, Maintenance of expression of tumor antigens in three-dimensional *in vitro* human tumor gel-supported histoculture, *Anticancer Res.* **11:**543–546.

Guadagni, F., Roselli, M., Schlom, J., and Greiner, J. W., 1996, *In vitro* and *in vivo* regulation of human tumor antigen expression by human recombinant interferons: A review, *Int. J. Biol. Mark.* **9:**53–60.

Guadagni, F., Schlom, J., Johnston, W. W., Szpak, C. A., Goldstein, D., Simpson, J. F., Borden, E. C., Pestka, S., and Greiner, J. W., 1989, Selective interferon-induced enhancement of tumor-associated antigens on a spectrum of freshly isolated human adenocarcinoma cells, *J. Natl. Cancer Inst.* **81:**502–512.

Guadagni, F., Schlom, J., Pothen, S., Pestka, S., and Greiner, J. W., 1988, Parameters involved in the enhancement of monoclonal antibody targeting *in vivo* with recombinant interferon, *Cancer Immunol. Immunother.* **26:**222–230.

Guadagni, F., Tortora, G., Clair, T., Cho-Chung, Y. S., Schlom, J., and Greiner, J. W., 1991, Regulation of carcinoembryonic antigen expression on human colorectal tumor cells by site-selective acyclic AMP analogues, *Int. J. Cancer* **48:**413–422.

Guadagni, F., Witt, P., Robbins, P., Schlom, J., and Greiner, J. W., 1990, Regulation of carcinoembryonic antigen expression in differential human colorectal tumor cells by interferon-gamma, *Cancer Res.* **50:**6248–6255.

Hagan, P. L., Halpern, S. E., Dillman, R. O., Shawler, D. L., Johnson, D. E., Chen, A., Krishnan, L., Frincke, J., Bartholemew, R. M., David, G. S., and Carlo D., 1986, Tumor size: Effect on monoclonal antibody uptake in tumor models, *J. Nucl. Med.* **27:**422–427.

Haisma, H. J., Vanmuijen, M., Scheffer, G., Scheper, R. J., Pinedo, H. M., and Boven, E., 1995, A monoclonal antibody against human beta-glucuronidase for application in antibody-directed enzyme prodrug therapy, *Hybridoma* **14:**377–382.

Hakomori, S. and Kannagi, R., 1983, Glycosphingolipids as tumor-associated and differentiation markers, *J. Natl. Cancer Inst.* **71:**231–251.

Hamada, Y., Yamamura, M., Hioki, K., Yamamoto, M., Nagura, H., and Watanabe, K., 1985, Immunohistochemical study of carcinoembryonic antigen in patients with colorectal cancer. Correlation with plasma carcinoembryonic antigen levels, *Cancer* **55:**136–141.

Hammarstrom, S., Shively, J. E., Paxton, R. J., Beatty, B. G., Larsson, A., Ghosh, R., Bormer, O., Buchegger, F., Mach, J. P., Burtin, P., Seguin, P., Darbouret, P., Degorce, F., Sertour, J., Jolu, J. P., Fuks, A., Kalthoff, H., Schmiegel, W., Arndt, R., Kloppel, G., von Kleist, S., Grunert, F., Schwarz, K., Matsuoka, Y., Kuroki, M., Wagener, C., Weber, T., Yachi, A., Imai, K., Hishikawa, N., and Tsujisaki, M., 1989, Antigenic sites in carcinoembryonic antigen, *Cancer Res.* **49:**4852–4858.

Hand, P. H., Calvo, B., Milenic, D., Yokota, T., Finch, M., Snoy, P., Garmestani, K., Gansow, O., Schlom, J., and Kashmiri, S. V. S., 1992, Comparative biological properties of a recombinant chimeric anticarcinoma Mab and a recombinant aglycosylated variant, *Cancer Immunol. Immunother.* **35:**165–174.

Hand, P. H., Colcher, D., Salomon, D., Ridge, J., Noguchi, P., and Schlom, J., 1985, Influence of spatial configuration of carcinoma cell populations on the expression of a tumor-associated glycoprotein, *Cancer Res.* **45:**833–840.

Hand, P. H., Nuti, M., Colcher, D., and Schlom, J., 1983, Definition of antigenic heterogeneity and modulation among human mammary carcinoma cell populations using monoclonal antibodies to tumour-associated antigens, *Cancer Res.* **43:**728–735.

Harwood, P. J., Britton, D. W., Southall, P. J., Boxer, G. M., Rawlins, G., and Rogers, G. T., 1986, Mapping epitope characteristics on carcinoembryonic antigen, *Br. J. Cancer* **54:**75–82.

Haugen, D. R., Akslen, L. A., Varhaug, J. E., and Lillehaug, J. R., 1996, Expression of c-erbB-3 and c-erbB-4 proteins in papillary thyroid carcinomas, *Cancer Res.* **56:**1184–1188.

Hernando, J. J., von Kleist, S., and Grunert, F., 1994, A repertoire of monoclonal antibodies reveals extensive epitope heterogeneity in CEA purified from neoplasms originating from different organs, *Int. J. Cancer* **56:**655–661.

Hilkens, J. and Buijs, F., 1988, Biosynthesis of MAM-6, an epithelial sialomucin. Evidence for involvement of a rare proteolytic cleavage step in the endoplasmic reticulum, *J. Biol. Chem.* **263:**4215–4222.

Hird, V., Maraveyas, A., Snook, D., Dhokia, B., Soutter, W. P., Meares, C., Stewart, J. S. W., Mason, P., Lambert, H. E., and Epenetos, A. A., 1993, Adjuvant therapy of ovarian cancer with radioactive monoclonal antibody, *Br. J. Cancer* **68:**403–406.

Hirohashi, S., Clausen, H., Yamada, T., Shimosato, Y., and Hakomori, S., 1985, Blood group A cross-reacting epitope defined by monoclonal antibodies NCC-LU-35 and -81 expressed in cancer of blood group O or B individuals: Its identification as Tn antigen, *Proc. Natl. Acad. Sci. USA* **82:** 7039–7043.

Hopkins, K., Chandler, C., Bullimore, J., Sandeman, D., Coakham, H., and Kemshead, J. T., 1995, A pilot study of the treatment of patients with recurrent malignant glioma with intratumoral yttrium-90 radioimmunoconjugates, *Radiother. Oncol.* **34:**121–131.

Horoszewicz, J. S., Kawinski, E., and Murphy, G. P., 1987, Monoclonal antibodies to a new antigenic marker in epithelial prostatic cells and serum of prostatic cancer patients, *Anticancer Res.* **7:** 927–935.

Houba, P. H. J., Boven, E., and Haisma, H. J., 1996, Improved characteristics of a human beta-glucuronidase-antibody conjugate after deglycosylation for use in antibody-directed enzyme prodrug therapy, *Bioconjugate Chem.* **7:**606–611.

Houghton, A. N., Eisinger, M., Albino, A. P., Cairncross, J. G., and Old, L. J., 1982, Surface antigens of melanocytes and melanomas. Markers of melanocyte differentiation and melanoma subsets, *J. Exp. Med.* **151:**1755–1766.

Houldsworth, J., Cordon Cardo, C., Ladanyi, M., Kelsen, D. P., and Chaganti, R. S., 1990, Gene amplification in gastric and esophageal adenocarcinomas, *Cancer Res.* **50:**6417–6422.

Humphrey, P. A., Wong, A. J., Vogelstein, B., Friedman, H. S., Werner, M. H., Bigner, D. D., and Bigner, S. H., 1988, Amplification and expression of the epidermal growth factor receptor gene in human glioma xenografts, *Cancer Res.* **48:**2231–2238.

Humphrey, P. A., Wong, A. J., Vogelstein, B., Zalutsky, M. R., Fuller, G. N., Archer, G. E., Friedman, H. S., Kwatra, M. M., Bigner, S. H., and Bigner, D. D., 1990, Anti-synthetic peptide antibody reacting at the fusion junction of deletion-mutant epidermal growth factor receptors in human glioblastoma. *Proc. Natl. Acad. Sci. USA* **87:**4207–4211.

Hwang, W. I., Sack, T., and Kim, Y. S., 1986, Effects of cyclic adenosine 3′,5′-monophosphate upon glycoprotein and carcinoembryonic antigen synthesis and release by human colon cancer cells, *Cancer Res.* **46:**3371–3374.

Iles, R. K., Ind, T. E., and Chard, T., 1994, Production of placental alkaline phosphatase (PLAP) and PLAP-like material by epithelial germ cell and non-germ cell tumours in vitro, *Br. J. Cancer* **69:** 274–278.

Iles, R. K., Purkis, P. E., Whitehead, P. C., Oliver, R. T. D., Leigh, I., and Chard, T., 1990, Expression of beta human chorionic gonadotrophin by non-trophoblastic nonendocrine 'normal' and malignant epithelial cells, *Br. J. Cancer* **61:**663–666.

Israeli, R. S., Powell, C. T., Fair, W. R., and Heston, W. D., 1993, Molecular cloning of a complementary DNA encoding a prostate-specific membrane antigen, *Cancer Res.* **53:**227–230.

Jain, R. K., 1988, Determinants of tumor blood flow: A review, *Cancer Res.* **48:**2641–2658.

Jain, R. K., 1996a, 1995 Whitaker lecture: Delivery of molecules, particles, and cells to solid tumors, *Ann. Biomed. Eng.* **24:**457–473.

Jain, R. K., 1996b, Transport of molecules in the tumour interstitium: A review, *Cancer Res.* **47:**3039–3051.

Jain, R. K. and Baxter, L. T., 1988, Mechanisms of heterogeneous distribution of monoclonal antibodies and other macromolecules in tumors: Significance of elevated interstitial pressure, *Cancer Res.* **48:**7022–7032.

Jemmerson, R., Klier, F. G., and Fishman, W. H., 1985, Clustered distribution of human placental alkaline phosphatase on the surface of both placental and cancer cells. Electron microscopic observations using gold-labeled antibodies, *J. Histochem. Cytochem.* **33:**1227–1234.

Jirtle, R. L., 1988, Chemical modification of tumour blood flow, *Int. J. Hyperthermia* **4:**355–371.

Juhl, H., Sievers, M., Baltzer, K., Helmig, F., Wolf, H., Brenner, W., and Kalthoff, H., 1995, A monoclonal antibody cobra venom factor conjugate increases the tumor-specific uptake of a Tc-99m-labeled anti-carcinoembryonic antigen antibody by a two-step approach, *Cancer Res.* **55:** S5749–S5755.

Kahn, D., Williams, R. D., Seldin, D. W., Libertino, J. A., Hirschhorn, M., Dreicer, R., Weiner, G. J., Bushnell, D., and Gulfo, J., 1994, Radioimmunoscintigraphy with 111-indium-labeled CYT-356 for the detection of occult prostate cancer recurrence. *J. Urol.* **152:**1490–1495.

Kalofonos, H. P., Kosmas, C., Hird, V., Snook, D. E., and Epenetos, A. A., 1994, Targeting of tumours with murine and reshaped human monoclonal antibodies against placental alkaline phosphatase: Immunolocalisation, pharmacokinetics and immune response, *Eur. J. Cancer* **30A:**1842–1850.

Kameda, T., Yasui, W., Yoshida, K., Tsujino, T., Nakayama, H., Ito, M., Ito, H., and Tahara, E., 1990, Expression of ERBB2 in human gastric carcinomas: Relationship between $p185^{ERBB2}$ expression and the gene amplification, *Cancer Res.* **50:**8002–8009.

Keenan, A. M., Weinstein, J. N., and Carasquillo, J. A., 1987, Immunolymphoscintigraphy and the dose dependence of In-111-labeled T101 monoclonal antibody in patients with cutaneous T-cell lymphoma, *Cancer Res.* **47:**6093–6099.

Kennel, S. J., 1992, Effects of target antigen competition on distribution of monoclonal antibody to solid tumors, *Crit. Rev. Biochem. Mol. Biol.* **52:**1284–1290.

Kerr, D. E., Garrigues, U. S., Wallace, P. M., Hellstrom, K. E., Hellstrom, I., and Senter, P. D., 1993, Application of monoclonal antibodies against cytosine deaminase for the *in vivo* clearance of a cytosine deaminase immunoconjugate, *Bioconjugate Chem.* **4:**353–357.

King, C. R., Kraus, M. H., and Aaronson, S. A., 1985, Amplification of a novel v-erbB-related gene in a human mammary carcinoma, *Science* **229:**974–976.

King, C. R., Swain, S. M., Porter, L., Steinberg, S. M., Lippman, M. E., and Gelmann, E. P., 1989, Heterogeneous expression of erbB-2 messenger RNA in human breast cancer, *Cancer Res.* **49:**4185–4191.

Kjeldsen, T., Clausen, H., Hirohashi, S., Ogawa, T., Iijima, H., and Hakomori, S., 1988, Preparation and characterization of monoclonal antibodies directed to the tumor-associated O-linked sialosyl-2 → 6 alpha-*N*-acetylgalactosaminyl (sialosyl-Tn) epitope, *Cance; Res.* **48:**2214–2220.

Klagsbrun, M. and D'Amore, P. A., 1991, Regulators of angiogenesis, *Annu. Rev. Physiol.* **53:**217–239.

Knox, R. J., Friedlos, F., Jarman, M., and Roberts, J. J., 1988, A new cytotoxic, DNA interstrand cross-linking agent, 5-(aziridin-1-yl)-4-hydroxylamino-2-nitrobenzamide, is formed from 5-(aziridin-1-yl)-2,4-dinitrobenzamide (CB1954) by a nitroreductase enzyme in Walker carcinoma cells, *Biochem. Pharmacol.* **37:**4661–4669.

Korc, M., Chandrasekar, B., Yamanaka, Y., Friess, H., Buchier, M., and Beger, H. G., 1992, Over-expression of the epidermal growth factor receptor in human pancreatic cancer is associated with concomitant increases in the levels of epidermal growth factor and transforming growth factor alpha, *J. Clin. Invest.* **90:**1352–1360.

Köhler, G. and Milstein, C., 1975, Continuous cultures of fused cells secreting antibody of predefined specificity, *Nature* **256:**495–497.

Kraus, M. H., Issing, W., Miki, T., Popescu, N. C., and Aaronson, S. A., 1989, Isolation and characterization of ERBB3, a third member of the ERBB/epidermal growth factor receptor family: Evidence for overexpression in a subset of human mammary tumors, *Proc. Natl. Acad. Sci. USA* **86:**9193–9197.

Krichevsky, A., Campbell-Acevedo, E. A., Tong, J. Y., and Acevedo, H. F., 1995, Immunological detection of membrane-associated human luteinizing hormone correlates with gene expression in cultured human cancer and fetal cells, *Endocrinology* **136:**1034–1039.

Kufe, D. W., Nadler, L., Sargent, L., Shapiro, H., Hand, P., Austin, F., Colcher, D., and Schlom, J., 1983, Biological behavior of human breast carcinoma-associated antigens expressed during cellular proliferation, *Cancer Res.* **43:**851–857.

Kumar, R., Shepard, H. M., and Mendelsohn, J., 1991, Regulation of phosphorylation of the c-erbB-2/HER2 gene product by a monoclonal antibody and serum growth factor(s) in human mammary carcinoma cells, *Mol. Cell Biol.* **11:**979–986.

Kuroki, M., Arakawa, F., Haruno, M., Murakami, M., Wakisaka, M., Higuchi, H., Oikawa, S., Nakazato, H., and Matsuoka, Y., 1992, Biochemical characterization of 25 distinct carcinoembryonic antigen (CEA) epitopes recognized by 57 monoclonal antibodies and categorized into 7 groups in terms of domain structure of the CEA molecule, *Hybridoma* **11:**391–407.

Kuroki, M., Fernsten, P. D., Wunderlich, D., Colcher, D., Simpson, J. F., Poole, D. J., and Schlom, J., 1990, Serological mapping of the TAG-72 tumor-associated antigen using 19 distinct monoclonal antibodies, *Cancer Res.* **50:**4872–4879.

Kuroki, M., Ichiki, S., and Matsuoka, Y., 1984, Identification and partial characterization of the unglycosylated peptide of carcinoembryonic antigen synthesized by human tumor cell lines in the presence of tunicamycin, *Mol. Immunol.* **21:**743–746.

Kyriakos, R. J., Shih, L. B., Ong, G. L., Patel, K., Goldenberg, D. M., and Mattes, M. J., 1992, The fate of antibodies bound to the surface of tumor cells *in vitro*, *Cancer Res.* **52:**835–842.

Lan, M. S., Batra, S. K., Qi, W. N., Metzgar, R. S., and Hollingsworth, M. A., 1990, Cloning and sequencing of a human pancreatic tumor mucin cDNA, *J. Biol. Chem.* **265:**15294–15299.

Leberthon, B., Khawli, L. A., Alauddin, M., Miller, G. K., Charak, B. S., Mazumder, A., and Epstein, A. L., 1991, Enhanced tumor uptake of macromolecules induced by a novel vasoactive interleukin-2 immunoconjugate, *Cancer Res.* **51:**2694–2698.

Leek, J., Lench, N., Maraj, B., Bailey, A., Carr, I. M., Andersen, S., Cross, J., Whelan, P., MacLennan, K. A., Meredith, D. M., and Markham, A. F., 1995, Prostate-specific membrane antigen: Evidence for the existence of a second related human gene, *Br. J. Cancer* **72:**583–588.

Lemoine, N. R., Barnes, D. M., Hollywood, D. P., Hughes, C. M., Smith, P., Dublin, E., Prigent, S. A., Gullick, W. J., and Hurst, H. C., 1992a, Expression of the ERBB3 gene product in breast cancer, *Br. J. Cancer* **66:**1116–1121.

Lemoine, N. R., Lobresco, M., Leung, H., Barton, C., Hughes, C. M., Prigent, S. A., Gullick, W. J., and Kloppel, G., 1992b, The erbB-3 gene in human pancreatic cancer, *J. Pathol.* **168:**269–273.

Less, J. R., Skalak, T. C., Sevick, E. M., and Jain, R. K., 1991, Microvascular architecture in a mammary carcinoma: Branching patterns and vessel dimensions, *Cancer Res.* **51:**265–271.

Liao, S. K., Meranda, C., Avner, B. P., Romano, T., Husseini, S., Kimbro, B., and Oldham, R. K., 1989, Immunohistochemical phenotyping of human solid tumors with monoclonal antibodies in devising biotherapeutic strategies, *Cancer Immunol. Immunother.* **28:**77–86.

Libermann, T. A., Razon, N., Bartal, A. D., Yarden, Y., Schlessinger, J., and Soreq, H., 1984, Expression of epidermal growth factor receptors in human brain tumors, *Cancer Res.* **44:**753–760.

Ligtenberg, M. J., Kruijshaar, L., Buijs, F., van Meijer, M., Litvinov, S. V., and Hilkens, J., 1992, Cell-associated episialin is a complex containing two proteins derived from a common precursor, *J. Biol. Chem.* **267:**6171–6177.

Ligtenberg, M. J., Vos, H. L., Gennissen, A. M., and Hilkens, J., 1990, Episialin, a carcinoma-associated mucin, is generated by a polymorphic gene encoding splice variants with alternative amino termini, *J. Biol. Chem.* **265:**5573–5578.

Lopes, A. D., Davis, W. L., Rosenstraus, M. J., Uveges, A. J., and Gilman, S. C., 1990, Immunohistochemical and pharmacokinetic characterization of the site-specific immunoconjugate CYT-356 derived from antiprostate monoclonal antibody 7E11-C5, *Cancer Res.* **50:**6423–6429.

Madersbacher, S., Kratzik, C., Gerth, R., Dirnhofer, S., and Berger, P., 1994, Human chorionic gonadotropin (hCG) and its free subunits in hydrocele fluids and neoplastic tissue of testicular cancer patients: Insights into the *in vivo* hCG-secretion pattern, *Cancer Res.* **54:**5096–5100.

Maier, L. A., Xu, F. J., Hester, S., Boyer, C. M., McKenzie, S., Bruskin, A. M., Argon, Y., and Bast, R. C. J., 1991, Requirements for the internalization of a murine monoclonal antibody directed against the HER-2/neu gene product c-erbB-2, *Cancer Res.* **51:**5361–5369.

Majuri, M. L., Hakkarainen, M., Paavonen, T., and Renkonen, R., 1994, Carcinoembryonic antigen is expressed on endothelial cells. A putative mediator of tumor cell extravasation and metastasis, *APMIS* **102:**432–438.

Malaitsi, J., Zorzos, J., Varvarigou, A. D., Archimandritis, S., Dassiou, C., Skarlos, D. V., Dimitriou, P., Likourinas, M., Zizi, A., and Proukakis, C., 1995, Immunolocalisation of transitional cell carcinoma of the bladder with intravesically administered technetium-99m labelled HMFG1 monoclonal antibody, *Eur. J. Nucl. Med.* **22:**25–31.

Mangili, F., Sassi, I., Di Rocco, M., Leone, B. E., Garancini, P., and Santambrogio, G., 1996, Breast carcinoma detection with a combination of radiolabeled monoclonal antibodies. Promising results from immunohistochemistry studies, *Cancer* **78:**2334–2339.

Marcillac, I., Troalen, F., Bidart, J. M., Ghillani, P., Ribrag, V., Escudier, B., Malassagne, B., Droz, J. P., Lhomme, C., Rougier, P., Duvillard, P., Prade, M., Lugagne, P. M., Richard, F., Poynard, T., Bohuan, C., Wands, J., and Bellet, D., 1992, Free human chorionic gonadotropin beta subunit in gonadal and nongonadal neoplasms, *Cancer Res.* **52:**3901–3907.

Mardirossian, G., Brill, A. B., Dwyer, K. M., Kahn, D., and Nelp, W., 1996, Radiation absorbed dose from indium-111-CYT-356, *J. Nucl. Med.* **37:**1583–1588.

Masuho, Y., Zalutsky, M., Knapp, R. C., and Bast, R., 1985, Interaction of monoclonal antibodies with cell surface antigens of human ovarian carcinomas, *Cancer Res.* **44:**2813–2819.

Mattes, M. J., Major, P. P., Goldenberg, D. M., Dion, A. S., Hutter, R. V., and Klein, K. M., 1990, Patterns of antigen distribution in human carcinomas, *Cancer Res.* **50:**880s–884s.

Matzku, S., Tilgen, W., Kalthoff, H., Schmiegel, W. H., and Brocker, E.-B., 1988, Dynamics of antibody transport and internalisation, *Int. J. Cancer Suppl.* **2:**11–14.

Mauger, A. B., Burke, P. J., Somani, H. H., Friedlos, F., and Knox, R. J., 1994, Self-immolative prodrugs: Candidates for antibody-directed enzyme prodrug therapy in conjunction with a nitroreductase enzyme, *J. Med. Chem.* **37:**3452–3458.

McDicken, I. W., Stamp, G. H., McLaughlin, P. J., and Johnson, P. M., 1983, Expression of human placental-type alkaline phosphatase in primary breast cancer, *Int. J. Cancer* **32:**205–209.

McLaughlin, P. J., Travers, P. J., McDicken, I. W., and Johnson, P. M., 1984, Demonstration of placental and placental-like alkaline phosphatase in non-malignant human tissue extracts using monoclonal antibodies in an enzyme immunoassay, *Clin. Chim. Acta* **137:**341–348.

Melton, R. G., Rowland, J. A., Pietersz, G. A., Sherwood, R. F., and McKenzie, I. F. C., 1993, Tumour necrosis factor increases tumour uptake of co-administered antibody-carboxypeptidase-G_2 conjugate, *Eur. J. Cancer* **29A:**1177–1183.

Melton, R. G., Searle, F., Sherwood, R. F., Bagshawe, K. D., and Boden, J. A., 1990, The potential of carboxypeptidase G_2:antibody conjugates as anti-tumour agents. II. *In vivo* localising and clearance properties in a choriocarcinoma model, *Br. J. Cancer* **61:**420–424.

Melton, R. G. and Sherwood, R. F., 1996, Antibody-enzyme conjugates for cancer therapy, *J. Natl. Cancer Inst.* **88:**153–165.

Miller, G. K., Naeve, G. S., Gaffar, S. A., and Epstein, A. L., 1993, Immunologic and biochemical analysis of TNT-1 and TNT-2 monoclonal antibody binding to histones, *Hybridoma* **12:**689–698.

Mulshine, J. L., 1990, Bronchial epithelial changes in the early detection of lung cancer: Application of monoclonal antibodies in the management of lung cancer, *J. Jpn. Soc. Bronchol.* **12:**569–575.

Muraro, R., Wunderlich, D., Thor, A., Lundy, J., Noguchi, P., Cunningham, R., and Schlom, J., 1985, Definition by monoclonal antibodies of a repertoire of epitopes on carcinoembryonic antigen differentially expressed in human colon carcinomas versus normal adult tissues, *Cancer Res.* **45:**5769–5780.

Murray, J. C., Randhawa, S., and Denekamp, J., 1987, The effect of melphalan and misonidazole on the vasculature of a murine sarcoma, *Br. J. Cancer* **55:**233–238.

Murray, J. L., Macey, D. J., Grant, E. J., Rosenblum, M. G., Kasi, L. P., Zhang, H. Z., Gatz, R. L., Rieger, P. T., Lebherz, D., Bhadkamkar, V., Greiner, J. W., Schlom, J., and Podoloff, D. A., 1995, Enhanced TAG-72 expression and tumor uptake of radiolabeled monoclonal antibody CC49 in metastatic breast cancer patients following alpha-interferon treatment, *Cancer Res.* **55:**S5925–S5928.

Nagy, J. A., Brown, L. F., Senger, D. R., Lanir, N., Van-de-Water, L., Dvorak, A. M., and Dvorak, H. F., 1988, Pathogenesis of tumor stroma generation: A critical role for leaky blood vessels and fibrin deposition, *Biochim. Biophys. Acta* **948:**305–326.

Nakasaki, H., Mitomo, T., Noto, T., Ogoshi, K., Hanaue, H., Tanaka, Y., Makuuchi, H., Clausen, H., and Hakamori, S., 1989, Mosaicism in the expression of tumor-associated carbohydrate antigens in human colonic and gastric cancers, *Cancer Res.* **49:**3662–3669.

Natali, P. G., Cavaliere, R., Bigotti, A., Nicotra, M. R., Russo, C., Ng, A. K., Giacomini, P., and Ferrone, S., 1983, Antigenic heterogeneity of surgically removed primary and autologous metastatic human melanoma lesions, *J. Immunol.* **130:**1462–1466.

Niehans, G. A., Singleton, T. P., Dykoski, D., and Kiang, D. T., 1993, Stability of HER-2/neu expression over time and at multiple metastatic sites, *J. Natl. Cancer Inst.* **85:**1230–1235.

Nouwen, E. J., Pollet, D. E., Schelstraete, J. B., Eerdekens, M. W., Hansch, C., Van de Voorde, A., and De Broe, M. E., 1985, Human placental alkaline phosphatase in benign and malignant ovarian neoplasia, *Cancer Res.* **45:**892–902.

Nuti, M., Teramoto, Y. A., Mariani Costantini, R., Hand, P. H., Colcher, D., and Schlom, J., 1982, A monoclonal antibody (B72.3) defines patterns of distribution of a novel tumor-associated antigen in human mammary carcinoma cell populations, *Int. J. Cancer* **29:**539–545.

Ogata, S., Ho, I., Chen, A., Dubois, D., Maklansky, J., Singhal, A., Hakomori, S., and Itzkowitz, S. H., 1995, Tumor-associated sialylated antigens are constitutively expressed in normal human colonic mucosa, *Cancer Res.* **55:**1869–1874.

Ohannesian, D. W., Lotan, D., and Lotan, R., 1994, Concomitant increases in galectin-1 and its glycoconjugate ligands (carcinoembryonic antigen, lamp-1, and lamp-2) in cultured human colon carcinoma cells by sodium butyrate, *Cancer Res.* **54:**5992–6000.

Oikawa, S., Imajo, S., Noguchi, T., Kosaki, G., and Nakazato, H., 1987, The carcinoembryonic antigen (CEA) contains multiple immunoglobulin-like domains, *Biochem. Biophys. Res. Commun.* **144:** 634–642.

Oikawa, S., Inuzuka, C., Kuroki, M., Matsuoka, Y., Kosaki, G., and Nakazato, H., 1989, Cell adhesion activity of non-specific cross-reacting antigen (NCA) and carcinoembryonic antigen (CEA) expressed on CHO cell surface: Homophilic and heterophilic adhesion, *Biochem. Biophys. Res. Commun.* **164:**39–45.

Oikawa, S., Nakazato, H., and Kosaki, G., 1987, Primary structure of human carcinoembryonic antigen (CEA) deduced from cDNA sequence, *Biochem. Biophys. Res. Commun.* **142:**511–518.

Oredipe, O. A., Barth, R. F., Rotaru, J. H., Hinkle, G. H., and Steplewski, Z., 1990, Alterations in monoclonal antibody affinity and antigenic receptor site expression on mycoplasma-infected human colorectal cancer cells, *Proc. Soc. Exp. Biol. Med.* **194:**301–307.

Owens, R. J., King, D. J., Howat, D., Lisle, H., Mountain, A., Harvey, A., Bergin, S., Turner, A., Pedley, B., Boden, J., Begent, R., and Yarranton, G. T., 1991, Tumor binding properties of B72.3 Fv fragments, *Antibody Immunoconjugate Radiopharm.* **4:**459–467.

Pai, L. H., Bookman, M. A., Ozols, R. F., Young, R. C., Smith, J. W., Longo, D. L., Gould, B., Frankel, A., McClay, E. F., Howell, S., Reed, E., Willingham, M. C., Fitzgerald, D. J., and Pastan, I., 1991, Clinical evaluation of intraperitoneal *Pseudomonas* exotoxin immunoconjugate OVB3-PE in patients with ovarian cancer. *J. Clin. Oncol.* **9:**2095–2103.

Paxton, R. J., Mooser, G., Pande, H., Lee, T. D., and Shively, J. E., 1987, Sequence analysis of carcinoembryonic antigen: Identification of glycosylation sites and homology with the immunoglobulin supergene family, *Proc. Natl. Acad. Sci. USA* **84:**920–924.

Pedley, R. B., Boden, J. A., Keep, P. A., Harwood, P. J., Green, A. J., and Rogers, G. T., 1987, Relationship between tumour size and uptake of a radiolabelled anti-CEA in a colon tumour xenograft, *Eur. J. Nucl. Med.* **13:**197–202.

Pervez, S., Kirkland, S. C., Epenetos, A. A., Mooi, W. J., Evans, D. J., and Krausz, T., 1989, Effect of polarity and differentiation on antibody localization in multicellular tumour spheroid and xenograft models and its potential importance for *in vivo* immunotargeting, *Int. J. Cancer* **44:**940–947.

Pesando, J. M., Hoffman, P., and Abed, M., 1986, Antibody-induced antigenic modulation is antigen dependent: Characterization of 22 proteins on a malignant human B cell line. *J. Immunol.* **137:**3689–3695.

Pierce, J. G., and Parsons, T. F., 1981, Glycoprotein hormones: Structure and function, *Annu. Rev. Biochem.* **50:** 465–495.

Pimm, M. V., 1990, An examination of the influence of vasoactive drugs on blood flow and localisation of a monoclonal antibody in human tumour xenografts, *Br. J. Cancer* **62:**69–71.

Pimm, M. V., 1995, Circulating antigen: Bad or good for immunoscintigraphy? *Nucl. Med. Biol.* **22:** 137–145.

Pimm, M. V., Gribben, S. J., and Morris, T. M., 1991, Influence of recombinant tumour necrosis factor-alpha on blood flow and antibody localisation in human tumour xenografts in nude mice, *J. Cancer Res. Clin. Oncol.* **117:**543–548.

Plowman, G. D., Whitney, G. S., Neubauer, M. G., Green, J. M., McDonald, V. L., Todaro, G. J., and Shoyab, M., 1990, Molecular cloning and expression of an additional epidermal growth factor receptor-related gene. *Proc. Natl. Acad. Sci. USA* **87:**4905–4909.

Press, M. F., Cordon Cardo, C., and Slamon, D. J., 1990, Expression of the HER-2/neu proto-oncogene in normal human adult and fetal tissues, *Oncogene* **5:**953–962.

Press, M. F., Hung, G., Godolphin, W., and Slamon, D. J., 1994, Sensitivity of HER-2/neu antibodies in archival tissue samples: Potential source of error in immunohistochemical studies of oncogene expression, *Cancer Res.* **54:**2771–2777.

Press, M. F., Pike, M. C., Chazin, V. R., Hung, G., Udove, J. A., Markowicz, M., Danyluk, J., Godolphin, W., Sliwkowski, M., Akita, R., Peterson, M. C., and Slamon, D. J., 1993, Her-2/neu expression in node-negative breast cancer: Direct tissue quantitation by computerized image analysis and association of overexpression with increased risk of recurrent disease, *Cancer Res.* **53:**4960–4970.

Prigent, S. A., Lemoine, N. R., Hughes, C. M., Plowman, G. D., Selden, C., and Gullick, W. J., 1992, Expression of the c-erbB-3 protein in normal human adult and fetal tissues, *Oncogene* **7:**1273–1278.

Rachwal, W. J., Bongiorno, P. F., Orringer, M. B., Whyte, R. I., Ethier, S. P., and Beer, D. G., 1995, Expression and activation of erbB-2 and epidermal growth factor receptor in lung adenocarcinomas, *Br. J. Cancer* **72:**56–64.

Rajkumar, T., Gooden, C. S., Lemoine, N. R., Gullick, W. J., and Gooden, C. S., 1993, Expression of the c-erbB-3 protein in gastrointestinal tract tumours determined by monoclonal antibody RTJ1, *J. Pathol.* **170:**271–278.

Rajkumar, T., Stamp, G. W. H., Pandha, H. S., Waxman, J., and Gullick, W. J., 1996, Expression of the type 1 tyrosine kinase growth factor receptors EGF receptor, c-erbB2 and c-erbB3 in bladder cancer, *J. Pathol.* **179:**381–385.

Reist, C. J., Archer, G. E., Kurpad, S. N., Wikstrand, C. J., Vaidyanathan, G., Willingham, M. C., Moscatello, D. K., Wong, A. J., Bigner, D. D., and Zalutsky, M. R., 1995, Tumor-specific anti-epidermal growth factor receptor variant III monoclonal antibodies: Use of the tyramine-cellobiose radioiodination method enhances cellular retention and uptake in tumor xenografts, *Cancer Res.* **55:**4375–4382.

Reist, C. J., Garg, P. K., Alston, K. L., Bigner, D. D., and Zalutsky, M. R., 1996, Radioiodination of internalizing monoclonal antibodies using *N*-succinimidyl 5-iodo-3-pyridinecarboxylate, *Cancer Res.* **56:**4970–4977.

Rippe, B. and Haraldsson, B., 1987, Fluid and protein fluxes across small and large pores in the microvasculature, *Acta Physiol. Scand.* **131:**411–428.

Riva, P., Arista, A., Sturiale, C., Lazzari, S., Franech, G., Moscatelli, G., Saro, G., Campori, F., Riva, N., Casi, N., and Gentile, R., 1995, Intralesional therapy of glioma, in: *New Antibody Technology and the Emergence of Useful Cancer Therapy.* (R. Begent and A. Hamblin, eds.), p. 2326.

Ro, J., North, S. M., Gallick, G. E., Hortobagyi, G. N., Gutterman, J. U., and Blick, M., 1988, Amplified and overexpressed epidermal growth factor receptor gene in uncultured primary human breast carcinoma. *Cancer Res.* **48:**161–164.

Robertson, K. W., Reeves, J. R., Smith, G., Keith, W. N., Ozanne, B. W., Cooke, T. G., and Stanton, P. D., 1996, Quantitative estimation of epidermal growth factor receptor and c-erbB-2 in human breast cancer, *Cancer Res.* **56:**3823–3830.

Rodrigues, M. L., Presta, L. G., Kotts, C. E., Wirth, C., Mordenti, J., Osaka, G., Wong, W. L. T., Nuijens, A., Blackburn, B., and Carter, P., 1995, Development of a humanized disulfide-stabilized anti-beta p185(HER2) Fv-beta-lactamase fusion protein for activation of a cephalosporin doxorubicin prodrug, *Cancer Res.* **55:**63–70.

Rogers, G. T., 1976, Heterogeneity of carcinoembryonic antigen. Implications on its role as a tumour marker substance, *Biochim. Biophys. Acta* **458:**355–373.

Rogers, G. T., Burke, P. J., Sharma, S. K., Koodie, R., and Boden, J. A., 1995, Plasma clearance of an antibody-enzyme conjugate in ADEPT by monoclonal anti-enzyme: Its effect on prodrug activation *in vitro*, *Br. J. Cancer* **72:**1357–1363.

Roselli, M., Guadagni, F., Buonomo, O., Belardi, A., Vittorini, V., Cavazzana, A., Spagnoli, L. G., Cipriani, C., and Casciani, C. U., 1996, Intraoperative radioimmunolocalization of an anti-CEA Mab $F(Ab')_2$ (FO23C5) in CEA serum-negative colorectal cancer patients, *Anticancer Res.* **16:**883–889.

Rowlinson-Busza, G., Bamias, A., Krausz, T., and Epenetos, A. A., 1991, Uptake and distribution of specific and control monoclonal antibodies in subcutaneous xenografts following intra-tumour injection, *Cancer Res.* **51:**3251–3256.

Rowlinson-Busza, G., Deonarain, M. P., and Epenetos, A. A., 1996, Comparison of intact monoclonal antibody, its $F(ab')_2$ and Fab fragments and recombinant single chain Fv in a human tumor xenograft model, *Tumor Targeting* **2:**37–48.

Rowlinson-Busza, G., Maraveyas, A., and Epenetos, A. A., 1995, Effect of tumour necrosis factor on the uptake of specific and control monoclonal antibodies in a human tumour xenograft model, *Br. J. Cancer* **71:**660–665.

Rowlinson, G., Balkwill, F., Snook, D., Hooker, G., and Epenetos, A. A., 1986, Enhancement by gamma-interferon of *in vivo* tumour radiolocalisation by a monoclonal antibody against HLA-DR antigen, *Cancer Res.* **46:**6413–6417.

Russell, S. M., Krauer, K. G., McKenzie, I. F. C., and Pietersz, G. A., 1990, Effect of tumour necrosis factor on the anti-tumour efficacy and toxicity of aminopterin-monoclonal antibody conjugates: Parameters for optimisation of therapy, *Cancer Res.* **50:**6028–6033.

Sands, H., Jones, P. L., Shah, S. A., Palme, D., Vessella, R. L., and Gallagher, B. M., 1988, Correlation of vascular permeability and blood flow with monoclonal antibody uptake by human Clouser and renal cell xenografts, *Cancer Res.* **48:**188–193.

Sanidas, E. E., Filipe, M. I., Linehan, J., Lemoine, N. R., Gullick, W. J., Rajkumar, T., and Levison, D. A., 1993, Expression of the c-erbB-3 gene product in gastric cancer, *Int. J. Cancer* **54:**935–940.

Schechter, A. L., Stern, D. F., Vaidyanathan, L., Decker, S. J., Drebin, J. A., Greene, M. I., and Weinberg, R. A., 1984, The neu oncogene: An erb-B-related gene encoding a 185,000-Mr tumour antigen, *Nature* **312:**513–516.

Schlom, J., Eggensperger, D., Colcher, D., Molinolo, A., Houchens, D., Miller, L. S., Hinkle, G., and Siler, K., 1992, Therapeutic advantage of high-affinity anticarcinoma radioimmunoconjugates, *Cancer Res.* **52:**1067–1072.

Schmidt, F., Florent, J. C., Monneret, C., Straub, R., Czech, J., Gerken, M., and Bosslet, K., 1997, Glucuronide prodrugs of hydroxy compounds for antibody directed enzyme prodrug therapy (ADEPT): A phenol nitrogen mustard carbamate, *Bioorg. Med. Chem. Lett.* **7:**1071–1076.

Schroff, R. W., Farrell, M. M., Klein, R. A., Oldham, R. K., and Foon, K. A., 1984, T65 antigen modulation in a phase I monoclonal antibody trial with chronic lymphocytic leukemia patients, *J. Immunol.* **133:**1641–1648.

Schultz, K. R., Badger, C. C., Dombi, G. W., Greenberg, P. D., and Bernstein, I. D., 1992, Effect of interleukin-2 on biodistribution of monoclonal antibody in tumor and normal tissues in mice bearing SL-2 thymoma, *J. Natl. Cancer Inst.* **84:**109–113.

Schumacher, U., Adam, E., Horny, H. P., and Dietl, J., 1996, Transplantation of a human ovarian cystadenocarcinoma into severe combined immunodeficient (SCID) mice—Formation of metastases without significant alteration of the tumour cell phenotype, *Int. J. Exp. Pathol.* **77:**219–227.

Seko, A., Ohkura, T., Kitamura, H., Yonezawa, S., Sato, E., and Yamashita, K., 1996, Quantitative differences in GlcNAc:beta1 → 3 and GlcNAc:beta1 → 4 galactosyltransferase activities between human colonic adenocarcinomas and normal colonic mucosa, *Cancer Res.* **56:**3468–3473.

Semba, K., Kamata, N., Toyoshima, K., and Yamamoto, T., 1985, A v-erbB-related protooncogene, c-erbB-2, is distinct from the c-erbB-1/epidermal growth factor-receptor gene and is amplified in a human salivary gland adenocarcinoma, *Proc. Natl. Acad. Sci. USA* **82:**6497–6501.

Sharkey, R. M., Juweid, M., Shevitz, J., Behr, T., Dunn, R., Swayne, L. C., Wong, G. Y., Blumenthal, R. D., Griffiths, G. L., Siegel, J. A., Leung, S., Hansen, H. J., and Goldenberg, D. M., 1995, Evaluation of a complementarity-determining region-grafted (humanized) anti-carcinoembryonic antigen monoclonal antibody in preclinical and clinical studies. *Cancer Res.* **55:**5935s–5945s.

Shih, C., Padhy, L. C., Murray, M., and Weinberg, R. A., 1981, Transforming genes of carcinomas and neuroblastomas introduced into mouse fibroblasts, *Nature* **290:**261–264.

Siemann, D. W., 1984, Modification of chemotherapy by nitroimidazoles, *Int. J. Radiat. Oncol. Biol. Phys.* **10:**1585–1594.

Silver, D. A., Pellicer, L., Fair, W. R., Heston, W. D. W., and Cordon-Cardo, C., 1997, Prostate-specific membrane antigen expression in normal and malignant human tissues, Clin. Can. Res. **3:**81:–85.

Simpson, B. J., Weatherill, J., Miller, E. P., Lessells, A. M., Langdon, S. P., and Miller, W. R., 1995, C-erbB-3 protein expression in ovarian tumours, *Br. J. Cancer* **71:**758–762.

Sivinski, C. L., Lindner, D. J., Borden, E. C., and Tempero, M. A., 1995, Modulation of tumor-associated antigen expression on human pancreatic and prostate carcinoma cells *in vitro* by alpha- and gamma-interferons, *J. Immunother.* **18:**156–165.

Slamon, D. J., Godolphin, W., Jones, L. A., Holt, J. A., Wong, S. G., Keith, D. E., Levin, W. J., Stuart, S. G., Udove, J., Ullrich, A., and Press, M. F., 1989, Studies of the HER-2/neu proto-oncogene in human breast and ovarian cancer, *Science* **244:**707–712.

Smyth, M. J., Pietersz, G. A., and McKenzie, I. F. C., 1987, The use of vasoactive agents to increase tumour perfusion and the anti-tumour efficiency of drug-monoclonal antibody conjugates, *J. Natl. Cancer Inst.* **79:**1367–1368.

Smyth, M. J., Pietersz, G. A., and McKenzie, I. F. C., 1988, Increased anti-tumor effect of immunoconjugates and tumour necrosis factor *in vivo*, *Cancer Res.* **48:**3607–3612.

Springer, C. J., Bagshawe, K. D., Sharma, S. K., Searle, F., Boden, J. A., Antoniw, P., Burke, P. J., Rogers, G. T., Sherwood, R. F., and Melton, R. G., 1991, Ablation of human choriocarcinoma xenografts in nude mice by antibody-directed enzyme prodrug therapy (ADEPT) with three novel compounds, *Eur. J. Cancer* **27:**1361–1366.

Springer, G. F., 1984, T and Tn, general carcinoma autoantigens, *Science* **224:**1198–1206.

Stramignoni, D., Bowen, R., Atkinson, B. F., and Schlom, J., 1983, Differential reactivity of monoclonal antibodies with human colon adenocarcinomas and adenomas, *Int. J. Cancer* **31:**543–552.

Su, S. L., Huang, I. P., Fair, W. R., Powell, C. T., and Heston, W. D., 1995, Alternatively spliced variants of prostate-specific membrane antigen RNA: Ratio of expression as a potential measurement of progression, *Cancer Res.* **55:**1441–1443.

Sutherland, R., Buchegger, F., Schreyer, M., Vacca, A., and Mach, J.-P., 1997, Penetration and binding of radiolabelled anti-carcinoembryonic antigen monoclonal antibodies and their antigen binding fragments in human colon cancer multicellular tumor spheroids, *Cancer Res.* **47:**1627–1633.

Swabb, E. A., Wei, J., and Gullino, P. M., 1974, Diffusion and convection in normal and neoplastic tissues, *Cancer Res.* **34:**2814–2822.

Swallow, D. M., Gendler, S., Griffiths, B., Corney, G., Taylor Papadimitriou, J., and Bramwell, M. E.,

1987, The human tumour-associated epithelial mucins are coded by an expressed hypervariable gene locus PUM, *Nature* **328:**82–84.

Tateishi, M., Ishida, T., Mitsudomi, T., Kaneko, S., and Sugimachi, K., 1990, Immunohistochemical evidence of autocrine growth factors in adenocarcinoma of the human lung, *Cancer Res.* **50:** 7077–7080.

Teglund, S., Olsen, A., Khan, W. N., Frangsmyr, L., and Hammarstrom, S., 1994, The pregnancy-specific glycoprotein (PSG) gene cluster on human chromosome 19: Fine structure of the 11 PSG genes and identification of 6 new genes forming a third subgroup within the carcinoembryonic antigen (CEA) family, *Genomics* **23:**669–684.

Thor, A., Ohuchi, N., Szpak, C. A., Johnston, W. W., and Schlom, J., 1986, Distribution of oncofetal antigen tumor-associated glycoprotein-72 defined by monoclonal antibody B72.3, *Cancer Res.* **46:**3118–3124.

Thor, A., Viglione, M. J., Muraro, R., Ohuchi, N., Schlom, J., and Gorstein, F., 1987, Monoclonal antibody B72.3 reactivity with human endometrium: A study of normal and malignant tissues, *Int. J. Gynecol. Pathol.* **6:**235–247.

Togo, S., Shimada, H., Kubota, T., Moossa, A. R., and Hoffman, R. M., 1995, Host organ specifically determines cancer progression, *Cancer Res.* **55:**681–684.

Travis, A., Pinder, S. E., Robertson, J. F., Bell, J. A., Wencyk, P., Gullick, W. J., Nicholson, R. I., Poller, D. N., Blamey, R. W., Elston, C. W., and Ellis, I. O., 1996, C-erbB-3 in human breast carcinoma: Expression and relation to prognosis and established prognostic indicators, *Br. J. Cancer* **74:** 229–233.

Tsuchida, T., Ravindranath, M. H., Saxton, R. E., and Irie, R. F., 1987, Gangliosides of human melanoma: Altered expression *in vivo* and *in vitro*, *Cancer Res.* **47:**1278–1281.

Ullrich, A. and Schlessinger, J., 1990, Signal transduction by receptors with tyrosine kinase activity, *Cell* **61:**203–212.

van Duk, J., Zegveld, S. Th., Fleuren, G. J., and Warnaar, S. O., 1991, Localisation of monoclonal antibody G250 and bispecific monoclonal antibody CD3/G250 in human renal-cell carcinoma xenografts: relative effects of size and affinity, *Int. J. Cancer* **48:**738–743.

Velders, M. P., Vanrhijn, C. M., Cornelissen, I. M. H. A., Vanmuijen, G. N. P., Briaire, I. H., Dohlsten, M., Fleuren, G. J., Warnaar, S. O., and Litvinov, S. V., 1996, The role of monoclonal antibody affinity in tumor immunotherapy evaluated in *in vivo* models for minimal residual disease, *J. Immunother.* **19:**245–256.

Vogel, C. W., 1991, Cobra venom factor: The complement-activating protein of cobra venom, in: *Handbook of Natural Toxins* (A. T. Tu, ed.), Marcel Dekker, New York, p. 147.

Wahl, R. L., Parker, C. W., and Philpott, G. W., 1983, Improved radioimaging and tumor localisation with monoclonal $F(ab')_2$, *J. Nucl. Med.* **24:**316–325.

Wang, X., Fu, X., Brown, P. D., Crimmin, M. J., and Hoffman, R. M., 1994, Matrix metalloproteinase inhibitor BB-94 (Batimastat) inhibits human colon tumour growth in a patient-like orthotopic model in nude mice, *Cancer Res.* **54:**4726–4728.

Wang, X., Fu, X., and Hoffman, R. M., 1992a, A new patient-like metastatic model of human lung cancer constructed orthotopically with intact tissue via thoracotomy in immuno-deficient mice, *Int. J. Cancer* **51:**992–995.

Wang, X., Fu, X., Kubota, T., and Hoffman, R. M., 1992b, A new patient-like metastatic model of human small-cell lung cancer constructed orthotopically with intact tissue via thoracotomy in nude mice, *Anticancer Res.* **12:**1403–1406.

Warren, B., 1979, The vascular morphology of tumours, in: *Tumour Blood Circulation*. (H. Peterson, ed.), CRC Press, Boca Raton, FL, p. 1.

Watanabe, H., Tokuyama, H., Ohta, H., Satomura, Y., Okai, T., Ooi, A., Mai, M., and Sawabu, N., 1990, Expression of placental alkaline phosphatase in gastric and colorectal cancers. An immunohistochemical study using the prepared monoclonal antibody, *Cancer* **66:**2575–2582.

Weinstein, J. N., 1990, Antibody lymphoscintigraphy, in: *Cancer Imaging with Radiolabelled Antibodies.* (D. M. Goldenberg, ed.), Kluwer, Boston, p. 365.

Weinstein, J. N. and van Osdol, W., 1992, Early intervention in cancer using monoclonal antibodies and other biological ligands: Micropharmacology and the "Binding Site Barrier," *Cancer Res.* **52:** 2747s–2751s.

Wikstrand, C. J., Hale, L. P., Batra, S. K., Hill, M. L., Humphrey, P. A., Kurpad, S. N., Mclendon, R. E., Moscatello, D., Pegram, C. N., Reist, C. J., Traweek, S. T., Wong, A. J., Zalutsky, M. R., and Bigner, D. J., 1995, Monoclonal antibodies against EGFRvIII are tumor specific and react with breast and lung carcinomas and malignant gliomas, *Cancer Res.* **55:**3140–3148.

Wong, A. J., Bigner, S. H., Bigner, D. D., Kinzler, K. W., Hamilton, S. R., and Vogelstein, B., 1987, Increased expression of the epidermal growth factor receptor gene in malignant gliomas is invariably associated with gene amplification, *Proc. Natl. Acad. Sci. USA* **84:**6899–6903.

Wong, J. Y. C., Williams, L. E., Yamauchi, D. M., Odom-Maryon, T., Esteban, J. M., Neumaier, M., Wu, A. M., Johnson, D. K., Primus, F. J., Shively, J. E., and Raubitschek, A. A., 1995, Initial experience evaluating ^{90}Yttrium-radiolabeled anti-carcinoembryonic antigen chimeric T84.66 in a Phase I radioimmunotherapy trial, *Cancer Res.* **55:**5929s–5934s.

Wright, G. L., Haley, C., Beckett, M. L., and Schellhammer, P. F., 1995, Expression of prostate-specific membrane antigen in normal, benign and malignant prostate tissues, *Urol. Oncol.* **1:**18–28.

Xu, M., Real, F. X., Welt, S., Schussler, M. H., Oettgen, H. F., and Old, L. J., 1989, Expression of TAG-72 in normal colon, transitional mucosa, and colon cancer, *Int. J. Cancer* **44:**985–989.

Yamamoto, T., Ikawa, S., Akiyama, T., Semba, K., Nomura, N., Miyajima, N., Saito, T., and Toyoshima, K., 1986, Similarity of protein encoded by the human c-erb-B-2 gene to epidermal growth factor receptor, *Nature* **319:**230–234.

Yamashita, K., Totani, K., Kuroki, M., Matsuoka, Y., Ueda, I., and Kobata, A., 1987, Structural studies of the carbohydrate moieties of carcinoembryonic antigens, *Cancer Res.* **47:**3451–3459.

Yokota, T., Milenic, D. E., Whitlow, M., and Schlom, J., 1992, Rapid tumour penetration of a single-chain Fv and comparison with other immunoglobulin forms. *Cancer Res.* **52:**3402–3408.

Yuan, F., Baxter, L. T., and Jain, R. K., 1991, Pharmacokinetic analysis of two-step approaches using bifunctional and enzyme-conjugated antibodies, *Cancer Res.* **51:**3119–3130.

Zaretsky, J. Z., Weiss, M., Tsarfaty, I., Hareuveni, M., Wreschner, D. H., and Keydar, I., 1990, Expression of genes coding for pS2, c-erbB2, estrogen receptor and the H23 breast tumor-associated antigen. A comparative analysis in breast cancer, *FEBS Lett.* **265:**46–50.

Zhu, X., Lai, C., Thomas, S., and Burden, S. J., 1995, Neuregulin receptors, erbB3 and erbB4, are localized at neuromuscular synapses, *EMBO J.* **14:**5842–5848.

Zotter, S., Hageman, P. C., Lossnitzer, A., Mooi, W. J., and Hilgers, J., 1988, Tissue and tumour distribution of human polymorphic epithelial mucin, *Cancer Res.* **11/12:**55–101.

Zwi, J. L., Baguley, B. C., Gavin, J. B., and Wilson, W. R., 1994, The morphological effects of the anti-tumour agents flavone acetic acid and 5,6-dimethylxanthenone acetic acid on the colon 38 mouse tumour, *Pathology* **24:**161–169.

4

Enzymes and Prodrugs Used for ADEPT

Richard J. Knox

1. INTRODUCTION

At present, chemotherapy is not very effective against common solid cancers especially once they have metastasized because the majority of the agents now used in cancer chemotherapy act by an antiproliferative mechanism. Thus, their action is on cells that are in cycle rather than by specific toxicity directed toward a particular type of cancer cell. Even for the more resistant solid cancers, cures could be achieved if the total dose given was about a hundredfold the equivalent of a normal clinical dose of present day agents. However, this cannot be achieved because of the toxic effect on the normal most rapidly dividing host tissues, such as bone marrow, gut mucosa, and the lymphatic system.

Many approaches have been examined to overcome the intrinsic problems associated with administering cytotoxic drugs. One of these is drug targeting. This came with the advent of monoclonal antibody technology in the 1970s and at the time was heralded as the fruition of the search for the "magic bullet" first proposed by Ehrlich in 1906 (Ehrlich 1960). This involves conjugating a cytotoxic agent to an antibody or other molecules that can specifically bind to antigen or receptor sites that are overexpressed on the target tumor cell. However, the goal has proved elusive, and a number of limitations are apparent in using this approach to treat solid tumors in humans. These include poor penetration of the tumor, lack of intertumoral accumulation of the cytotoxic component and the heterogeneity of tumor-associated antigens (Pimm, 1988; Shockley *et al.*, 1992; Wawrzynczak,

---

Richard J. Knox • Department of Medical Oncology, Charing Cross Hospital, London, W6 8RP, United Kingdom.

Enzyme-Prodrug Strategies for Cancer Therapy, edited by Melton and Knox. Kluwer Academic/Plenum Publishers, New York, 1999.

1991). However, Bagshawe (Bagshawe, 1987; Bagshawe *et al.*, 1988) suggested how the disappointing clinical results of drug targeting could be overcome. Activating enzymes, not drugs, could be targeted to human tumors using tumor-associated monoclonal antibodies before administering a prodrug. This approach has been termed ADEPT (antibody-directed enzyme prodrug therapy). For ADEPT, an enzyme of nonhuman or nonmammalian origin can be used, and this could metabolize substrates not normally activated in humans. The enzyme is linked to a tumor-associated antibody (this can be done chemically or by using recombinant DNA techniques) and allowed to localize to the tumor (Bagshawe, 1987, 1994, 1995; Bagshawe *et al.*, 1988). Thus, ADEPT creates a tumoral environment with a high concentration of an enzyme that would convert a normally inert substrate to a highly reactive metabolite. The unlocalized conjugate is given time to be eliminated from the body or is hastened on its way by a "clearance" antibody. Then a nontoxic prodrug is administered and activated at the site of the tumor by the bound conjugate. The active drug can be of low molecular weight and therefore can rapidly diffuse and reach tumoral regions inaccessible to the antibody-enzyme conjugate (Bagshawe 1987, 1994, 1995; Bagshawe *et al.*, 1988). The targeted antibody-enzyme conjugate remains extracellular. It should not be internalized because the conjugate would be expected to be rapidly degraded in the lysosomal compartment. Further, the prodrugs can exploit the external enzyme by being charged and thus excluded from the cell until activated. If secreted antigen accumulates within the interstitial spaces of the tumor, it can also be exploited to achieve higher conjugate levels at the tumor. A major advantage of the ADEPT approach (over, for example, drug targeting or immunotoxins) is that it is catalytic and a single enzyme molecule can, in theory, generate many hundreds of active molecules per second from the prodrug. Further, because the enzyme-conjugate is, by itself, inactive, it can be allowed to clear before administering the prodrug. Not all tumor cells would be required to bind the antibody because the drug released by the enzyme could diffuse to neighboring cells that are not expressing the antigen. It is fundamental to the concept of ADEPT that the prodrug is not activated by normal human enzymes. Therefore the enzyme must have little equivalent activity in humans, particularly in serum because the prodrug may be excluded from entering cells. Further, the enzyme must be active under physiological conditions and remain active when conjugated to an antibody.

This chapter describes the prodrugs and, in particular, the enzymes that have been proposed for ADEPT.

2. ENZYMES AND PRODRUGS USED FOR ADEPT

The prodrug is an integral component of ADEPT systems and requires careful design in its own right. An ideal prodrug is one that has a large differential in cytotoxicity between drug and prodrug, is a good substrate for the enzyme under physiological conditions, and has no mammalian homologue capable of performing the same reaction. Equal cytotoxicity of the released active drug toward proliferating and quiescent cells is also desirable if residual deposits of viable but nonproliferating cells with the potential for outgrowth are to be eradicated. Once formed, it would be desirable for the drug to have a very short half-life to limit the possibility of active drug escaping back into the circulation and of access to healthy tissue. Development of drug resistance may limit the effects of conventional active drugs produced from prodrugs. Thus the nitrogen mustard group of compounds are the most commonly used active drugs for ADEPT because they are not cell-cycle-specific, they kill well-oxygenated and hypoxic cells, and cells develop only low levels of resistance to them (Frei *et al.*, 1988).

With respect to the prodrug activating enzyme, it is possible to use a known enzyme and identify or design prodrugs for it or, alternatively to design a prodrug (based on a proven active agent) and search for a suitable enzyme. Both approaches have been used. It is even becoming possible to design the required specificity into an enzyme by protein engineering. However, the following criteria required for using an enzyme in ADEPT remain the same (although in no examples were they are all totally fulfilled!):

1. No equivalent activity in humans
2. High specificity (low K_m) and high k_{cat} for the prodrug
3. Active at neutral pH
4. Nonreversible reaction
5. Stable at 37°C
6. Stable to conjugation to an antibody
7. Available in large amounts (recombinant expression)
8. No cosubstrate requirements
9. Monomeric form
10. No inhibition by common substances

A nonmammalian, usually microbial, enzyme with no mammalian equivalent activity is the ideal choice to meet the criteria for high specificity with minimal endogenous activation of the prodrug. However, such foreign enzymes may raise immunogenic responses and thus preclude repetitive therapies. Human enzymes may circumvent the immune response but risk activation of the prodrug in nontarget tissues. This could be avoided if the enzyme were normally highly compartmentalized or if its substrate specificity were engineered for the prodrug, although, in these cases, the enzyme may still be recognized as "foreign."

2.1. Glucose Oxidase and Xanthine Oxidase

Although not designed for ADEPT, mention should also be made of two radical-generating systems. Glucose oxidase was the first example of an antibody-

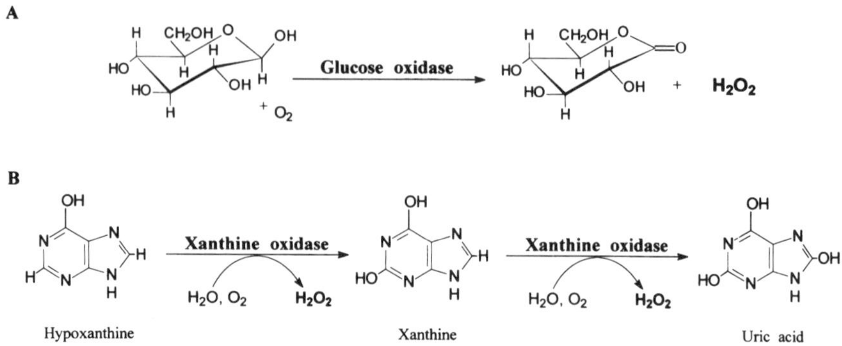

Figure 1. The formation of hydrogen peroxide from either (A) glucose by the enzyme glucose oxidase or (B) from hypoxanthine or xanthine by the enzyme xanthine oxidase.

targeted enzyme (Muzykantov *et al.*, 1988, 1990; Stanislawski *et al.*, 1989) and was made before the advent of monoclonal antibody technology. The enzyme generates hydrogen peroxide upon oxidation of glucose (Fig. 1A). However, the system was not very cytotoxic and glucose would be a poor choice of a prodrug for ADEPT because it is normally present in serum. Another radical-generating system is xanthine oxidase oxidizing either xanthine or hypoxanthine (Fig. 1B) (Ito *et al.*, 1989, 1990). Xanthine oxidase is a fairly ubiquitous enzyme and is expressed in many mammalian tissues, such as the liver. However, *in vivo*, the liver enzyme exists mainly in the dehydrogenase form and thus produces NADH from NAD^+ and not hydrogen peroxide (the dehydrogenase is converted to the oxidase by the action of proteolytic enzymes or simply by storage at −20°C and this happens readily *in vitro*). Radical-generating systems may not be very suitable for ADEPT, but such conjugates may prove useful for applications requiring a high degree of selectivity, for example in purging T cells in bone marrow transplantation (Ito *et al.*, 1989, 1990).

2.2. Carboxypeptidase G2

The first ADEPT system proposed by Bagshawe in 1987 (Bagshawe, 1987) used carboxypeptidase G2 (CPG2) to cleave a deactivating glutamate moiety from a benzoic acid mustard (Springer *et al.*, 1990). CPG2 is a dimeric protein (Mr 83 kDa) consisting of two identical, zinc-containing, subunits and was originally isolated from *Pseudomonas sp* RS-16 and subsequently cloned in *E. coli* (Minton *et al.*, 1983). This class of enzymes has an antitumor effect in their own right by depleting the reduced folate pool (Romanini *et al.*, 1989) but was actually prepared in large quantities for methotrexate rescue protocols (DeAngelis *et al.*, 1996). CPG2 remains the best characterized ADEPT system and is the only system to date for which clinical studies have commenced (Bagshawe, 1995; Bagshawe *et al.*, 1991). The initial prodrug synthesized was 4-[*N,N*-bis(2-chloroethyl)amino] benzoyl-L-glutamic acid from which CPG2 cleaves the glutamic acid to yield 4-[*N,N*-bis(2-chloroethyl)amino]benzoic acid (Fig. 2). This compound is an excellent substrate for CPG2 (K_m = 3 μM, k_{cat} = 700 s^{-1}). Subsequently, modified versions were developed with one or both of the 2-chloroethyl arms replaced by more reactive 2-mesyloxyethyl moieties. The mono-(2-mesyloxyethyl) compound (MMCl) was more cytotoxic than the bis-(2-chloroethyl) compound, but the bis-(2-mesyloxyethyl) form was too reactive and rapidly hydrolyzed in the absence of enzyme (Springer *et al.*, 1990, 1995). MMC1 also remained a good substrate for CPG2 (K_m = 3.4 μM, k_{cat} = 580 s^{-1}). The enzyme conjugate/prodrug combination was initially tested *in vitro* using the JAR choriocarcinoma cell line and gave greater than 100-fold differential toxicity between drug or prodrug plus enzyme (IC_{50} 20 μM) and prodrug (IC_{50} > 800 μM) (Bagshawe *et al.*, 1988). In

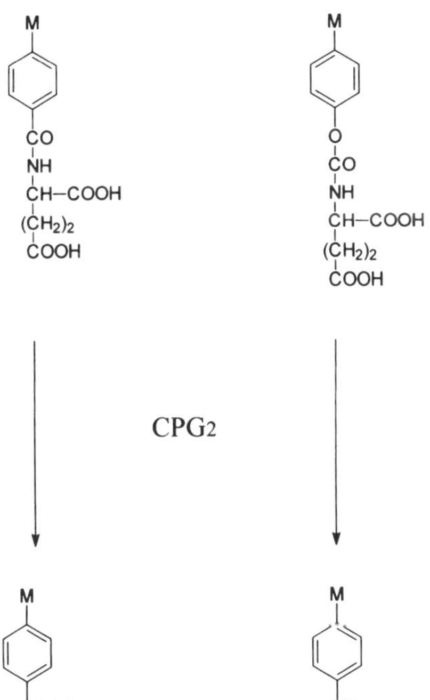

Figure 2. The activation of glutamate prodrugs to active mustards by the enzyme carboxypeptidase G2.

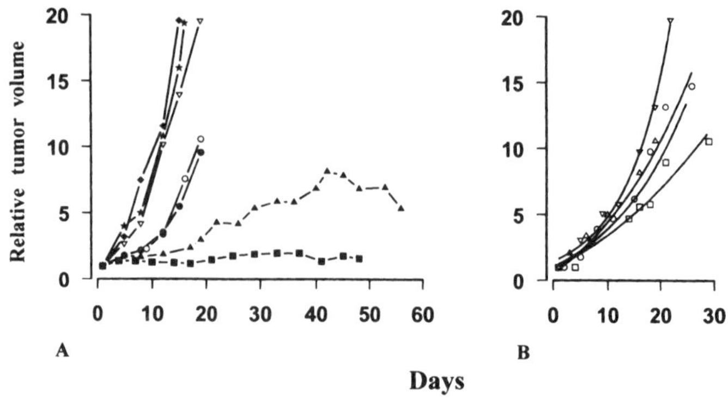

Figure 3. The effect of ADEPT therapy on the growth of a xenograft *in vivo*. Groups of mice bearing the CC3 choriocarcinoma xenograft were treated (A) intravenously with a W14 F(ab′)2:Carboxypeptidase G2 conjugate followed 56, 72, and 80h later with either 5 mg/kg (▲) or 10 mg/kg (■) of p-N-bis(2-chloroethyl)aminobenzoyl glutamic acid prodrug. This can be compared with conventional therapy with either 5 mg/kg methotrexate (▽), 50 mg/kg hydroxyurea (◆), 7.5 μg/ml actinomycin D (●), 20 mg/kg cyclophosphamide (★), or 20 mg/kg ara-C (○). (B) with either saline (△), conjugate alone (○), prodrug alone (□), or active drug alone (▽) (both at their maximum tolerated doses). (Data from Bagshawe *et al.*, 1988; Bagshawe, 1989).

antitumor studies using the CC3 human choriocarcinoma xenograft model in nude mice, the combination of W14 F(ab′)2-CPG2 conjugate followed and the MMCl prodrug resulted in a dramatic growth delay of greater than 50 days (Fig. 3A). The prodrug was administered as a three-times divided dose of 22.5 mg/kg, 72 hr after administration of the conjugate (Fig. 3A) (Bagshawe, 1989; Bagshawe *et al.*, 1988). In contrast, weekly intravenous injections of divided doses of methotrexate (5 mg/kg), hydroxyurea (50 mg/kg), actinomycin D (7.5 μg/kg) or cytosine arabinoside (20 mg/kg) failed to retard tumor growth (Fig. 3B) (Bagshawe, 1989; Bagshawe *et al.*, 1988). Similar encouraging results have also been reported with a conjugate in which the antibody recognizes the c-erbB2 antigen (Eccles *et al.*, 1994). However, when the same two-phase system was applied to the carcinoembryonic antigen (CEA)-expressing LS174T colon carcinoma model, using conjugate constructed with F(ab′)2 fragment of A5B7 anti-CEA antibody, it proved less effective, although growth delays of 10–15 days were still seen (Blakey *et al.*, 1993) probably because of the much slower clearance of the conjugate from the blood in this model compared to the choriocarcinoma system. In the latter case, high levels of circulating antigen probably form immunocomplexes leading to accelerated clearance. However, much lower levels of antigen are present in the circulation of CEA-expressing tumors, and thus the clearing effect of free antigen is not apparent. In view of this, a second galactosylated clearing antibody, SB43, which inactivates CPG2, has been used to remove unbound conjugate from the circulation, making the system a three-phase approach (Sharma *et al.*, 1990, 1991, 1994a,b). When the three-phase system was tested against the same model, prolonged growth delays of LS174T xenografts were seen using the MMCl compound and, to a lesser extent, the bischloro compound as prodrugs (Sharma *et al.*, 1991, 1994a,b).

The preclinical studies using the colon carcinoma xenograft model and the MMCl compound were sufficiently encouraging to justify a limited clinical trial on patients suffering from advanced metastatic, poorly differentiated colon or rectal cancer (Bagshawe *et al.*, 1991, 1994). Mylosuppression was observed and this was ascribed to the fact that active drug was reaching the bone marrow and indeed the active drug could be detected in the clinical serum samples (Springer *et al.*, 1993). This suggested that the release of a more reactive drug would be beneficial. With the demonstration that carbamates were also reasonable substrates for CPG2 ($K_m = 1\ \mu M$, $k_{cat} = 30\ s^{-1}$), the way was open for a series of derivatives that would be converted by the enzyme to an aniline derivative rather than the acid (Fig. 2) (Blakey *et al.*, 1995; Dowell *et al.*, 1996). Aniline mustards are highly toxic and reactive chemicals that would probably hydrolyze before reaching sensitive host tissues. Prodrugs of this structure will probably be used in future ADEPT clinical trials with this enzyme (Springer *et al.*, 1995).

2.3. Carboxypeptidase A and B

Like CPG2, these enzymes are zinc-containing exopeptidases but are very different from CPG2. Carboxypeptidase A has a much wider substrate specificity than CPG2 and a preference for substrates with either a C-terminal phenylalanine or alanine. As isolated from bovine pancreas, it is a monomer of about 34 kDa. Carboxypeptidase B (isolated from porcine pancreas) is also a 34 kDa monomer but has a preference for a C-terminal arginine or lysine. The work reported concentrated on Carboxypeptidase A that had been conjugated to a monoclonal antibody (KS1/4) which was raised against a lung adenocarcinoma cell line (UCLA-P3) and was used to activate a prodrug. This prodrug, based on methotrexate (Fig. 4), is deactivated (about 200×) by adding an alanine moiety but is not rapidly turned over by the enzyme ($k_{cat} = 0.37\ s^{-1}$) compared to the prodrugs activated by CPG2 (Esswein *et al.*, 1991; Haenseler *et al.*, 1992; Kuefner *et al.*, 1989). The results reported have been disappointing. The prodrug was cytotoxic on prolonged exposure to the cells (perhaps due to cell surface peptidases), and the conjugate reactivated the prodrug only by a factor of about 6 (Haenseler *et al.*, 1992). Improved prodrugs have been made by substituting the alanine moiety with another amino acid, such as tyrosine or phenylalanine (Perron and Page, 1996; Vitols *et al.*, 1995), that are the best substrates for carboxypeptidase A. Thus, as

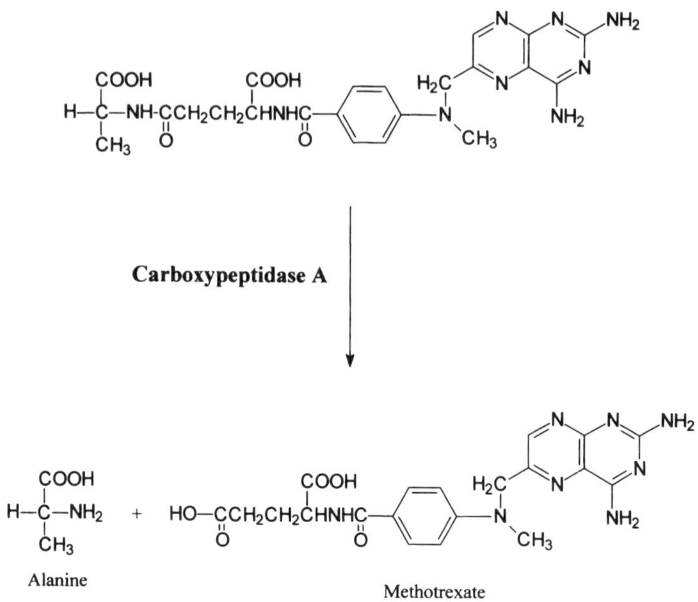

Figure 4. The generation of methotrexate from a prodrug by the enzyme carboxypeptidase A.

expected, greater than 10-fold more enzyme is required to activate methotrexate-α-alanine than to make methotrexate-α-phenylalanine equicytotoxic to methotrexate (Vitols *et al.*, 1995). Methotrexate-α-phenylalanine was the best substrate for CPA compared with 10 other methotrexate-α-peptides and showed moderate activity in an *in vitro* assay (Perron and Page, 1996). Prodrugs of quinazoline antifolates have also been synthesized (Springer *et al.*, 1996). The α-carboxyl of the glutamic acid of the thymidylate synthase inhibitors [*N*-[5-[*N*-(3,4-dihydro-2-methyl-4-oxoquinazolin-6-ylmethyl)-*N*-methylamino]-2-thenoyl]-L-glutamic acid] (ZD1694) and {*N*-[4-[*N*-[(2-methyl-3,4-dihydro-4-oxo-6-quinazolinyl)-methyl]-*N*-prop-2-ynylamino]benzoyl]-L-glutamic acid} (ICI-198583) has been linked through an amide bond to an L-alanine or an L-glutamic acid. These dipeptides had greatly decreased cytotoxicity against the LS174T human colorectal tumor cell line, compared to the parent drug, about 100× for the ZD1694 prodrugs and 20–200× for the ICI-198583 prodrugs. The L-glutamic acid based dipeptides were only poorly activated by adding CPA. However, activation of the α-linked L-alanine dipeptides with CPA led to a 10-fold increase in the cytotoxicity of the ZD1694 based prodrug and a 100-fold increase in the cytotoxicity of the ICI-198583 prodrug (Springer *et al.*, 1996).

Human prepro-CPA homologues have also been cloned and their kinetics studied to evaluate them as potential candidates for use in a human-based ADEPT approach. The kinetics are similar to the porcine enzyme (Laethem *et al.*, 1996).

No animal studies have been reported for any carboxypeptidase A system.

2.4. Aminopeptidase

This is a group of exopeptidases that cleave either N-terminal α-amino acids or pyroglutamic acid from peptides. A series of prodrugs of methotrexate has been synthesized in which the 2-amino group of the pteridine ring is acylated with various α-aminoacids or L-pyroglutamic acid. A free amino group is essential for tightly binding methotrexate to its target, dihydrofolate reductase. Thus these prodrugs have a greatly reduced inhibitory effect on this enzyme, and an aminopeptidase, with the appropriate substrate specificity, hydrolyzes the prodrug to regenerate methotrexate (Fig. 5). Compounds have been proposed for use in either prodrug therapy or ADEPT. For prodrug therapy the compounds are activated by endogenous aminopeptidase N that is overexpressed on some tumors. For ADEPT, the best prodrug candidate was 2-L-pyroglutamyl-methotrexate. This prodrug is

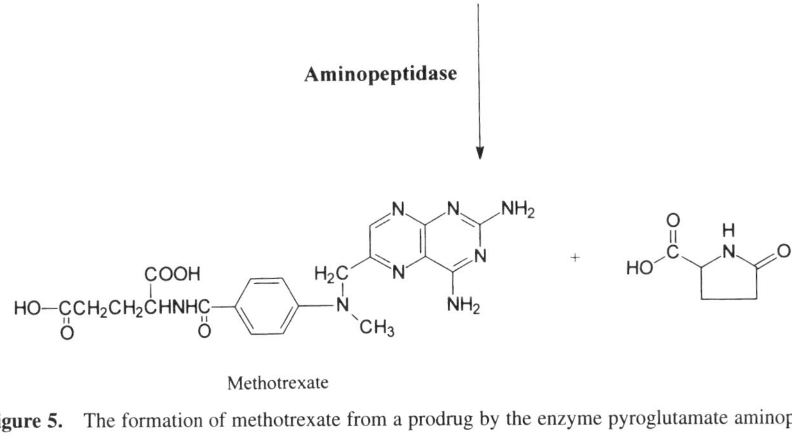

Methotrexate

Figure 5. The formation of methotrexate from a prodrug by the enzyme pyroglutamate aminopeptidase.

hydrolyzed to methotrexate by pyroglutamate aminopeptidase isolated from *Bacillus amyloliquefaciens* or D-aminopeptidase from *Ochrobactrum anthropi* (Smal *et al.*, 1995). No cytotoxicity studies have been reported.

2.5. Alkaline Phosphatase

Alkaline phosphatase, a ubiquitous enzyme that has a relative lack of substrate specificity, liberates phosphate ions from a wide range of organic phosphates. A number of groups have presented evidence that alkaline phosphatase activates a range of prodrugs (Fig. 6). Human alkaline phosphatase, in conjunction with a humanized or human antibody, may result in a conjugate that has reduced immunogenicity compared with bacterial enzyme conjugates. Calf intestinal alkaline phosphatase (a dimeric protein, Mr 140 kDa) has been used to cleave phosphate groups from etoposide phosphate (k_{cat} = 144 s^{-1}) (Haisma *et al.*, 1992b; Senter *et al.*, 1988, 1989), mitomycin c phosphate (k_{cat} = 98 s^{-1}) (Sahin *et al.*, 1990; Senter *et al.*, 1989), doxorubicin phosphate (Senter, 1990), and phenol mustard phosphate (k_{cat} = 3 s^{-1}) (Wallace and Senter, 1991). Using a human colon carcinoma cell line (H3347), etoposide phosphate was more than a hundred times less cytotoxic than the released drug, etoposide (Haisma *et al.*, 1992b). A conjugate of alkaline phosphatase with the L6 antibody showed antigen-specific cytotoxicity *in vitro* and growth delays *in vivo* using human colon carcinoma (H3347) xenograft models and the etoposide phosphate prodrug. An interesting approach, which has been tested with alkaline phosphate is using a bispecific antibody (Sahin *et al.*, 1990). In this case the antibody (expressed from a hybrid hybridoma) recognized

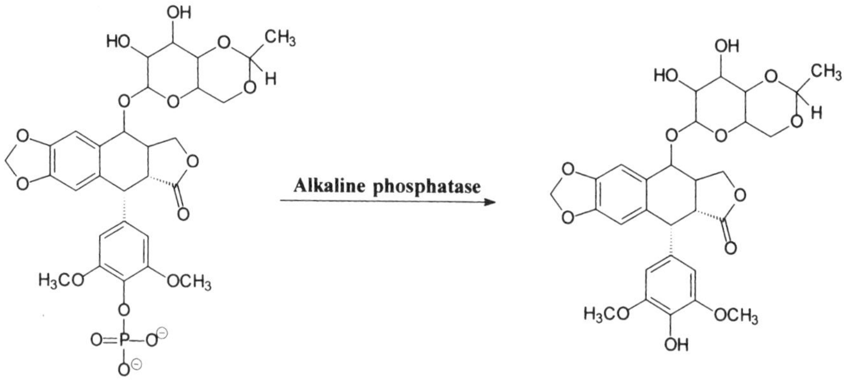

Figure 6. The formation of etoposide from etoposide phosphate by the enzyme alkaline phosphatase.

both the antigen (CD-30) and the enzyme alkaline phosphate. The antibody was incubated with alkaline phosphate and then tested on a CD-30 expressing Hodgkin's lymphoma-derived cell line, where it activated a mitomycin phosphate prodrug. Immunological coupling eliminates the need for chemically cross-linking the components of a conjugate. It is simpler and avoids the loss of enzyme activity that occurs by chemical conjugation methods, but the antibody must not significantly inhibit the enzyme when it binds.

Alkaline phosphatase may not represent an ideal ADEPT enzyme because it is widely distributed in the blood and its selectivity for drug release at the tumor site is limited by the endogenous enzyme that activates prodrugs in normal tissues. For example, etopofos (etoposide phosphate) is itself undergoing development as an orally administered prodrug of etoposide relying on systemic activation (Chabot *et al.*, 1996; Doyle and Vyas, 1990; Schacter *et al.*, 1994).

2.6. β-Glucuronidase

Endogenous β-glucuronidase was proposed as a target for glucuronide-deactivated prodrug therapy more than 30 years ago (Connors and Whisson, 1966). However, it was found clinically unfeasible because high enzyme activity was rare and not predictable by tumor type (Young *et al.*, 1976). Now systems have been proposed that involve targeting β-glucuronidase to the tumor. The *E. coli* form of the enzyme is a 70 kDa monomer with a pH optimum of 6.8. In contrast the human placental (lysosomal) is a 300 kDa tetramer with an optimum around pH 5. The *E. coli* enzyme has been used in conjunction with the prodrug form of aniline mustard (Fig. 7) (Roffler *et al.*, 1991; Wang *et al.*, 1992) and a glucuronide of epirubicin (4′-deoxydoxorubicin) (Haisma *et al.*, 1992b). This latter compound is actually isolated as a urinary metabolite from patients treated with epirubicin and is reported to be 100–1000× less cytotoxic than the parent compound (Fig. 8). Thus it has been suggested that the active drug formed from either of these prodrugs will be metabolized in the liver back to the glucuronide, thus minimizing circulating levels of the active compound and damage to normal tissue (Roffler *et al.*, 1991). The basis for the difference in cytotoxicity between the drug and the glucuronide prodrug is that the glucuronide is charged and thus excluded from entering cells (Fig. 9). Chemical conjugates of β-glucuronidase have been made with monoclonal antibodies that recognize a pan-carcinoma membrane glycoprotein (Haisma *et al.*, 1992a) or an epitope present on colon carcinoma (Roffler *et al.*, 1991) and activation of both prodrugs has been demonstrated. The rate of internalization of the conjugate was studied and was low (90% of bound conjugate remains on the cell surface after 24 hr incubation at 37°C). However, it was also found that the enzyme is inhibited by serum proteins and

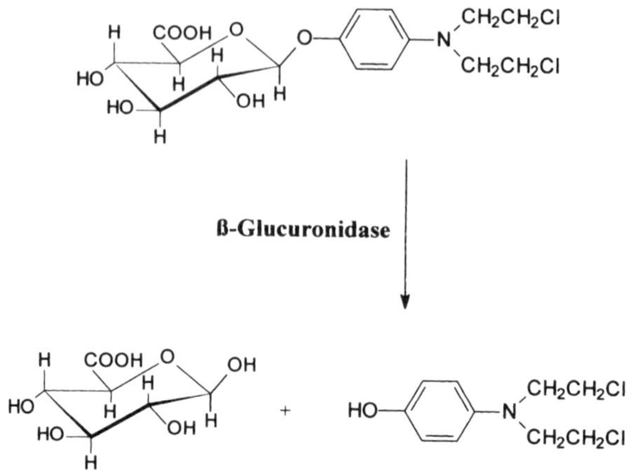

Figure 7. The generation of aniline mustard from its glucuronide by the enzyme β-glucuronidase.

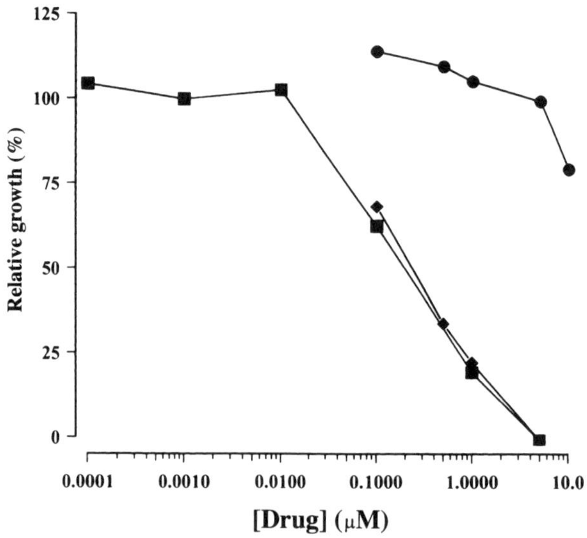

Figure 8. The growth inhibition of human MCF-7 cells by either epirubicin (■), epirubicin-glucuronide (●) or by epirubicin-glucuronide after the cells were preincubated with a β-glucuronidase-anti-pan carcinoma monoclonal antibody conjugate (◆) Unbound conjugate was removed by washing before drug treatment. The cells were exposed to the drugs for 24 hr and growth was measured at 72 hr. (Data from Haisma *et al.*, 1992a).

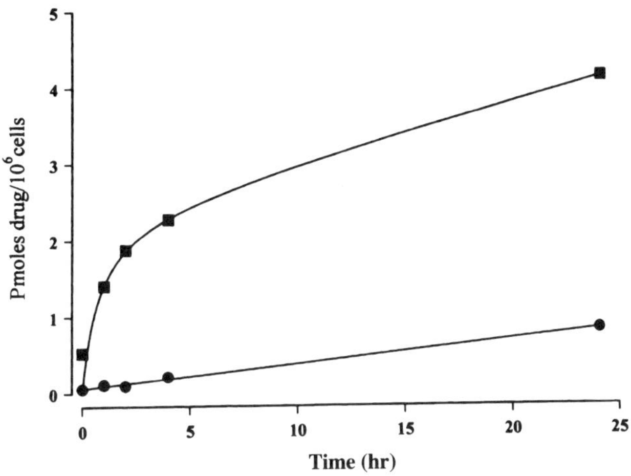

Figure 9. The cellular uptake of either 10 μM epirubicin (■) or epirubicin-glucuronide (●) into human OVCAR-3 cells. (Data from Haisma *et al.*, 1992a).

loses about 50% of its activity after 24 hr incubation (Senter, 1990). These are important but rarely addressed considerations.

Epirubicin-glucuronide is poorly turned over by β-glucuronidase (K_m = 10 μM, but k_{cat} = 0.046 s^{-1}), but hydrolysis rates are improved when daunorubicin glucuronides is synthesized by coupling a glucuronide group to daunorubicin via an aliphatic or an aromatic carbamate spacer (Houba *et al.*, 1996). A further class of glucuronidyl prodrugs based on anthracycline compounds have been synthesized. These prodrugs use an benzylcarbamate self-immolative linker between a glucuronide (or galactose) and the primary amine group of daunosamine, the carbohydrate constituent of daunorubicin or doxorubicin. However, although the prodrug form shows markedly reduced cytotoxicity against L1210 cells (IC_{50} for prodrug >1 μg/ml compared to daunorubicin (IC_{50} 0.02 μg/ml), it is not yet known whether these compounds are substrates for glucuronidase (or galactosidase) (Andrianomenjanahary *et al.*, 1992). A similar approach has been used to produce a phenol nitrogen mustard carbamate (Schmidt *et al.*, 1997). It is more stable than the prodrug produced by directly attaching the mustard to β-glucuronic acid and is a substrate for *E. coli* β-glucuronidase (Schmidt *et al.*, 1997), although the kinetics have not been published. The cytotoxicity was measured against LoVo cells. The prodrug has an IC_{50} > 830 μM, but this decreases to 10.5 μM in the presence of *E. coli* β-glucuronidase, equal to that of the free drug (Schmidt *et al.*, 1997).

A fusion protein has been constructed from the humanized Fab fragment of the anti-CEA antibody BW431 and human glucuronidase. The enzyme component of the fusion protein is presumably a monomer, but the fusion protein had antigenic and enzymatic activity ($K_m = 1.3$ mM, $k_{cat} = 0.8\ s^{-1}$). Maximum activity was seen at pH 4.5 and activity fell to 10–20% of this at pH 7, which is expected from the human enzyme. The fusion protein was tested *in vivo* using the prodrug *N*-(4-β-glucuronyl-3-nitrobenzyloxycarbonyl)-doxorubicin. Therapeutic responses were seen in CEA-expressing xenografts, and 4- to 12-fold higher doxorubicin concentrations were found in the tumors of the mice given the ADEPT treatment compared to those given drug alone (Bosslet *et al.*, 1992, 1994). Fusion proteins, such as this, address the anticipated immune response against nonhuman monoclonal antibodies and protein.

2.7. β-Glucosidase

This enzyme, isolated from sweet almonds (*Amygdalae dulces*), catalyzes the conversion of the disaccharide amygdalin to benzaldehyde and hydrogen cyanide (Fig. 10). Amygdalin is the main component of Laetrile®. Laetrile was not effec-

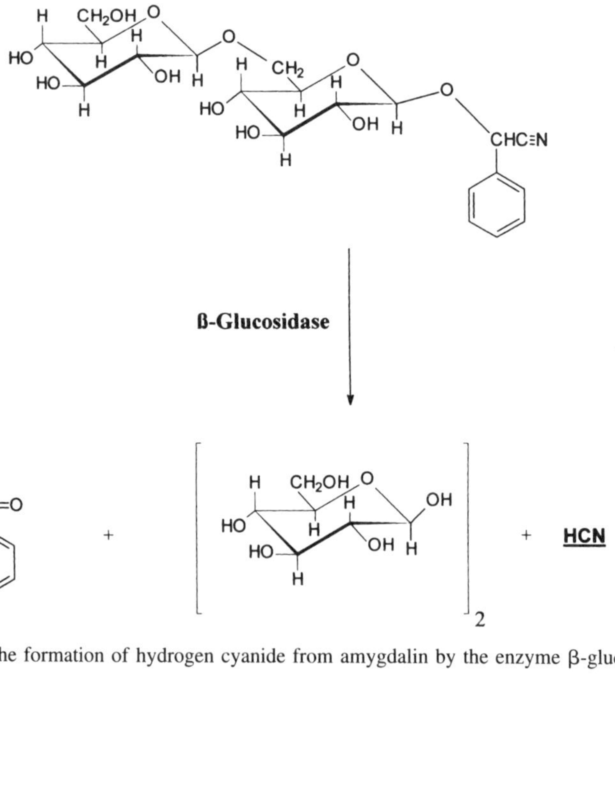

Figure 10. The formation of hydrogen cyanide from amygdalin by the enzyme β-glucosidase.

tive as an anticancer treatment because amygdalin is not a substrate for human tumor glucosidase. The almond enzyme has a much broader range of specificity and efficiently hydrolyzes this disaccharide. A system known as AGENT (Antibody-Guided Enzyme Nitrile Therapy) has been proposed that uses β-glucosidase conjugated to a tumor-specific antibody. *In vitro*, a cell-bound conjugate (antiplacental alkaline phosphatase) increases the cytotoxicity of amygdalin about a 1000-fold (Rowlinson-Busza *et al.*, 1992; Rowlinson-Busza and Epenetos, 1992). The combination of a specific conjugate and amygdalin is almost as effective as potassium cyanide, whereas a nonspecific conjugate plus amygdalin is much less effective. Contrary to what most people think, cyanide is not particularly cytotoxic, and its IC_{50} of 200 μM reported in this system is much higher than expected from the other active agents generated in ADEPT systems. However, the cytotoxicity of amygdalin is very low, and the ability to give a very large dose may generate sufficient cyanide at a tumor site.

2.8. β-Lactamase

The β-lactamases are a group of enzymes that are widely distributed in both gram-negative and gram-positive bacteria. They are vary in specificity, but all are capable of hydrolyzing β-lactams to substituted β-amino acids and thereby confer resistance to β-lactams containing antibiotics (penicillins and cephalosporins). Some act more readily on penicillins whereas other have greater activity against cephalosporins. β-Lactamase activity is not endogenous to the mammalian systems, and therefore it is subject to minimal interference from inhibitors, enzyme substrates, or endogenous enzyme systems. The β-lactamases are monomers. The molecular mass of enzyme isolated from *Enterobacter cloacae* is 43 kDa whereas that from *E. coli* is 30 kDa and has high specific activity against its natural substrates. Activity is optimal at neutral pH and 37°C. The synthesis of a range of 3′-substituted-cephalosporin derivatives of various mustards has been described (Alexander *et al.*, 1991; Svensson *et al.*, 1992), and these are relatively good substrates for the enzyme ($K_m = 7–108$ μM, $k_{cat} = 120–510\ s^{-1}$). Prodrugs based on doxorubucin have also been described (Vrudhula *et al.*, 1995). The enzyme opens the lactam ring and expels the 3′-subtituent. *In vivo*, an immunological specific antitumor effect was observed with a mustard prodrug when used with a β-lactamase (from *Enterobacter cloacae* 265A) conjugate which was targeted toward a tumor antigen (Svensson *et al.*, 1994; Vrudhula *et al.*, 1993b). The antitumor activity of a F(ab′)-β-lactamase conjugate (again the enzyme was isolated from *Enterobacter cloacae* 265A) to activate a cephalosporin derivative of the vinca alkaloid, 4-desacetylvinblastine-3-carboxhydrazide (DAVLBHYD) (Fig. 11), has

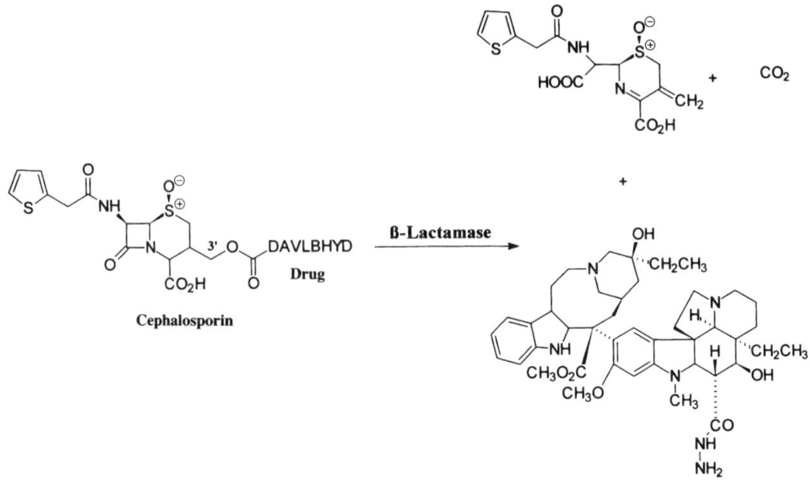

Figure 11. The activation of a prodrug to 4-desacetylvinblastine-3-carboxhydrazide (DAVLBHYD) by β-lactamase.

also been investigated (Meyer *et al.*, 1992, 1993, 1995; Shepherd *et al.*, 1991). This is an excellent substrate (K_m = 9 μM, k_{cat} = 3300 s^{-1}) that rapidly releases the active vinca alkaloid. Conjugates were made that targeted either CEA or KS1/4 and were used to treat the LS174T tumor xenograft *in vivo*. The prodrug, the active drug, and an irrelevant conjugate had a mild antitumor effect. However, the ADEPT treatment resulted in tumor regression and long-term stabilization (Fig. 12). Importantly, this study has shown that single-binding arm [F(ab′)] conjugates mediate an ADEPT effect. Such conjugates should facilitate rapid clearance from the circulation and should be easier to produce by recombinant techniques. A *B. cereus* β-lactamase II, single-chain antibody fusion protein has also been constructed. The L6 antibody used recognizes an antigen expressed on breast, colon, lung, and ovarian carcinomas. The fusion protein is produced by *E. coli* in an active form that does not require refolding and retains immunological and enzymatic properties similar to an equivalent chemically linked conjugate (Bosslet *et al.*, 1992, 1994). A disulfide-linked fusion protein, also been described, consists of *E. coli* β-lactamase and the Fv region of a humanized antibody (A64D5-8) raised against the *c-erb*B2 human breast antigen (Rodrigues *et al.*, 1995). Such fusion proteins are a good alternative to chemical conjugation. In this context, a bifunctional chimeric antibody has been made in which one antigen-binding arm of the antibody (raised against human placental alkaline phosphatase) is replaced by *E. coli* β-lactamase (De-Sutter and Fiers, 1994).

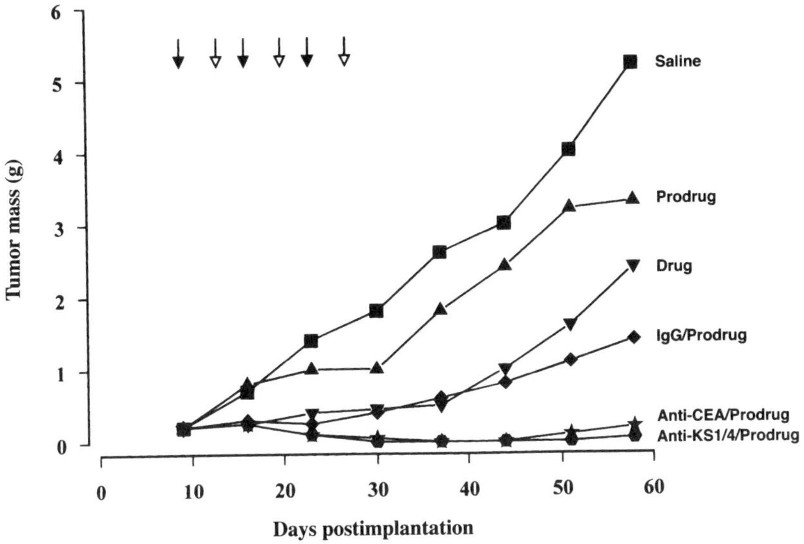

Figure 12. The effect of ADEPT therapy on the growth of the LS174T tumor xenograft *in vivo*. A β-lactamase conjugate (35 μg) was administered at the times indicated by the solid arrows, and this was followed after 96 hr by the prodrug of DAVLBHYD (4 mg/kg, open arrows). The control groups were treated with saline followed by the prodrug or the equivalent dose of DAVLBHYD. IgG was used as an irrelevant antibody (Data from Meyer *et al.*, 1993).

2.9. Penicillin Amidase

This enzyme is isolated from a wide range of bacteria or fungi and hydrolyzes various penicillins to produce a carboxylic acid and 6-aminopenicillanate. It also hydrolyzes a variety of phenylacetamides unrelated to penicillin. This lack of substrate specificity has been exploited by developing phenylacetyl-substituted derivatives of anticancer agents. Two forms of the enzyme have been investigated for ADEPT. The first is penicillin-V (phenoxymethyl penicillin) amidase from *Fusarium oxysporum*. This enzyme is readily available because it is used industrially to produce synthetic penicillins. The second is penicillin G (benzylpenicillin) amidase from *E. coli*. Both are momomeric, zinc-containing enzymes of about 70 kDa. Although they have a good temperature and pH profile, their specific activity is rather low. Penicillin G amidase has the advantage for ADEPT of not being glycosolysated (unlike penicillin-V amidase) and thus is not likely to have an accelerated clearance rate from the body. Penicillin V amidase was used to activate prodrugs prepared by coupling its substrate (*p*-hydroxyphenooxyacetic acid) with either doxorubicin or melphalan (Fig. 13). These phenylacetyl deriva-

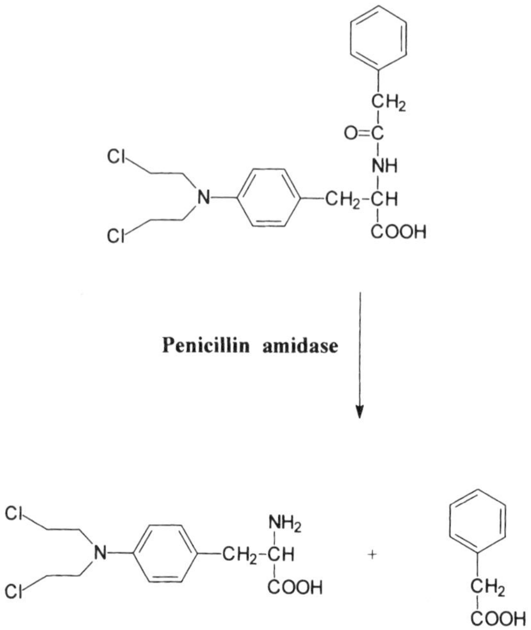

Figure 13. The hydrolysis of the phenylacetyl derivative of melphalan by the enzyme penicillin amidase to generate melphalan.

tives are substrates for penicillin-V amidase although of low specific activity (doxorubicin phenoxyacetamide, $k_{cat} = 0.3\ s^{-1}$; melphalan phenoxyacetamide, $k_{cat} = 0.01\ s^{-1}$ (Kerr *et al.*, 1990). Depending on the cell line, the doxorubicin prodrug was 20–80× less cytotoxic than the parent compound whereas the melphalan compound was 100–1000× less active. Activation of the doxorubicin prodrug occurs in an antigen-specific manner with cell-bound conjugate (about a 10-fold increase in cytotoxicity). However, there was no significant increase in cytotoxicity when the melphalan prodrug was used, and it was concluded that penicillin-V amidase could not hydrolyze the prodrug fast enough to achieve a cytotoxic concentration (Kerr *et al.*, 1990). Later studies have shown that penicillin G amidase conjugates activate either of these prodrugs or a water soluble analog [*N*-(4-aminophenylacetyl) doxorubicin] of the doxorubicin-based prodrug in an antigen-specific manner (Vrudhula *et al.*, 1993a). Cells were treated with the conjugate and unbound material was washed off. In the case of the phenylacetamido derivative of melphalan, cells subsequently treated with the prodrug were as cytotoxically affected as those cells treated with melphalan alone (Fig. 14) (Vrudhula *et al.*, 1993a). Thus, in this case, the prodrug is efficiently activated by antigen-bound conjugate.

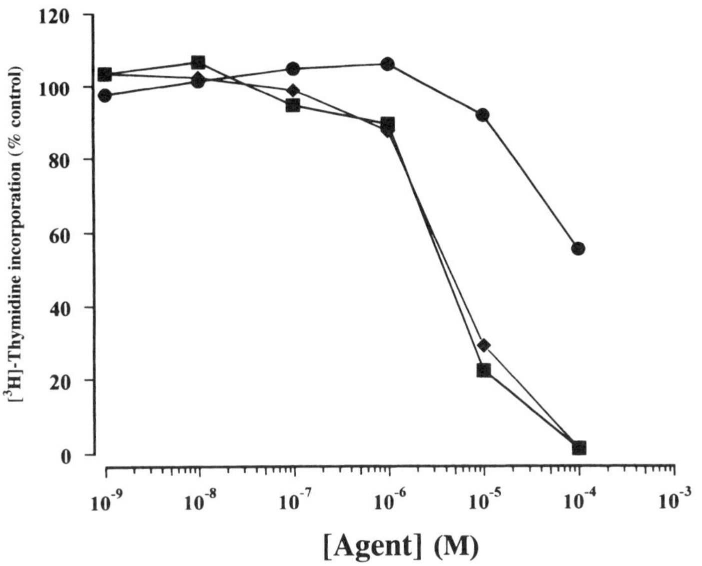

Figure 14. The *in vitro* cytotoxicity of the phenylacetyl derivative of melphalan on L6 antigen positive lung cells. Cells were treated with either medium or a L6-penicillin G amidase conjugate (0.1 mg/mL). The prodrug was added to the cells after unbound material was washed off. (●) Prodrug alone (no antibody). (♦)melphalan, (■) prodrug + conjugate. (Data from Vrudhula *et al.*, 1993a).

Penicillin G amidase has also been used to activate the 4′-hydroxyphenyl-acetamide derivative of palytoxin (which was about 100× less cytotoxic than palytoxin itself) (Bignami *et al.*, 1992). Palytoxin is a large, naturally occurring, highly cytotoxic compound that was isolated from soft coral (*Palythoa tuberculosa*). It is active at picomolar concentrations (compounds, such as mustard or doxorubicin, are active at micromolar concentrations) and acts at the cell surface by binding to a receptor and creating a pore in the cell membrane. Thus, unlike the other drugs released in ADEPT systems, it does not need to enter a cell to exert its effects. Further, it is unlikely to elicit drug-resistant cells. That such a large and complex molecule as the prodrug of palytoxin is a substrate for penicillin amidase illustrates the diverse specificity of this enzyme.

2.10. Cytosine Deaminase

This enzyme catalyzes the conversion of cytosine to uracil but also converts the clinically used antifungal agent 5-fluorocytosine (5-FC) into the known antitumor drug 5-fluorouracil (5-FU) (Fig. 15). This enzyme prodrug/prodrug combi-

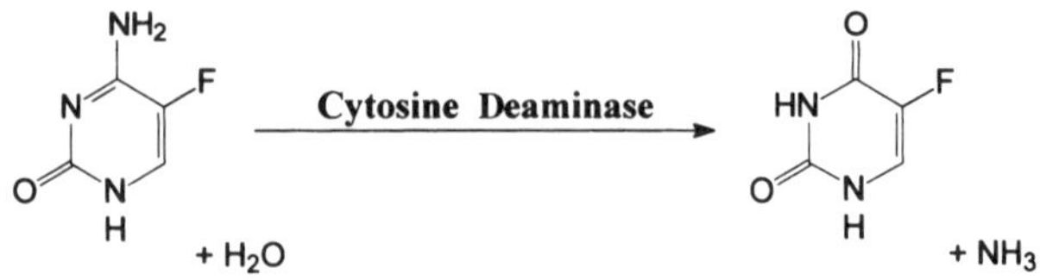

Figure 15. The formation of 5-fluorouracil from 5-fluorocytosine by the enzyme cytosine deaminase.

nation has been extensively studied for gene-directed enzyme prodrug therapy (GDEPT) (see Chapter 9) but has also been evaluated for ADEPT. The enzyme from baker's yeast (*Saccharomyces cerevisiae*), a 32 kDa dimer, converts 5-FC to 5-FU at a reasonable rate (K_m = 1.3 mM, k_{cat} = 38 s^{-1}). It has been chemically conjugated to an antibody (Senter *et al.*, 1991). In an *in vitro* cytotoxicity assay against H2981 colon cancer cells, 5-FC plus an L6 (H2981-specific) antibody-cytosine deaminase conjugate showed identical cytotoxicity to 5-FU, giving an IC_{50} of 20 μM. In contrast, 5-FC alone or in combination with a nonspecific antibody-cytosine deaminase conjugate shows little or no toxicity at concentrations up to 20 μM. *In vivo*, studies were undertaken to measure how much 5-FU is generated by the conjugate in tumor-bearing mice. By using a clearing antibody, high tumor to blood ratios are obtained (42:1). In blood and normal tissue there was little conversion of 5-fluorocytosine too 5-FU whereas 5-FU was formed in the tumor. It was estimated that the intratumoral levels of 5-FU obtained were 17-fold more than with conventional, systemic administration of 5-FU (Senter *et al.*, 1991; Wallace *et al.*, 1994).

2.11. Nitroreductases

Interest in this class of enzymes resulted from work with the prodrug CB 1954 [5-(aziridin-1-yl)-2,4-dinitrobenzamide]. In the cases of the enzymes described previously, the activating chemical event of the prodrug is hydrolysis to form the active agent. However, nitro reduction also activates prodrugs. In the case of CB 1954, it is converted from a weak monofunctional alkylating agent to a potent difunctional agent (which can form DNA interstrand cross-links) through reduction of its 4-nitro group. The active agent formed is 5-(aziridin-1-yl)-4-hydroxylamino-2-nitrobenzamide, which reacts chemically with cellular thioesters (such as *S*-acetylthiocholine) to form the highly reactive, proximal, cross-linking species that is probably the acetoxy derivative (Fig. 16) (Knox *et al.*, 1988b, 1991). There is a massive difference between the prodrug and the active form (10,000) of CB 1954 (Roberts *et al.*, 1986), which makes this system very

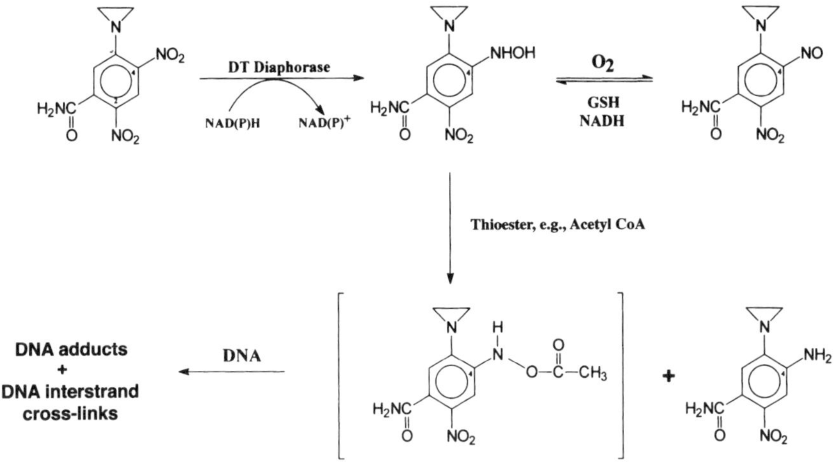

Figure 16. The bioactivation of CB 1954. The initial step is the reduction of CB 1954 by the enzyme DT diaphorase to form 5-(aziridin-1-yl)-4-hydroxylamino-2-nitrobenzamide. This hydroxylamine derivative reacts with thioesters to produce DNA reactive species. It is postulated that this is the N-acetoxy derivative. The major product of this reaction, 4-amino-5-(aziridin-1-yl)-2-nitrobenzamide, does not react readily with DNA. Formation of 4-amino-5-(aziridin-1-yl)-2-nitrobenzamide competes with the production of DNA binding products.

attractive. In rat cells the reduction of CB 1954 is mediated by the enzyme DT diaphorase (Figs. 16 and 17) which is better characterized as a quinone reductase (Knox *et al.*, 1988a). DT diaphorase is a flavin adenine dinucleotide (FAD)-containing flavoenzyme consisting of two homodimers of 33 kDa each. In fact, CB 1954 is a very poor substrate for DT diaphorase ($K_m = 0.83$ mM, $k_{cat} = 0.07\ s^{-1}$) compared to the quinone, menadione ($K_m = 1.3\ \mu M$, $k_{cat} = 1083\ s^{-1}$). The human enzyme reduces CB 1954 at a very slow rate ($K_m = 1.4$ mM, $k_{cat} = 0.009\ s^{-1}$), and human cells (even those expressing high levels of human DT diaphorase) are resistant to CB 1954 (Boland *et al.*, 1991). Thus, rat DT diaphorase was proposed as an ADEPT enzyme in combination with CB 1954 (Knox *et al.*, 1988b; Sunters *et al.*, 1991) although no work with antibody conjugates has been reported.

DT diaphorase was superseded as a potential ADEPT enzyme by the isolation of a nitroreductase enzyme from *E. coli* B (NR) that also activates CB 1954 but at a much faster rate ($K_m = 1.4$ mM, $k_{cat} = 6.0\ s^{-1}$). NR is also a flavoenzyme but contains FMN (flavin mononucleotide) and is a 23 kDa monomer (Anlezark *et al.*, 1995; Knox *et al.*, 1992). In contrast to DT diaphorase, NR reduces either (but not both) nitro group of CB 1954 and forms a mixture of the 2- and 4-hydroxylamines in equal proportions (Fig. 17) (Knox *et al.*, 1992). 5-(Aziridin-1-yl)-2-hydroxylamino-4-nitrobenzamide is not as cytotoxic as the 4-hydroxylamine and

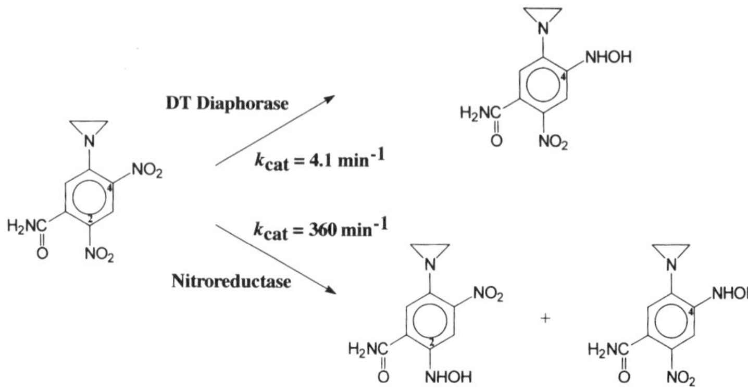

Figure 17. The reduction of CB 1954 by either DT diaphorase or *E. coli* nitroreductase. DT diaphorase forms only 5-(aziridin-1-yl)-4-hydroxylamino-2-nitrobenzamide, but the nitroreductase generates both the 2- and 4-hydroxylamino forms in equal proportions. The 2-hydroxylamine is less cytotoxic than the 4-hydroxylamine, but it is still much more cytotoxic than CB 1954 itself. CB 1954 is intrinsically reduced about 90-fold faster by the nitroreductase than by rat DT diaphorase.

is a less potent DNA cross-linking agent. However, it is still much more cytotoxic than the prodrug CB 1954 (Knox *et al.*, 1988b). NR activates CB 1954 to a cytotoxic form *in vitro* (Knox *et al.*, 1992) and is stable to conjugation (unpublished work). No data has been presented with a conjugate either *in vivo* nor *in vitro*. Other potential prodrugs have been proposed for use with the nitroreductase for ADEPT. A series of 2,4-dinitrobenzamide mustard analogs of CB 1954 has been synthesized (Anlezark *et al.*, 1995; Atwell *et al.*, 1996). These compounds were not reduced at all by DT diaphorase (Anlezark *et al.*, 1995) which is an advantage for ADEPT. In contrast to CB 1954 in which either nitro group is reducible to the corresponding hydroxylamine, SN 23862, the bischloroethylamino analog of CB 1954, is reduced by the nitroreductase to form only the 2-hydroxylamine. However, this hydroxylamine reacts with *S*-acetylthiocholine to form a species that produces interstrand cross-links in naked DNA (Anlezark *et al.*, 1995). SN 23862 is reduced more rapidly than CB 1954 (K_m = 2.5 mM, k_{cat} = 26.4 s^{-1}) and it was more cytotoxic than CB 1954 in an *in vitro* assay when incubated with UV4 cells in the presence of NR and NADH. Other compounds also showed advantageous properties. SN 25261 (with a dihydroxypropylcarboxamide ring substituent) is 10× more water soluble than SN 23862 (and CB 1954) but retains similar kinetic characteristics (K_m = 3.1 mM, k_{cat} = 32.5 s^{-1}) and cytotoxic potency. In SN 25084, a change in the position of the carboxamide group relative to the mustard results in an increased rate of reduction (K_m = 1.2 mM, k_{cat} = 75.7 s^{-1}) and greater cytotoxicity than SN 23862 when incubated with NR. An analog (SN 25507) incorporating both these structural changes has an enhanced k_{cat} of 576 s^{-1}

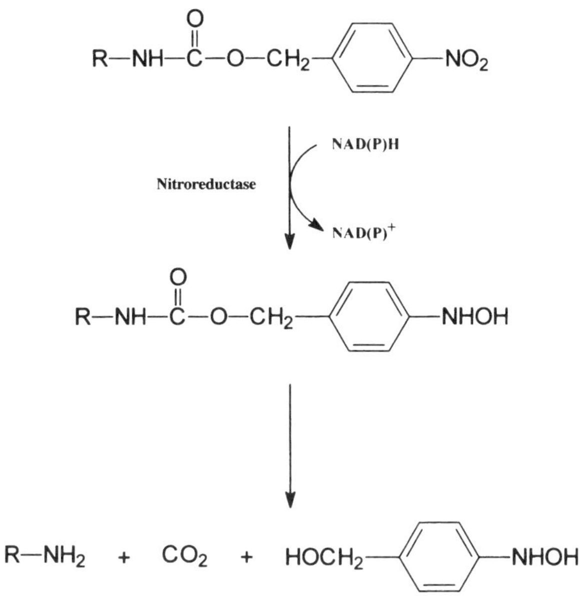

Figure 18. The generation of an active drug (R-NH2) by self-immolation of a prodrug after nitro-reduction. Carbamates and also carbonates undergo reduction and self-immolation to form active drugs (R-OH).

but is less cytotoxic than SN 23862 when activated by NR. Another class of prodrugs consists of compounds activated by a self-immolative mechanism (Fig. 18) which can form active drugs, such as mustards, actinomycins, mitomycin C (Fig. 19) (Mauger *et al.*, 1994), enediynes (Hay *et al.*, 1995), seco-CI alkylating agents (Tercel *et al.*, 1996), or tallimustine (Lee *et al.*, 1997). The prodrugs are 4-nitrobenzyloxycarbonyl derivatives of these drugs, which upon enzymatic reduction, generate the drug through self-immolation of the 4-(hydroxylamino)-benzyloxycarbonyl group. In the case of actinomycin D, the ratio of the dose required between drug and prodrug to give the same cytotoxicity *in vitro* is greater than 100. The prodrug is also much less toxic (~100×) than actinomycin D to mice *in vivo* (Mauger *et al.*, 1994).

However, for a nitroreductase enzyme to be used in ADEPT, a cofactor also has to be present to supply a source of reducing equivalents. The biogenic cofactors for reductases, NADH and NADPH, are unsuitable in this respect because they are rapidly oxidized and degraded by serum enzymes (Friedlos and Knox, 1992). A novel approach to this problem has been the development of enzyme-selective cofactors, and it has been shown that very simple reduced pyridinium compounds are still good cofactors for the nitroreductase (Knox *et al.*,

1.

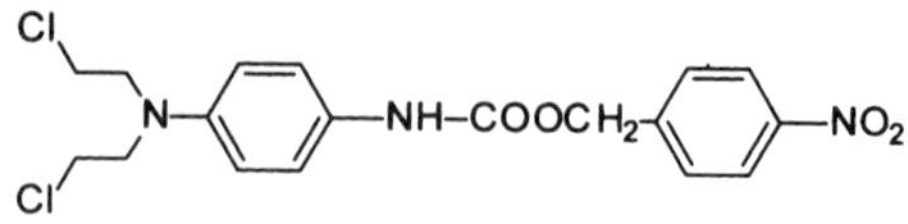

2.

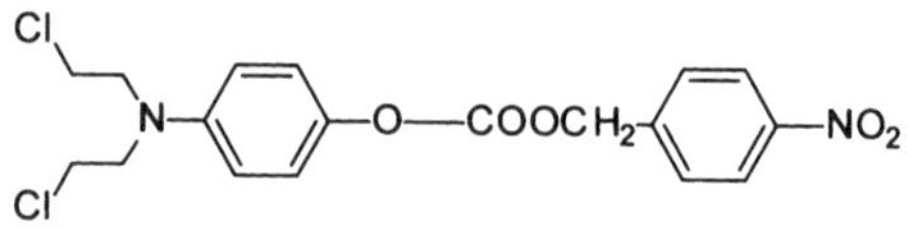

3.

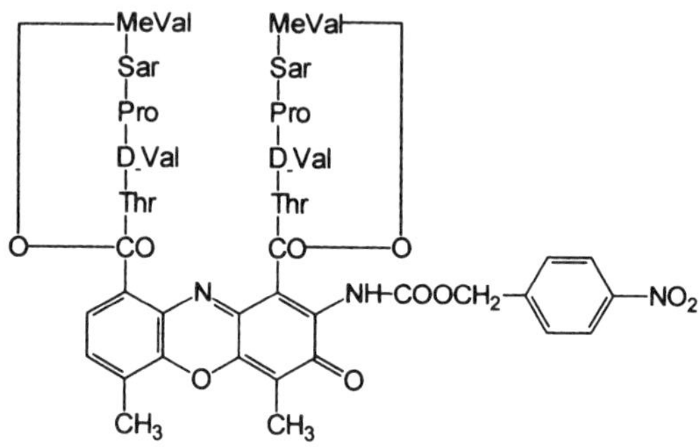

4.

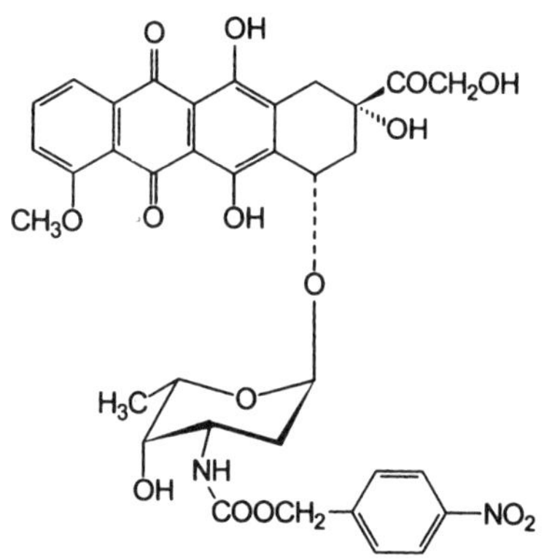

5.

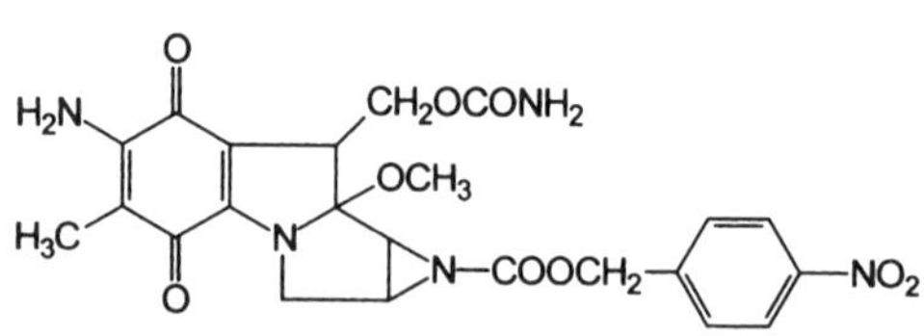

Figure 19. Prodrugs that could be activated by a self-immolative mechanism following nitroreduction: (1) 4-Nitrobenzyl 4′-[bis(2-chloroethyl)amino]phenylcarbamate; (2) 4-[bis(2-chloroethyl)-amino]phenyl 4′-nitrobenzyl carbonate; (3) *N*-4-nitrobenzyloxycarbonyl-AMD; (4) *N*-4-nitrobenzyloxycarbonyl-doxorubicin; (5) *N*-4-nitrobenzyloxycarbonyl-mitomycin C.

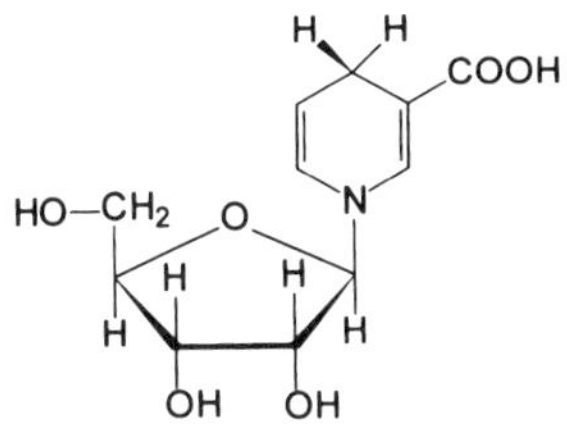

Figure 20. The structure of the reduced form of nicotinic acid riboside. An example of a synthetic, enzyme-selective cofactor.

1995). Further, they are not substrates for the serum enzymes that metabolize NAD(P)H and therefore are serum-stable (Knox *et al.*, 1995). Such a compound is the reduced form of nicotinic acid riboside (Fig. 20). This compound can totally substitute for NADH in the reduction of CB 1954 (Fig. 21) although there is a decreased affinity of nitroreductase for this compound (K_m = 97 μM) compared to NADH (K_m = <10 μM) (Knox *et al.*, 1995). Interestingly, this compound is enzyme-selective and is not used by other oxidoreductases, such as DT diaphorase

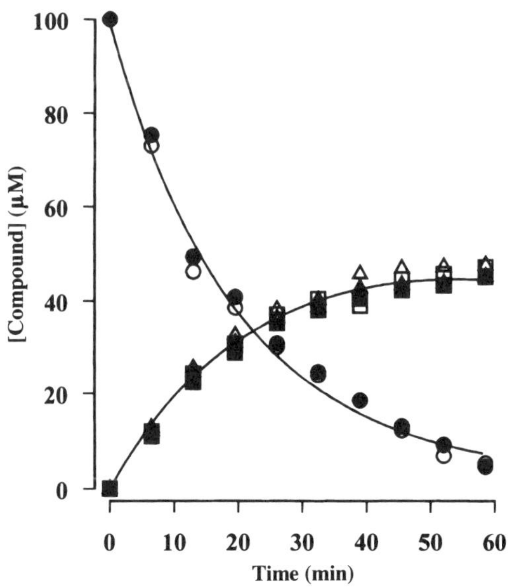

Figure 21. The aerobic reduction of CB 1954 (●) and the resulting formation of 5-(aziridin-1-yl)-4-hydroxylamino-2-nitrobenzamide (■) and 5-(aziridin-1-yl)-2-hydroxylamino-4-nitrobenzamide (▲) by the *E. coli* nitroreductase using either NADH (solid symbols) or NARH (open symbols) as a cofactor. The enzyme concentration was 2 μg/mL and the initial cofactor concentration was 500 μM. (Data from Knox *et al.*, 1995).

(Knox *et al.*, 1995). These observations suggest that the pharmacological limitations of NADH need not be an obstacle to using bioreductive enzymes for ADEPT therapy. However, the major application of nitroreductases is in GDEPT (Bailey *et al.*, 1996; Bridgewater *et al.*, 1995). Here, the cofactor requirement is not a problem because the enzyme is expressed intracellularly. It may even be an advantage because any enzyme that escapes (for example, from dying cells) will be unable to activate circulating prodrug because of the lack of a cofactor.

3. CONCLUSIONS

There is a rich diversity of potential enzyme and prodrug combinations for use in ADEPT applications, and it is highly probable that additional examples will materialize. ADEPT offers a novel field of opportunities for the enzymologist and the chemist. Antibody delivery combined with amplification provided by the enzymatic activation of prodrugs should enable adequate selection between tumor and normal tissue. In ADEPT the enzyme must be conjugated to an antibody. This can be done chemically but it is anticipated that recombinant DNA technology will be used increasingly. The first fusion protein developed specifically for an ADEPT application consisted of human β-glucuronidase fused to the CH domain of BW431, a high avidity anti-CEA antibody, humanized by complementarity-determining region (CDR) grafting. The protein thus produced has antigen affinity very similar to that of the parent antibody and retains enzymatic activity (Bosslet *et al.*, 1992, 1994). It is interesting that human β-glucuronidase is a tetramer but in this situation is active as a monomer. A *B. cereus* β-lactamase II, single-chain antibody fusion protein has also been constructed. The L6 antibody used recognizes an antigen expressed on breast, colon, lung, and ovarian carcinomas. The fusion protein is produced by *E. coli* in an active form that does not require refolding and retains immunological and enzymatic properties similar to an equivalent chemically linked conjugate (Goshorn *et al.*, 1993). The ultimate refinement now possible by protein engineering techniques is to make a bifunctional antibody in which one arm possesses an enzyme-mimicking catalytic function. A catalytic antibody capable of mimicking cytosine deaminase by converting 5-FC to 5-FU has been described (Campbell *et al.*, 1994). The turnover capacity of this catalytic antibody was very low ($k_{cat} = 0.00052\ s^{-1}$) and is not of direct value in ADEPT. However, it seems probable that systems capable of matching current chemically linked conjugates will be developed in due course as the technology matures. Indeed, a catalytic antibody (termed an "abzyme") has been made which, *in vitro*, produces cytotoxicity (Wentworth *et al.*, 1996). The abzyme hydrolyzes the carbamate prodrug 4-[*N*,*N*-bis(2-chloroethyl)]aminophenyl-*N*-[(1*S*)-(1,3-dicarboxy)-propyl]carbamate to generate the corresponding nitrogen mustard. *In vitro* studies with this abzyme and this prodrug lead to a marked reduction in viability of cultured human colonic carcinoma (LoVo) cells relative to appropriate controls. However, the kinetics of the abzyme for the prodrug (K_m = 201 μM, k_{cat} = 0.031 s^{-1}) was still poor relative to most of the conventional systems discussed.

The most common type of active drugs released are the nitrogen mustard type of DNA alkylating agent. As described previously, this type of agent is not cell-cycle-specific and does not require continuous exposure to mitigate its cytotoxicity. However, releasing a clinically used alkylating agent, such as melphalan, from a prodrug is not the right approach. These agents are not reactive enough and can diffuse away from the site of activation and cause unwanted toxicity. This was demonstrated in the initial clinical trial of a CGG2 conjugate and a mustard prodrug. Active drug was seen in the patients' serum and mylosupression was observed. Future clinical trials with this system will probably generate a highly reactive aniline mustard with a chemical half-life measured in seconds (melphalan has a chemical half-life of about 30 min). Generation of highly cytotoxic species is not without its disadvantages. Even if the prodrug is well deactivated, it may be cytotoxic at doses below the K_m of the enzyme for the prodrug. Thus activation may be suboptimal. This is an important consideration when using prodrugs based on very highly cytotoxic compounds, such as palytoxin.

It was found with the CPG2 system that residual circulating conjugate could limit ADEPT therapy by producing toxicity by activating the prodrug in the serum, not specifically at the tumor site. It has been suggested that this could be minimized if the kinetics of the enzyme for the prodrug were not optimal and a high K_m may be desired (Bagshawe *et al.*, 1994). However, such an approach requires very detailed pharmacokinetic modeling, and it will probably be much more effective to use effective clearance systems for removing residual circulating conjugate.

What is the best enzyme/prodrug system for ADEPT? None of the present systems fully meet the criteria listed at the beginning of this chapter, but impressive results have been achieved. The CPG2/mustard prodrug system is the most fully developed. Results from its initial clinical trial have raised a number of issues that need to be addressed. First, the need for the release of a more reactive drug discussed previously. Secondly, patients rapidly raised an antibody response against both the enzyme and antibody components of the conjugate. These antibody responses developed within 10–12 days effectively limited the treatment to a single cycle. This trial used a murine antibody, but now techniques, such as complementarity-determining region (CDR) grafting, exist to humanize them (see Chapter 6) and to eliminate the chances of developing antibodies against them. The enzyme is more of a problem! One approach is simply to use immunosuppressive drugs, such as cyclosporine (Bagshawe and Sharma, 1996; Sharma *et al.*, 1992). This may provide a partial remedy, but a less aggressive solution is desirable. It would also be possible to use a different activating enzyme for each cycle of treatment to circumvent the antibody response. However, this is complicated and would also require a different clearance system for each cycle. Alternatively, a human enzyme could be employed. The use of a human enzyme has been best exemplified by β-glucuronidase. This enzyme forms part of a fusion protein with a humanized antibody that has been tested *in vivo* (Bosslet *et al.*, 1994) but only in nude mice. The fundamental problem of using a human enzyme is activation of the prodrug by endogenous enzyme. This can be overcome by design of the prodrug. For example, glucuronides are charged at physiological pH. Therefore, they do not readily enter cells, and prodrugs exploiting this principle will not be activated by endogenous enzymes that are intracellular. A more elegant approach is modifying a human enzyme to have a new substrate specificity. This approach carries the risk that the "new" enzyme will be recognized as foreign by the immune system. However, very subtle changes can have profound effects on prodrug activating enzymes. For example, human DT diaphorase cannot effectively reduce the prodrug CB 1954 when compared to the rat enzyme. It has now been shown that a single amino acid residue at position 104, which is Tyr in the rat enzyme but Gln in the human, is mostly responsible for this difference. A mutant human enzyme that substitutes the rat residue for the human residue at this position effectively reduces CB 1954 (Chen *et al.*, 1997).

Very recently a mutated version of human ribonuclease has been reported (Blakey *et al.*, 1997). Human ribonuclease A is a low molecular weight (14 kDa) monomer with a pH optimum of 6.5 that cleaves RNA into ribonucleotides. Ribonucleases are renowned for their robustness. In the mutated enzyme the substrate specificity has been altered so that a prodrug is converted to an active drug by the mutant enzyme but is a relatively poor substrate for the native form. This was accomplished by mutating the positively charged lysine to a negatively charged glutamate at residue 66 of the enzyme. A basic substituent was incorporated into the prodrug so that it could interact with this mutated residue and be hydrolyzed to realize an active mustard (Fig. 22) which is 30-fold more cytotoxic. The mutant enzyme has both a greater affinity (K_m = 4.2 mM) and turns over the prodrug faster (k_{cat} = 15 s^{-1}) than the native enzyme (K_m = 17 mM, k_{cat} = 3.4 s^{-1}). A similar approach has been taken with a mutant of human carboxypeptidase A1 (Smith *et al.*, 1997). Based on a computer model of the human enzyme (built from the known crystal structure of bovine carboxypeptidase A), bulky phenylalanine- and tyrosine-based prodrugs of methotrexate that were metabolically stable *in vivo* and were not substrates for wild type human carboxypeptidase A were designed and synthesized. In addition, the model was used to design a mutant of human carboxypeptidase A1, changed at position 268 from the wild-type threonine to a glycine. Unlike the wild-type, this novel enzyme is activates the synthesized prodrugs. The mutant enzyme could use these new prodrugs as efficiently as the wild-type used its best substrates, such as methotrexate-α-phenylalanine. The k_{cat}/K_m value for the wild-type enzyme with methotrexate-α-phenylalanine is 0.44

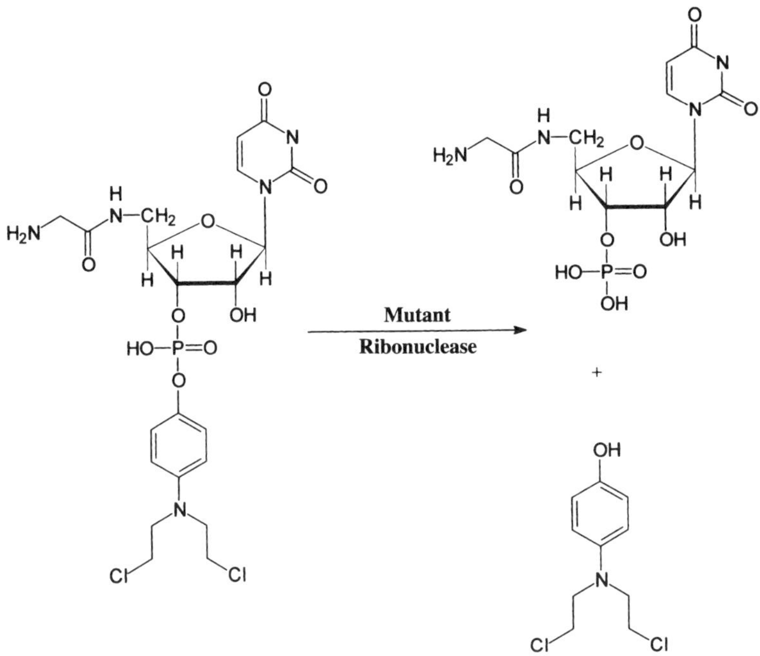

Figure 22. The generation of an active mustard from a prodrug by a mutant human ribonuclease A. The polarity for substrate recognition is reversed in the mutant. Therefore, the prodrug is a poor substrate for the native human enzyme.

$\mu M^{-1} s^{-1}$, and k_{cat}/K_m values for mutant enzyme with methotrexate-α-3-cyclobutylphenylalanine and methotrexate-α-3-cyclopentyltyrosine are 1.8 and 0.16 $\mu M^{-1} s^{-1}$, respectively. *In vitro*, a conjugate of the mutant enzyme with ING-1 (an antibody that binds to the tumor antigen Ep-Cam) kills human HT-29 colon adenocarcinoma cells (which express Ep-Cam) in the presence of either methotrexate-α-3-cyclobutylphenylalanine or methotrexate-α-3-cyclopentyltyrosine as efficiently as the active drug methotrexate. This effect was not seen when the conjugate was made with an irrelevant antibody (Smith *et al.*, 1997).

Such studies demonstrate the potential for making nonimmunogenic systems to activate prodrugs that are not substrates for native enzymes.

In summary, a diverse range enzymes and prodrugs have been proposed for ADEPT. Most have shown very encouraging results *in vitro*, and some excellent results have been obtained using *in vivo* models. A clinical trial has highlighted potential problems, and these are being addressed. Rapidly developing protein engineering techniques hold the promise of nonimmunogenic enzymes and conjugates. The future for ADEPT is in these developments.

REFERENCES

Alexander, R. P., Beeley, N. R. A., O'Driscoll, M., O'Neill, F. P., Millican, T. A., Pratt, A. J., and Willenbrock, F. W., 1991, Cephalosporin nitrogen mustard carbamate prodrugs for "ADEPT", *Tetrahedron Lett.* **32:**3269–3272.

Andrianomenjanahary, S., Dong, X., Florent, J. C., Gaudel, G., Gesson, J. P., Jacquesy, J. C., Koch, M., Michel, S., and Mondon, M., 1992, Synthesis of novel targeted pro-prodrugs of anthracyclines potentially activated by a monoclonal antibody galactosidase conjugate, *Bioorg. Medicinal Chem. Lett.* **2:**1093–1096.

Anlezark, G. M., Melton, R. G., Sherwood, R. F., Wilson, W. R., Denny, W. A., Palmer, B. D., Knox, R. J., Friedlos, F., and Williams, A., 1995, Bioactivation of dinitrobenzamide mustards by an *E. coli* B nitroreductase, *Biochem. Pharmacol.* **50:**609–618.

Atwell, G. J., Boyd, M., Palmer, B. D., Anderson, R. F., Pullen, S. M., Wilson, W. R., and Denny, W. A., 1996, Synthesis and evaluation of 4-substituted analogues of 5-[*N,N*-bis (2-chloroethyl)-amino]-2-nitrobenzamide as bioreductively activated prodrugs using an *Escherichia coli* nitroreductase, *Anticancer Drug Des.* **11:**553–567.

Bagshawe, K. D., 1987, Antibody directed enzymes revive anti-cancer prodrugs concept, *Br. J. Cancer* **56:**531–532.

Bagshawe, K. D., 1989, Towards generating cytotoxic agents at cancer sites, *Br. J. Cancer* **60:**275–281.

Bagshawe, K. D., 1994, ADEPT and related concepts, *Cell Biophys.* **25:**83–91.

Bagshawe, K. D., 1995, Antibody-directed enzyme prodrug therapy—a review, *Drug Development Res.* **34:**220–230.

Bagshawe, K. D., and Sharma, S. K., 1996, Cyclosporine delays host immune response to antibody enzyme conjugate in ADEPT, *Transplant. Proc.* **28:**3156–3158.

Bagshawe, K. D., Sharma, S. K., Springer, C. J., Antoniw, P., Boden, J. A., Rogers, G. T., Burke, P. J., Melton, R. G., and Sherwood, R. F., 1991, Antibody directed enzyme prodrug therapy (ADEPT): Clinical report, *Dis. Markers* **9:**233–238.

Bagshawe, K. D., Sharma, S. K., Springer, C. J., and Rogers, G. T., 1994, Antibody directed enzyme prodrug therapy (ADEPT). A review of some theoretical, experimental and clinical aspects, *Ann. Oncol.* **5:**879–891.

Bagshawe, K. D., Springer, C. J., Searle, F., Antoniw, P., Sharma, S. K., Melton, R. G., and Sherwood, R. F., 1988, A cytotoxic agent can be generated selectively at cancer sites, *Br. J. Cancer* **58:** 700–703.

Bailey, S. M., Knox, R. J., Hobbs, S. M., Jenkins, T. C., Mauger, A. B., Melton, R. G., Burke, P. J., Connors, T. A., and Hart, I. R., 1996, Investigation of alternative prodrugs for use with E. coli nitroreductase in 'suicide gene' approaches to cancer therapy, *Gene Ther.* **3:**1143–1150.

Bignami, G. S., Senter, P. D., Grothaus, P. G., Fischer, K. J., Humphreys, T., and Wallace, P. M., 1992, *N*-(4′-hydroxyphenylacetyl)palytoxin: A palytoxin prodrug that can be activated by a monoclonal antibody-penicillin G amidase conjugate, *Cancer Res.* **52:**5759–5764.

Blakey, D. C., Boyle, F. T., Davies, D. H., Dowell, R. I., Heaton, D. W., Rose, M. S., Slater, A. M., Eggelte, H. J., Tarragona-Fiol, A., and Taylorson, C. J., 1997, Mutant human ribonuclease for use in antibody-directed enzyme prodrug therapy (ADEPT), *Br. J. Cancer* **75(Suppl 1):**36 (Abstract).

Blakey, D. C., Davies, D. H., Dowell, R. I., East, S. J., Burke, P. J., Sharma, S. K., Springer, C. J., Mauger, A. B., and Melton, R. G., 1995, Anti-tumor effects of an antibody-carboxypeptidase G2 conjugate in combination with phenol mustard prodrugs, *Br. J. Cancer* **72:**1083–1088.

Blakey, D. C., Valcaccia, B. E., East, S., Wright, A. F., Boyle, F. T., Springer, C. J., Burke, P. J., Melton, R. G., and Bagshawe, K. D., 1993, Antitumor effects of an antibody-carboxypeptidase G2 conjugate in combination with a benzoic acid mustard prodrug, *Cell Biophys.* **22:**1–8.

Boland, M. P., Knox, R. J., and Roberts, J. J., 1991, The differences in kinetics of rat and human DT diaphorase result in a differential sensitivity of derived cell lines to CB 1954 (5-aziridin-1-yl)-2,4-dinitrobenzamide), *Biochem. Pharmacol.* **41:**867–875.

Bosslet, K., Czech, J., Lorenz, P., Sedlacek, H. H., Schuermann, M., and Seemann, G., 1992, Molecular and functional characterisation of a fusion protein suited for tumour specific prodrug activation, *Br. J. Cancer* **65:**234–238.

Bosslet, K., Czech, J., Seemann, G., Monneret, C., and Hoffmann, D., 1994, Fusion protein mediated prodrug activation (FMPA) *in vivo*, *Cell Biophys.* **25:**51–63.

Bridgewater, J. A., Springer, C. J., Knox, R. J., Minton, N. P., Michael, N. P., and Collins, M. K., 1995, Expression of the bacterial nitroreductase enzyme in mammalian cells renders them selectively sensitive to killing by the prodrug CB1954, *Eur. J. Cancer* **31a:**2362–2370.

Campbell, D. A., Gong, B., Kochersperger, L. M., Yonkovich, S., Gallop, M. A., and Schultz, P. G., 1994, Antibody-catalyzed prodrug activation, *J. Am. Chem. Soc.* **116:**2165–2166.

Chabot, G. G., Armand, J. P., Terret, C., de Forni, M., Abigerges, D., Winograd, B., Igwemezie, L., Schacter, L., Kaul, S., Ropers, J., and Bonnay, M., 1996, Etoposide bioavailability after oral administration of the prodrug etoposide phosphate in cancer patients during a phase I study, *J. Clin. Oncol.* **14:**2020–2030.

Chen, S., Knox, R., Wu, K., Deng, P. S. K., Zhou, D., Bianchet, M. A., and Amzel, L. M., 1997, Molecular basis of the catalytic differences among DT-diaphorase of human, rat, and mouse, *J. Biol. Chem.* **272:**1437–1439.

Connors, T. A., and Whisson, M. E., 1966, Cure of mice bearing advanced plasma cell tumours with aniline mustard: the relationship between glucuronidase activity and tumour sensitivity, *Nature* **210:**866–867.

DeAngelis, L. M., Tong, W. P., Lin, S., Fleisher, M., and Bertino, J. R., 1996, Carboxypeptidase G2 rescue after high-dose methotrexate, *J. Clin. Oncol.* **14:**2145–2149.

De-Sutter, K., and Fiers, W., 1994, A bifunctional murine: human chimeric antibody with one antigen-binding arm replaced by bacterial β-lactamase, *Mol. Immunol.* **31:**261–267.

Dowell, R. I., Springer, C. J., Davies, D. H., Hadley, E. M., Burke, P. J., Boyle, F. T., Melton, R. G., Connors, T. A., Blakey, D. C., and Mauger, A. B., 1996, New mustard prodrugs for antibody-directed enzyme prodrug therapy: Alternatives to the amide link, *J. Med. Chem.* **39:**1100–1105.

Doyle, T. M., and Vyas, D. M., 1990, Second generation analogs of etoposide and mitomycin C, *Cancer Treat. Rev.* **17:**127–131.

Eccles, S. A., Court, W. J., Box, G. A., Dean, C. J., Melton, R. G., and Springer, C. J., 1994, Regression of established breast carcinoma xenografts with antibody-directed enzyme prodrug therapy against c-erbB2 p185, *Cancer Res.* **54:**5171–5177.

Ehrlich, P., 1960, *The collected papers of P. Ehrlich*, Pergamon Press, London.

Esswein, A., Haenseler, E., Montejano, Y., Vitols, K. S., and Huennekens, F. M., 1991, Construction and chemotherapeutic potential of carboxypeptidase-A/monoclonal antibody conjugate, *Adv. Enzyme Regul.* **31:**3–12.

Frei, E., Teicher, B. A., Holden, S. A., Cathcart, K. N., and Wang, Y. Y., 1988, Preclinical studies and clinical correlation of the effect of alkylating dose, *Cancer Res.* **48:**6417–6423.

Friedlos, F., and Knox, R. J., 1992, Metabolism of NAD(P)H by blood components. Relevance to bioreductively activated prodrugs in a targeted enzyme therapy system, *Biochem. Pharmacol.* **44:**631–635.

Goshorn, S. C., Svensson, H. P., Kerr, D. E., Somerville, J. E., Senter, P. D., and Fell, H. P., 1993, Genetic construction, expression, and characterization of a single chain anti-carcinoma antibody fused to β-lactamase, *Cancer Res.* **53:**2123–2127.

Haenseler, E., Esswein, A., Vitols, K. S., Montejano, Y., Mueller, B. M., Reisfeld, R. A., and Hunnekens, F. M., 1992, Activation of methotrexate-α-alanine by carboxypeptidase A monoclonal antibody conjugate, *Biochemistry* **31:**891–897.

Haisma, H. J., Boven, E., van Muijen, M., de Jong, J., van der Vijgh, W. J., and Pinedo, H. M., 1992a,

A monoclonal antibody-β-glucuronidase conjugate as activator of the prodrug epirubicin-glucuronide for specific treatment of cancer, *Br. J. Cancer* **66:**474–478.

Haisma, H. J., Boven, E., van Muijen, M., De Vries, R., and Pinedo, H. M., 1992b, Analysis of a conjugate between anti-carcinoembryonic antigen monoclonal antibody and alkaline phosphatase for specific activation of the prodrug etoposide phosphate, *Cancer Immunol. Immunother.* **34:**343–348.

Hay, M. P., Wilson, W. R., and Denny, W. A., 1995, A novel enediyne prodrug for antibody-directed enzyme prodrug therapy (ADEPT) using *E. coli* B nitroreductase, *Bioorg. Medicinal Chem. Lett.* **5:**2829–2834.

Houba, P. H., Leenders, R. G., Boven, E., Scheeren, J. W., Pinedo, H. M., and Haisma, H. J., 1996, Characterization of novel anthracycline prodrugs activated by human β-glucuronidase for use in antibody-directed enzyme prodrug therapy, *Biochem. Pharmacol.* **52:**455–463.

Ito, H., Morizet, J., Coulombel, L., Goavec, M., Rousseau, V., Bernard, A., and Stanislawski, M., 1989, An immunotoxin system intended for bone marrow purging composed of glucose oxidase and lactoperoxidase coupled to monoclonal antibody 097, *Bone Marrow Transplant.* **4:**519–527.

Ito, H., Morizet, J., Coulombel, L., and Stanislawski, M., 1990, T cell depletion of human bone marrow using an oxidase-peroxidase enzyme immunotoxin, *Bone Marrow Transplant.* **6:**395–398.

Kerr, D. E., Senter, P. D., Burnett, W. V., Hirschberg, D. L., Hellstrom, I., and Hellstrom, K. E., 1990, Antibody-penicillin-V-amidase conjugates kill antigen-positive tumor cells when combined with doxorubicin phenoxyacetamide, *Cancer Immunol. Immunother.* **31:**202–206.

Knox, R. J., Boland, M. P., Friedlos, F., Coles, B., Southan, C., and Roberts, J. J., 1988a, The nitroreductase enzyme in Walker cells that activates 5-(aziridin-1-yl)-2,4-dinitrobenzamide (CB 1954) to 5-(aziridin-1-yl)-4-hydroxylamino-2-nitrobenzamide is a form of NAD(P)H dehydrogenase (quinone) (EC 1.6.99.2), *Biochem. Pharmacol.* **37:**4671–4677.

Knox, R. J., Friedlos, F., Jarman, M., Davies, L. C., Goddard, P., Anlezark, G. M., Melton, R. G., and Sherwood, R. F., 1995, Virtual cofactors for an *Escherichia coli* nitroreductase enzyme—relevance to reductively activated prodrugs in antibody-directed enzyme prodrug therapy (ADEPT), *Biochem. Pharmacol.* **49:**1641–1647.

Knox, R. J., Friedlos, F., Jarman, M., and Roberts, J. J., 1988b, A new cytotoxic, DNA interstrand cross-linking agent, 5-(aziridin-1-yl)-4-hydroxylamino-2-nitrobenzamide, is formed from 5-(aziridin-1-yl)-2,4-dinitrobenzamide (CB 1954) by a nitroreductase enzyme in Walker carcinoma cells, *Biochem. Pharmacol.* **37:**4661–4669.

Knox, R. J., Friedlos, F., Marchbank, T., and Roberts, J. J., 1991, Bioactivation of CB 1954: Reaction of the active 4-hydroxylamino derivative with thioesters to form the ultimate DNA–DNA interstrand cross-linking species, *Biochem. Pharmacol.* **42:**1691–1697.

Knox, R. J., Friedlos, F., Sherwood, R. F., Melton, R. G., and Anlezark, G. M., 1992, The bioactivation of 5-(aziridin-1-yl)-2,4-dinitrobenzamide (CB1954)—II. A comparison of an *Escherichia coli* nitroreductase and Walker DT diaphorase, *Biochem. Pharmacol.* **44:**2297–2301.

Kuefner, U., Lohrmann, U., Montejano, Y. D., Vitols, K. S., and Hunnekens, F. M., 1989, Carboxypeptidase-mediated release of methotrexate from methotrexate-α-peptides, *Biochemistry* **28:**2288–2297.

Laethem, R. M., Blumenkopf, T. A., Cory, M., Elwell, L., Moxham, C. P., Ray, P. H., Walton, L. M., and Smith, G. K., 1996, Expression and characterization of human pancreatic preprocarboxypeptidase A1 and preprocarboxypeptidase A2, *Arch. Biochem. Biophys.* **332:**8–18.

Lee, M., Simpson, J. E., Woo, S., Kaenzig, C., Anlezark, G. M., EnoAmooquaye, E., and Burke, P. J., 1997, Synthesis of an aminopropyl, analog of the experimental anticancer drug tallimustine, and activation of its 4-nitrobenzylcarbamoyl prodrug by nitroreductase and NADH, *Bioorg. Medicinal Chem. Lett.* **7:**1065–1070.

Mauger, A. B., Burke, P. J., Somani, H. H., Friedlos, F., and Knox, R. J., 1994, Self-immolative prodrugs: Candidates for antibody-directed enzyme prodrug therapy in conjunction with a nitroreductase enzyme, *J. Med. Chem.* **37:**3452–3458.

Meyer, D. L., Jungheim, L. N., Law, K. L., Mikolajczyk, S. D., Shepherd, T. A., Mackensen, D. G., Briggs, S. L., and Starling, J. J., 1993, Site-specific prodrug activation by antibody-β-lactamase conjugates: Regression and long-term growth inhibition of human colon carcinoma xenograft models, *Cancer Res.* **53:**3956–3963.

Meyer, D. L., Jungheim, L. N., Mikolajczyk, S. D., Shepherd, T. A., Starling, J. J., and Ahlem, C. N., 1992, Preparation and characterization of a β-lactamase-Fab′ conjugate for the site-specific activation of oncolytic agents, *Bioconjugate Chem.* **3:**42–48.

Meyer, D. L., Law, K. L., Payne, J. K., Mikolajczyk, S. D., Zarrinmayeh, H., Jungheim, L. N., Kling, J. K., Shepherd, T. A., and Starling, J. J., 1995, Site-specific prodrug activation by antibody-β-lactamase conjugates: Preclinical investigation of the efficacy and toxicity of doxorubicin delivered by antibody directed catalysis, *Bioconjugate Chem.* **6:**440–446.

Minton, N. P., Atkinson, T., and Sherwood, R. F., 1983, Molecular cloning of the *Pseudomonas* carboxypeptidase G2 gene and its expression in *Escherichia coli* and *Pseudomonas putida*, *J. Bacteriol.* **156:**1222–1227.

Muzykantov, V. R., Saharov, D. V., Sinitsyn, V. V., Domogatsky, S. P., Goncharov, N. V., and Danilov, S. M., 1988, Specific killing of human endothelial cells by antibody-conjugated glucose oxidase, *Anal. Chem.* **169:**383–389.

Muzykantov, V. R., Trubetskaya, O. V, Puchnina, E. A., Sakharov, D. V., and Domogatsky, 1990, Cytotoxicity of glucose oxidase conjugated with antibodies to target cells: Killing efficiency depends on conjugate internalization, *Biochim. Biophys. Acta* **1053:**27–31.

Perron, M. J., and Page, M., 1996, Activation of methotrexate-phenylalanine by monoclonal antibody-carboxypeptidase A conjugate for the specific treatment of ovarian cancer in vitro, *Br. J. Cancer* **73:**281–287.

Pimm, M. V., 1988, Drug-monoclonal antibody conjugates for cancer therapy: Potential and limitations, *CRC Crit. Rev. Ther. Drug Carrier Syst.* **5:**189–227.

Roberts, J. J., Friedlos, F., and Knox, R. J., 1986, CB 1954 (2,4-dinitro-5-aziridinyl benzamide) becomes a DNA interstrand cross-linking agent in Walker tumour cells, *Biochem. Biophys. Res. Commun.* **140:**1073–1078.

Rodrigues, M. L., Presta, L. G., Kotts, C. E., Wirth, C., Mordenti, J., Osaka, G., Wong, W. L., Nuijens, A., Blackburn, B., and Carter, P., 1995, Development of a humanized disulfide-stabilized anti-p185HER2 Fv-β-lactamase fusion protein for activation of a cephalosporin doxorubicin prodrug, *Cancer Res.* **55:**63–70.

Roffler, S. R., Wang, S. M., Chern, J. W., Yeh, M. Y., and Tung, E., 1991, Anti-neoplastic glucuronide prodrug treatment of human tumor cells targeted with a monoclonal antibody-enzyme conjugate, *Biochem. Pharmacol.* **42:**2062–2065.

Romanini, A., Sobrero, A. F., Chou, T. C., Sherwood, R. F., and Bertino, J. R., 1989, Enhancement of trimetrexate cytotoxicity in vitro and in vivo by carboxypeptidase G2, *Cancer Res.* **49:**6019–6023.

Rowlinson-Busza, G., Bamias, A., Krausz, T., and Epenetos, A. A., 1992, Antibody-guided enzyme nitrile therapy (Agent): *In vitro* cytotoxicity and *in vivo* tumour localisation, in: *Monoclonal Antibodies Applications in Clinical Oncology* (Epenetos, A. A., ed.), Chapman and Hall, London, pp. 111–118.

Rowlinson-Busza, G., and Epenetos, A. A., 1992, Targeted delivery of biologic and other antineoplastic agents, *Curr. Opinion Oncol.* **4:**1142–1148.

Sahin, U., Hartmann, F., Senter, P., Pohl, C., Engert, A., Diehl, V., and Pfreundschuh, M., 1990, Specific activation of the prodrug mitomycin phosphate by a bispecific anti-CD30/anti-alkaline phosphatase monoclonal antibody, *Cancer Res.* **50:**6944–6948.

Schacter, L. P., Igwemezie, L. N., Seyedsadr, M., Morgenthien, E., Randolph, J., Albert, E., and Santabarbara, P., 1994, Clinical and pharmacokinetic overview of parenteral etoposide phosphate, *Cancer Chemother. Pharmacol.* **34:**S58–63.

Schmidt, F., Florent, J. C., Monneret, C., Straub, R., Czech, J., Gerken, M., and Bosslet, K., 1997,

Glucuronide prodrugs of hydroxy compounds for antibody directed enzyme prodrug therapy (ADEPT): A phenol nitrogen mustard carbamate, *Bioorg. Medicinal Chem. Lett.* **7:**1071–1076.

Senter, P. D., 1990, Activation of prodrugs by antibody-enzyme conjugates: A new approach to cancer therapy, *FASEB J.* **4:**188–193.

Senter, P. D., Saulnier, M. G., Schreiber, G. J., Hirschberg, D. L., Brown, J. P., Hellstrom, I., and Hellstrom, K. E., 1988, Anti-tumor effects of antibody-alkaline phosphatase conjugates in combination with etoposide phosphate, *Proc. Natl. Acad. Sci. USA* **85:**4842–4846.

Senter, P. D., Schreiber, G. J., Hirschberg, D. L., Ashe, S. A., Hellstrom, K. E., and Hellstrom, I., 1989, Enhancement of the *in vitro* and *in vivo* antitumor activities of phosphorylated mitomycin C and etoposide derivatives by monoclonal antibody-alkaline phosphatase conjugates, *Cancer Res.* **49:**5789–5792.

Senter, P. D., Su, P. C. D., Katsuragi, T., Sakai, T., Cosand, W. L, Hellstrom, I., and Hellstrom, K. E., 1991, Generation of 5-fluorouracil from 5-fluorocytosine by monoclonal antibody-cytosine deaminase conjugates, *Bioconjugate Chem.* **2:**447–451.

Sharma, S. K., Bagshawe, K. D., Burke, P. J., Boden, J. A., Rogers, G. T., Springer, C. J., Melton, R. G., and Sherwood, R. F., 1994a, Galactosylated antibodies and antibody-enzyme conjugates in antibody-directed enzyme prodrug therapy, *Cancer* **73:**1114–1120.

Sharma, S. K., Bagshawe, K. D., Burke, P. J., Boden, R. W., and Rogers, G. T., 1990, Inactivation and clearance of an anti-CEA carboxypeptidase G2 conjugate in blood after localisation in a xenograft model, *Br. J. Cancer* **61:**659–662.

Sharma, S. K., Bagshawe, K. D., Melton, R. G., and Sherwood, R. F., 1992, Human immune response to monoclonal antibody-enzyme conjugates in ADEPT pilot clinical trial, *Cell Biophys.* **21:** 109–120.

Sharma, S. K., Bagshawe, K. D., Springer, C. J., Burke, P. J., Rogers, G. T., Boden, J. A., Antoniw, P., Melton, R. G., and Sherwood, R. F., 1991, Antibody directed enzyme prodrug therapy (ADEPT): A three phase system, *Dis. Markers* **9:**225–231.

Sharma, S. K., Boden, J. A., Springer, C. J., Burke, P. J., and Bagshawe, K. D., 1994b, Antibody-directed enzyme prodrug therapy (ADEPT). A three-phase study in ovarian tumor xenografts, *Cell Biophys.* **25:**219–228.

Shepherd, T. A., Jungheim, L. N., Meyer, D. M., and Starling, J. J., 1991, A novel targeted delivery system utilizing a cephalosporin-oncolytic prodrug activated by an antibody B-lactamase conjugate for the treatment of cancer, *Bioorg. Medicinal Chem. Lett.* **1:**21–26.

Shockley, T. R., Lin, K., Nagy, J. A., Tompkins, R. G., Yarmush, M. L., and Dvorak, H. F., 1992, Spatial distribution of tumor-specific monoclonal antibodies in human melanoma xenografts, *Cancer Res.* **52:**367–376.

Smal, M. A., Dong, Z., Cheung, H. T., Asano, Y., Escoffier, L., Costello, M., and Tattersall, M. H., 1995, Activation and cytotoxicity of 2-α-aminoacyl prodrugs of methotrexate, *Biochem. Pharmacol.* **49:**567–574.

Smith, G. K., Banks, S., Blumenkopf, T. A., Cory, M., Humphreys, J., Laethem, R. M., Miller, J., Moxham, C. P., Mullin, R., Ray, P. H., Walton, L. M., and Wolfe, L. A., 1997, Toward antibody-directed enzyme prodrug therapy with the T268G mutant of human carboxypeptidase A1 and novel *in vivo* stable prodrugs of methotrexate, *J. Biol. Chem.* **272:**15804–15816.

Springer, C. J., Antoniw, P., Bagshawe, K. D., Searle, F., Bisset, G. M., and Jarman, M., 1990, Novel prodrugs which are activated to cytotoxic alkylating agents by carboxypeptidase G2, *J. Med. Chem.* **33:**677–681.

Springer, C. J., Bavetsias, V., Jackman, A. L., Boyle, F. T., Marshall, D., Pedley, R. B., and Bisset, G. M., 1996, Prodrugs of thymidylate synthase inhibitors: Potential for antibody directed enzyme prodrug therapy (ADEPT), *Anticancer Drug Des.* **11:**625–636.

Springer, C. J., Dowell, R., Burke, P. J., Hadley, E., Davis, D. H., Blakey, D. C., Melton, R. G., and Niculescu Duvaz, I., 1995, Optimization of alkylating agent prodrugs derived from phenol and aniline mustards: A new clinical candidate prodrug (ZD2767) for antibody-directed enzyme prodrug therapy (ADEPT), *J. Med. Chem.* **38:**5051–5065.

Springer, C. J., Poon, G. K., Sharma, S. K., and Bagshawe, K. D., 1993, Identification of prodrug, active drug, and metabolites in an ADEPT clinical study, *Cell Biophys.* **22:**9–26.

Stanislawski, M., Rousseau, V., Goavec, M., and Ito, H., 1989, Immunotoxins containing glucose oxidase and lactoperoxidase with tumoricidal properties: *In vitro* killing effectiveness in a mouse plasmacytoma cell model, *Cancer Res.* **49:**5497–5504.

Sunters, A., Baer, J., and Bagshawe, K. D., 1991, Cytotoxicity and activation of CB1954 in a human tumour cell line, *Biochem. Pharmacol.* **41:**1293–1298.

Svensson, H. P., Kadow, J. F., Vrudhula, V. M., Wallace, P. M., and Senter, P. D., 1992, Monoclonal antibody-β-lactamase conjugates for the activation of a cephalosporin mustard prodrug, *Bioconjugate Chem.* **3:**176–181.

Svensson, H. P., Wallace, P. M., and Senter, P. D., 1994, Synthesis and characterization of monoclonal antibody-β-lactamase conjugates, *Bioconjugate Chem.* **5:**262–267.

Tercel, M., Denny, W. A., and Wilson, W. R., 1996, A novel nitro-substituted *seco*-CI: Application as a reductively activated ADEPT prodrug, *Bioorg. Medicinal Chem. Lett.* **6:**2741–2744.

Vitols, K. S., Haag Zeino, B., Baer, T., Montejano, Y. D., and Huennekens, F. M., 1995, Methotrexate-α-phenylalanine: Optimization of methotrexate prodrug for activation by carboxypeptidase A monoclonal antibody conjugate, *Cancer Res.* **55:**478–481.

Vrudhula, V. M., Senter, P. D., Fischer, K. J., and Wallace, P. M., 1993a, Prodrugs of doxorubicin and melphalan and their activation by a monoclonal antibody-penicillin-G amidase conjugate, *J. Med. Chem.* **36:**919–923.

Vrudhula, V. M., Svensson, H. P., Kennedy, K. A., Senter, P. D., and Wallace, P. M., 1993b, Antitumor activities of a cephalosporin prodrug in combination with monoclonal antibody-β-lactamase conjugates, *Bioconjugate Chem.* **4:**334–340.

Vrudhula, V. M., Svensson, H. P., and Senter, P. D., 1995, Cephalosporin derivatives of doxorubicin as prodrugs for activation by monoclonal antibody-β-lactamase conjugates, *J. Med. Chem.* **38:**1380–1385.

Wallace, P. M., MacMaster, J. F., Smith, V. F., Kerr, D. E., Senter, P. D., and Cosand, W. L., 1994, Intratumoral generation of 5-fluorouracil mediated by an antibody-cytosine deaminase conjugate in combination with 5-fluorocytosine, *Cancer Res.* **54:**2719–2723.

Wallace, P. M., and Senter, P. D., 1991, *In vitro* and *in vivo* activities of monoclonal antibody-alkaline phosphatase conjugates in combination with phenol mustard phosphate, *Bioconjugate Chem.* **2:**349–352.

Wang, S. M., Chern, J. W., Yeh, M. Y., Ng, J. C., Tung, E., and Roffler, S. R., 1992, Specific activation of glucuronide prodrugs by antibody-targeted enzyme conjugates for cancer therapy, *Cancer Res.* **52:**4484–4491.

Wawrzynczak, E. J., 1991, Systemic immunotoxin therapy of cancer: Advances and prospects, *Br. J. Cancer* **64:**624–630.

Wentworth, P., Datta, A., Blakey, D., Boyle, T., Partridge, L. J., and Blackburn, G. M., 1996, Toward antibody-directed "abzyme" prodrug therapy, ADAPT: Carbamate prodrug activation by a catalytic antibody and its *in vitro* application to human tumor cell killing, *Proc. Natl. Acad. Sci. USA* **93:**799–803.

Young, C. W., Yagoda, A., Bittar, E. S., Smith, S. W., Grabstald, H., and Whitmore, W., 1976, Therapeutic trial of aniline mustard in patients with advanced cancer. Comparison of therapeutic response with cytochemical assessment of tumor cell β-glucuronidase activity, *Cancer* **38:**1887–1895.

5

The Design and Synthesis of Prodrugs for Antibody-Directed Enzyme Prodrug Therapy (ADEPT)

P. J. Burke

1. INTRODUCTION

Antibody-directed enzyme prodrug therapy (ADEPT) is a strategy for treating cancer that combines antibody specificity with enzymatic catalysis to deliver drugs selectively to tumor sites. In this approach a conjugate consisting of an enzyme chemically linked to a tumor-selective antibody (mAb) is administered to the host and allowed to localize at the tumor site. After a suitable interval to allow clearance of the conjugate from the blood, a non-toxic prodrug is administered which releases a therapeutic drug on contact with the enzyme. This two-step approach has the potential to overcome many of the limitations of current chemotherapeutic strategies, allow the generation of high intratumoral concentrations of drug, and minimize damage to normal tissues.

The initial stage of prodrug design is usually selection of an appropriate drug. Preferably, it should be of low molecular weight and relatively lipophilic so that it can diffuse into target cells after extracellular generation by the enzyme. It must have a functional group that can be modified to accommodate the enzyme's substrate moiety and, ideally, this functional group should be essential for the cytotoxic activity of the drug. Further considerations regarding the selection of the

P. J. Burke • Department of Medical Oncology, Charing Cross Hospital, London W6 8RP, United Kingdom.

Enzyme-Prodrug Strategies for Cancer Therapy, edited by Melton and Knox. Kluwer Academic/Plenum Publishers, New York, 1999.

drug are the choice of target, availability of materials, accessible pharmacokinetic data, etc.

The prodrug is formed by covalently linking the drug to the enzyme's substrate moiety. The composition of the latter is determined by the choice of enzyme and may be a specific chemical grouping or a broad range of species. It may be possible to chemically modify the substrate moiety to improve substrate specificity, reduce cytotoxicity, or modify the prodrug pharmacokinetics, but such endeavours are probably best left until "second generation" compounds are evaluated. The nature of the linkage between the drug and the substrate moiety is critical to the prodrug's success. It may be a simple chemical bond between the two components or additional atoms may be added to facilitate the cleavage process. Whatever the nature of the bond, it must be stable under physiological conditions and must be capable of rapid cleavage by the enzyme to release the drug.

Prodrugs synthesized to date have usually been evaluated by determining the cytotoxicity of the prodrug alone and in the presence of enzyme or conjugate by measuring the enzyme kinetics of drug release and, in many cases, by antitumor studies in mouse xenograft models after administering an appropriate conjugate. A number of potential therapeutic systems have been identified, and one has undergone a pilot-scale clinical trial with encouraging results.

This chapter reviews the ADEPT systems described in the literature but concentrates on the prodrug/drug combinations rather than on the biological components. For convenience the prodrugs are classified according to the enzyme for which they have been designed.

2. PRODRUGS USED IN ADEPT

2.1. Prodrugs for Carboxypeptidases

2.1.1. CARBOXYPEPTIDASE G2

Prodrugs derived from benzoic acid mustards have been synthesized by Bagshawe and colleagues for use with an ADEPT system utilizing the bacterial enzyme carboxypeptidase G2 (CPG2) (Bagshawe, 1987; Bagshawe *et al.*, 1988; Blakey *et al.*, 1993; Springer *et al.*, 1991). This enzyme cleaves the amide bond of folic acid and several of its analogs to release L-glutamic acid and the appropriate pteroate. Prodrugs **1–3** were designed to release benzoic acid mustards upon activation by CPG2. The prodrugs were expected to be less cytotoxic than their parent drugs because the mustard arms would be deactivated by formation of the amide bond. It was also thought that the increased hydrophilicity imparted by the

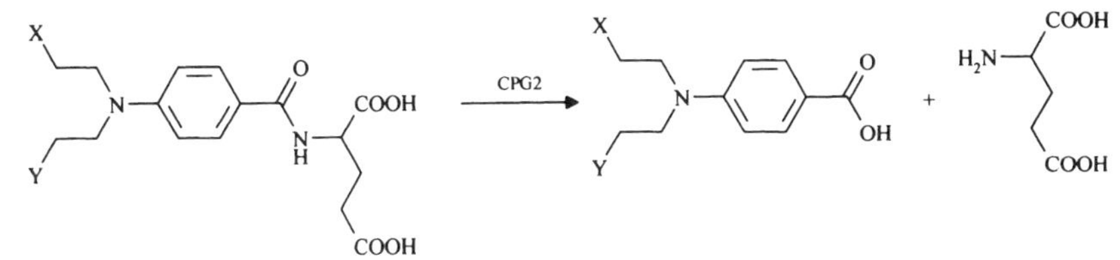

1 X=Y=Cl

2 X=Cl, Y=OSO_2CH_3

3 X=Y=OSO_2CH_3

glutamate carboxylic acids would lead to a reduced rate of cell uptake, thus reducing cytotoxicity.

Prodrugs **1–3** were prepared by reacting carboxy-protected L-glutamic acid with 4-nitrobenzoyl chloride followed by reduction of the nitro group with ammonium formate. The resulting amine **4** was alkylated with excess ethylene oxide to give di-*t*-butyl-4-[bis(2-hydroxyethyl)amino]benzoyl-L-glutamate **5**.

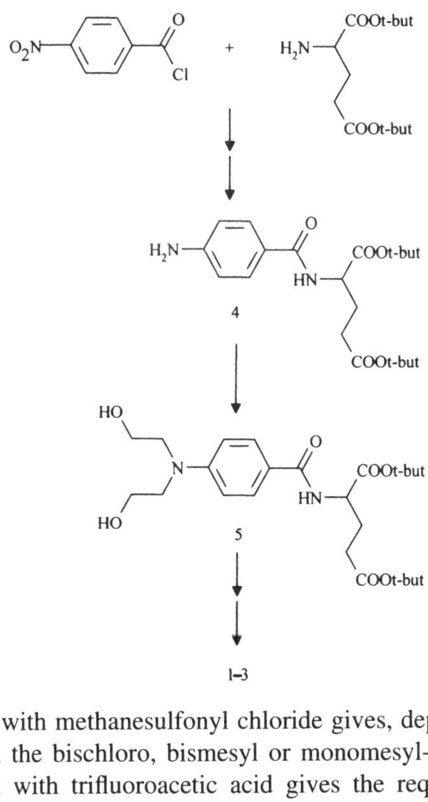

Reaction of **5** with methanesulfonyl chloride gives, depending on the temperature employed, the bischloro, bismesyl or monomesyl-monochloro derivatives. Deprotection with trifluoroacetic acid gives the required prodrugs **1–3**

(Springer *et al.*, 1990). Treatment of **1–3** with CPG2 rapidly regenerates the corresponding parent mustard drugs (Springer *et al.*, 1991). All three prodrugs are less toxic than their parent drugs and, in combination with an appropriate antibody-CPG2 conjugate, give rise to significant antitumor effects in mouse xenograft models (Bagshawe *et al.*, 1988; Blakey *et al.*, 1993; Springer *et al.*, 1991). A pilot-scale clinical trial using antibody-CPG2 and prodrug **2** has demonstrated the feasibility of using an ADEPT protocol in the clinic (Bagshawe *et al.*, 1995).

A series of prodrugs based on phenol or phenylenediamine mustards has also been evaluated for use with CPG2 (Blakey *et al.*, 1995, 1996; Dowell *et al.*, 1996; Springer *et al.*, 1996). In these compounds the aromatic ring bearing the haloalkyl groups is linked to L-glutamic acid by a carbamate or a urethane bond. A number of synthetic routes were employed to prepare the prodrugs and these are briefly outlined following. In general, prodrugs derived from phenol mustards were synthesized *via* the 4-nitrophenylcarbonate derivative **6** which was reacted with dibenzyl-L-glutamic acid to give the protected carbamates **7**. Hydrogenation with

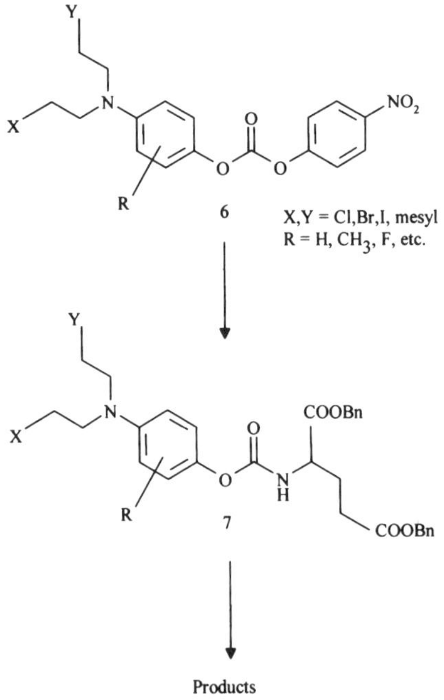

Pd/C gives the required products (Springer *et al.*, 1996). Other phenol mustard derivatives are obtained by reacting substituted 4-nitrophenols with $COCl_2$ and di-*t*-butyl L-glutamic acid and then introducing the mustard arms *via* reduction of the nitro group and hydroxyethylation using standard methods. Prodrugs derived from phenylenediamine mustards were synthesized from substituted 4-phenylenediamine mustards by reaction with $COCl_2$ and dibenzyl L-glutamic acid, followed by hydrogenation (Springer *et al.*, 1996).

The most promising compounds to emerge from these studies were the phenol mustard glutamate **8** and its di-iodo derivative **9**. Both **8** and **9** are less

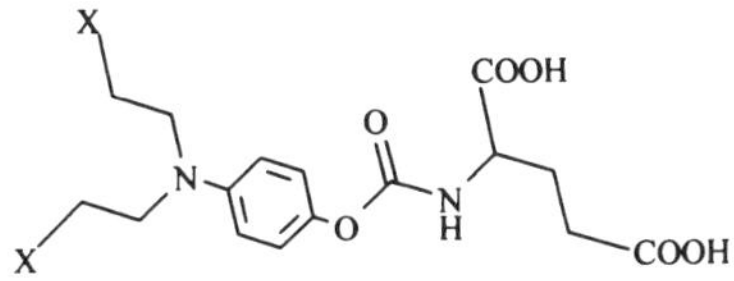

8 X = Cl
9 X = I

cytotoxic than their parent drugs against colon carcinoma wells and are surprisingly good substrates for CPG2. Both compounds gave tumor regressions and growth delays when tested *in vivo* in a mouse xenograft model (Blakey *et al.*, 1995, 1996). Compound **9** is currently under clinical development for treating colon cancer.

2.1.2. Carboxypeptidase A

Haenseler *et al.* reported prodrugs of methotrexate in which the α-carboxyl of the L-glutamate was linked to a peptide (Haenseler *et al.*, 1992; Vitols *et al.*, 1995). The rationale for this approach was that methotrexate α-peptides are known to be less cytotoxic than methotrexate, presumably because they are taken up into cells at a reduced rate. The optimum prodrug in terms of cytotoxicity and substrate specificity is the phenylalanine derivative **10** which was synthesized by reacting

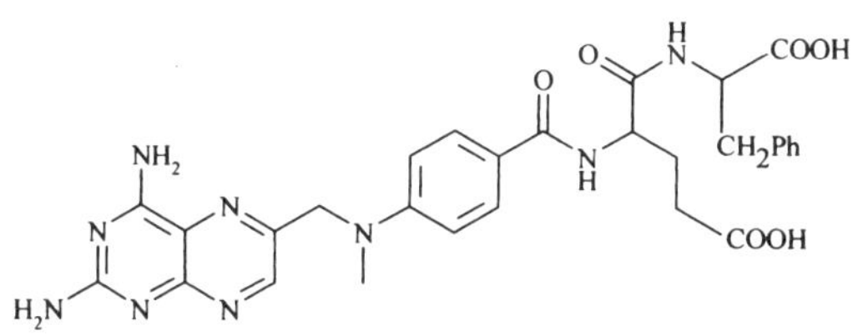

10

the 4-nitrophenyl ester of 4-amino 4 deoxy-10-methylpteroic acid with L-glutamyl-α-L-phenylalanine. The reaction conditions were carefully controlled to optimize the yield of the L,L-diastereoisomer because the D,L-stereoisomer is not a substrate for the enzyme. Prodrug **10** is considerably less cytotoxic than methotrexate but in the presence of carboxypeptidase A is as cytotoxic as the parent drug. To date no *in vivo* studies have been reported for this system.

In a similar vein, Springer *et al.* evaluated α-peptides of two potent thymidylate synthase inhibitors for activation by carboxypeptidase A (Springer *et al.*, 1996). α-L-Alanine and α-L-glutamic acid derivatives of ICI198583 and ZD1694 (**11–14**) were synthesized by coupling di-*t*-butyl protected esters of the amino acids with the appropriate quinazoline derivative, using diethylcyanophosphoridate as coupling agent. Deprotection with TFA gives the required prodrugs **11–14**.

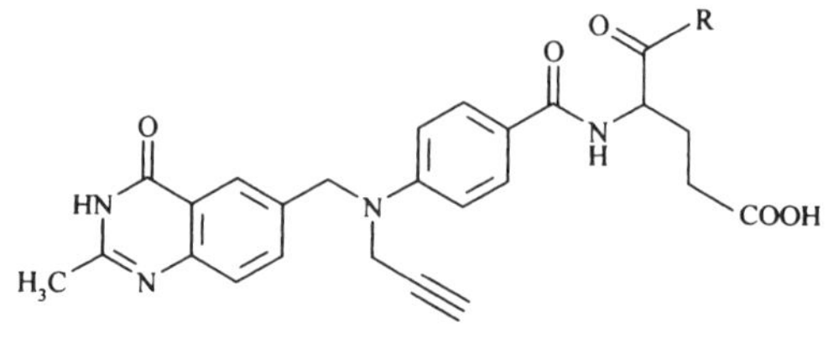

11, R = L-Ala
12, R = L-Glu

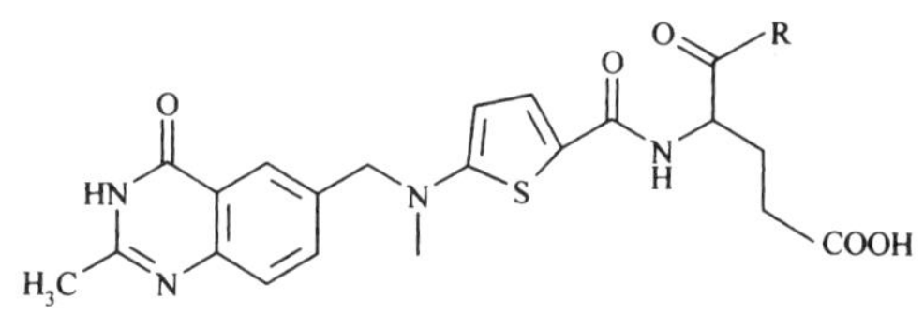

13, R = L-Ala
14, R = L-Glu

All four prodrugs are less cytotoxic than their parent drugs against LS174T colon carcinoma cells. Compound **11** shows a particularly favourable deactivation (>200). In the presence of CPA, the α-L-alanine dipeptide prodrugs **11** and **13** are activated to a more potent agent. This was most marked for **11** that increased in potency by 100-fold (Springer *et al.*, 1996).

2.2. Prodrugs for Alkaline Phosphatase

The synthesis of prodrugs for activation by alkaline phosphatases is relatively attractive because adding a phosphate group to a drug is likely to reduce the rate of cell uptake significantly relative to the parent drug, thus considerably reducing the cytotoxicity. In addition, such a prodrug will be water-soluble and will avoid many of the formulation problems encountered in administration to humans. Prodrugs **15–18**, which are based upon etoposide, doxorubicin, phenol mustard, and mitomycin respectively, have been synthesized for use with calf

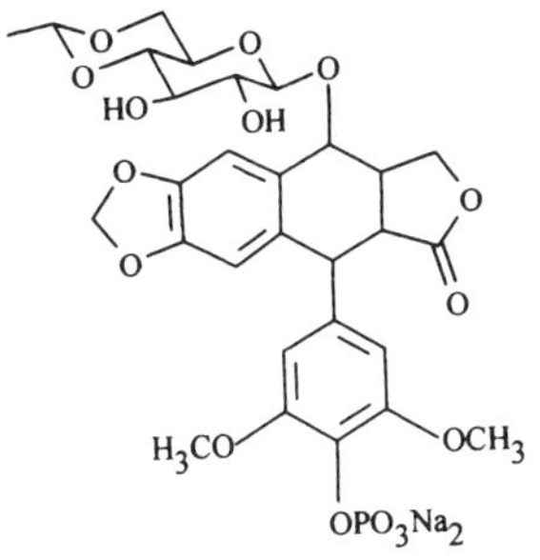

15

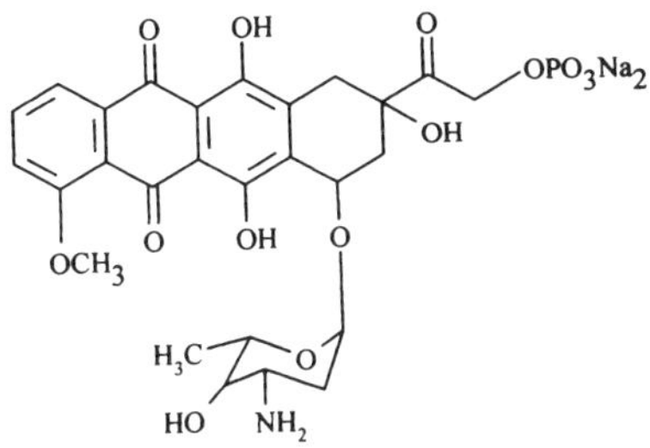

16

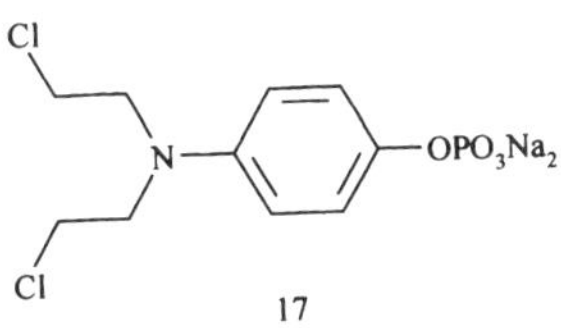

17

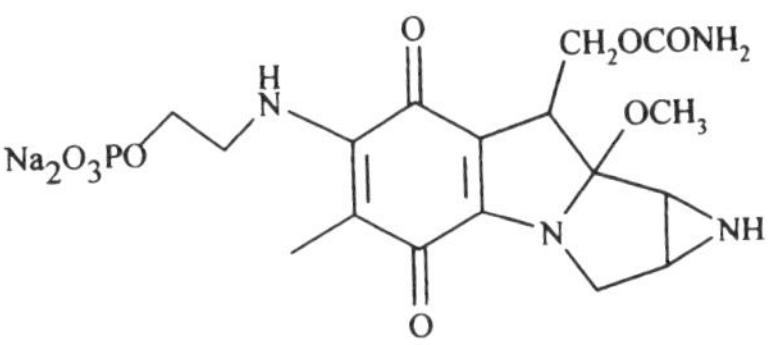

18

intestinal alkaline phosphatase (Senter *et al.*, 1989, 1990; Wallace *et al.*, 1991). Conversion of the parent drugs to prodrugs **15–17** is achieved simply by reaction with phosphoryl chloride followed by basic hydrolysis. The mitomycin analog **18** is obtained by displacing the 7-methoxy group in mitomycin A with 2-aminoethylphosphate. Where reported, the prodrugs were considerably less cytotoxic than their parent drugs and when tested *in vivo*, in combination with an antibody-alkaline phosphatase conjugate, gave significant antitumor responses.

An advantage of targeting an alkaline phosphatase is that the enzyme can activate several different prodrugs that could be used together in combination chemotherapy. A potential disadvantage of this system is the ubiquity of alkaline phosphatases in mammals that is likely to lead to nonspecific conversion of the prodrugs and increased toxicity.

2.3. Prodrugs for Glycosidases

2.3.1. β-Glucuronidase

Prodrugs containing the glucuronic acid moiety are attractive candidates for ADEPT because the ionizable carboxylic acid can reduce cell uptake and cytotoxicity. Several groups have examined prodrugs of this type. Roffler *et al.* (1991)

evaluated the phenol mustard glucuronide **19** for use with β-glucuronidase from *E. coli*. (Wang *et al.*, 1992). The compound was synthesized from phenol mustard, as previously reported (Bukhari *et al.*, 1972). Prodrug **19** is considerably less toxic than phenol mustard when tested against a variety of human cancer cell lines but in the presence of enzyme it is as cytotoxic as the parent drug.

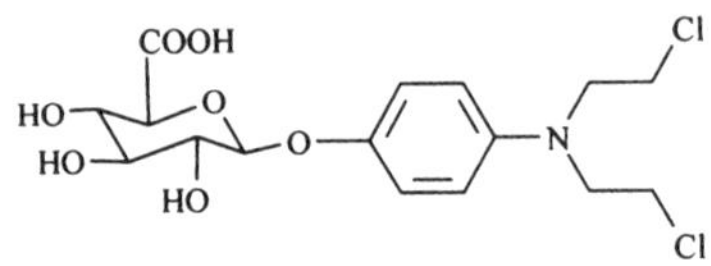

19

In an attempt to synthesize a more stable prodrug of phenol mustard, Schmidt *et al.* synthesized glucuronide **20** in which the mustard is separated from the sugar by a carbamate spacer (Schmidt *et al.*, 1997). Cleavage of the glucuronic acid in this molecule by β-glucuronidase results in liberating a phenolate ion that is free to cyclize and cause expulsion of phenol mustard. The synthesis of **20** is

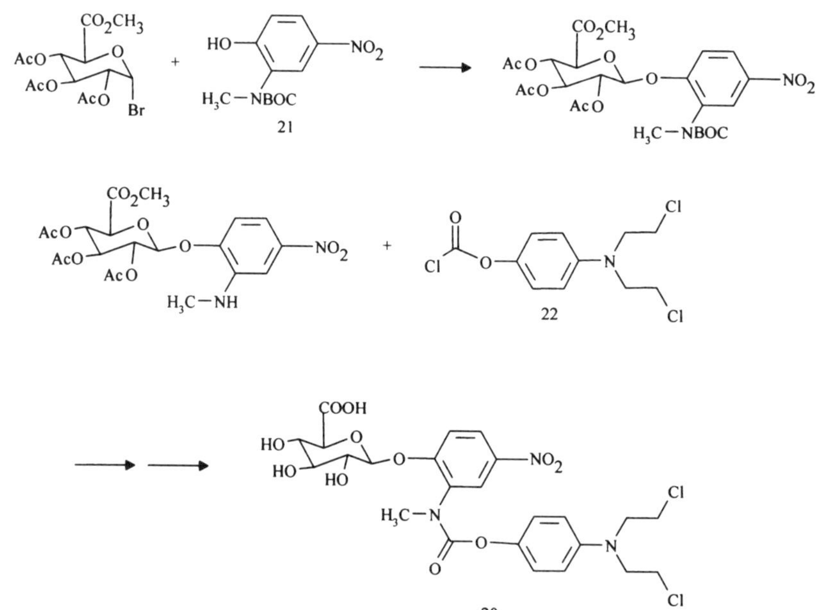

20

accomplished by coupling protected α-D-glucopyranosyl bromide uronate with phenol **21** followed by removal of the BOC-protecting group and condensation of the amine function with phenol mustard chloroformate **22**. Deprotection of the sugar by standard methods gives prodrug **20** in good yield. The compound is a good substrate for *E. coli* β-glucuronidase, and the mixture rapidly releases phenol mustard. The cytotoxicity of **20** is considerably less than phenol mustard, and in the presence of enzyme, the cytotoxicity of **20** equals that of the parent drug. In phosphate buffer (pH 7.2) the half-life of **20** is 44 min. These preliminary studies indicated the suitability of **20** as a potential prodrug for use in ADEPT.

Prodrug **23**, based upon doxorubicin, was evaluated by Bosslet *et al.* (1994) for activation by human β-glucuronidase. In this case the glucuronic acid is

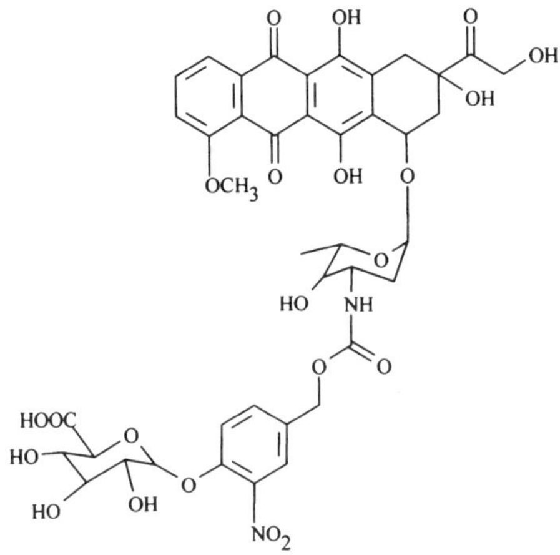

23

separated from the parent drug by a nitrobenzyl carbamate spacer attached to the amino group of the drug carbohydrate. After cleavage of the glucuronic acid by the enzyme the phenol moiety remaining spontaneously decomposes to release the parent drug. Prodrug **23** is less toxic than doxorubicin *in vivo* and in combination with a β-glucuronidase fusion protein shows good antitumor effects in a human tumor xenograft model (Bosslet *et al.*, 1994).

As with the phosphatase system, a potential limitation of glucuronidase ADEPT is the susceptibility of the drug to activation by endogenous enzyme. However, the concentration of β-glucuronidase in human serum is very low, indicating that glucuronide prodrugs should be stable in the blood after IV administration. In addition, it has been shown that prodrug **19** is relatively non-toxic to a number of cell lines that express high levels of endogenous β-glucuronidase (Wang *et al.*, 1992). This result probably reflects a low rate of intracellular transport due to the charge carried by the prodrug and suggests that endogenous glucuronidase may not compromise an ADEPT targeting strategy.

2.3.2. Galactosidases

Based on the assumption that incorporating any carbohydrate should reduce the penetration of prodrugs into cells, other groups examined galactose as a potential enzyme substrate moiety. Prodrugs **24**, **25**, and **26**, derived from daunorubicin, were synthesized for activation by α-galactosidase *via* spontaneous decomposition of the spacer (Gesson *et al.*, 1994). Compound **24** is synthesized

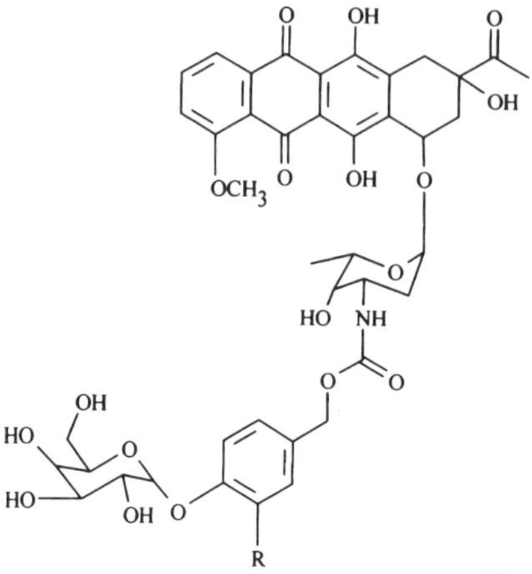

24, R = Cl
25, R = NO_2

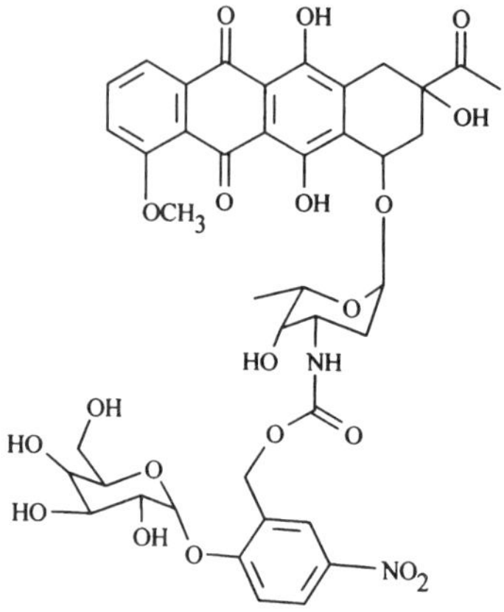

26

from tetra-*O*-acetyl-α-D-galactopyranose which is coupled to 2-chloro-4-methylphenol in the presence of zinc chloride. After benzylic bromination and hydrolysis with silver nitrate, the resulting alcohol is activated with *N,N′*-disuccinimidyl carbonate (DSC) and reacted with daunorubicin. Removal of the acetyl protecting groups by transesterification yields **24**. Prodrug **25** is prepared similarly from *p*-cresol and per-*O*-acetyl-D-galactose although in this case a nitration step is included after bromination. Prodrug **26** is prepared from 2-methyl-4-nitrophenol and per-*O*-acetyl-α-D-galactopyranose but the hydroxy derivative is activated with 4-nitrophenylchloroformate rather than DSC. All three prodrugs are considerably less cytotoxic than daunorubicin and are reasonably stable in plasma. In the presence of α-galactosidase, prodrugs **24** and **25** rapidly decompose to release the parent drug. Release of daunorubicin from **26** is somewhat slower (Gesson *et al.*, 1994).

A similar prodrug, based upon doxorubicin, has also been synthesized for activation by α-galactosidase (Azoulay *et al.*, 1995). In prodrug **27** the α-galactopyranosyl residue is directly linked to the carbohydrate amino group of doxorubicin through a self-immolative 4-aminobenzyloxycarbonyl spacer. **27** is synthesized *via* intermediate **28** that is formed by condensing *p*-tolyl isocyanate with tetra-*O*-acetyl-α-D-galactopyranose. After bromination with NBS and treatment

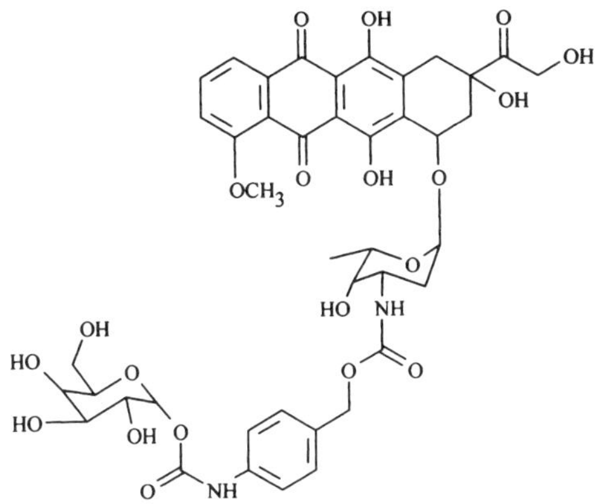

27

with $AgNO_3$, the required alcohol is activated with 4-nitrophenylchloroformate to give **29** which is reacted directly with doxorubicin and deprotected to give the required product. Prodrug **27** is essentially nontoxic against L1210 murine leukemia cells and is reasonably stable in plasma. In the presence of α-galactosidase, **27** is almost completely hydrolyzed within 150 min to afford free doxorubicin.

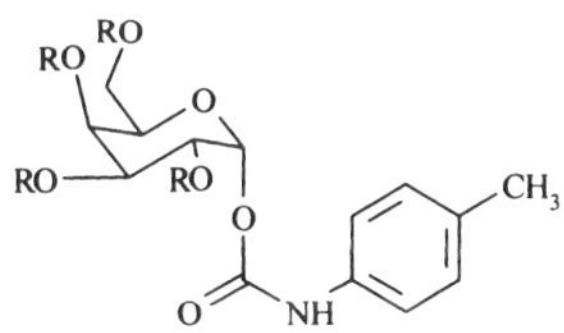

28 R = Ac

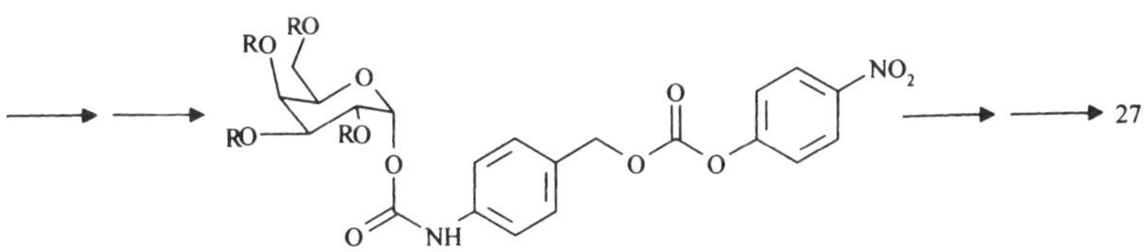

29

2.4. Prodrugs for β-Lactamases

A number of prodrugs based on the cephalosporin nucleus have been designed for activation by β-lactamases. Cephalosporins are readily hydrolyzed by β-lactamases with concurrent expulsion of the C-3′ substituent, so it should be possible to attach a variety of cytotoxic drugs to this position without compromising the substrate specificity of the enzyme. In addition, appropriate modification of the cephalosporin nucleus and/or the linking atoms should allow varying the pharmacokinetic properties, solubility, cytotoxicity, etc. according to requirements. The viability of the cephalosporin/lactamase system was initially demonstrated using prodrug **30** which was based on a derivative of the potent drug desacetylvinblastine (Shepherd *et al.*, 1991).The compound is synthesized from

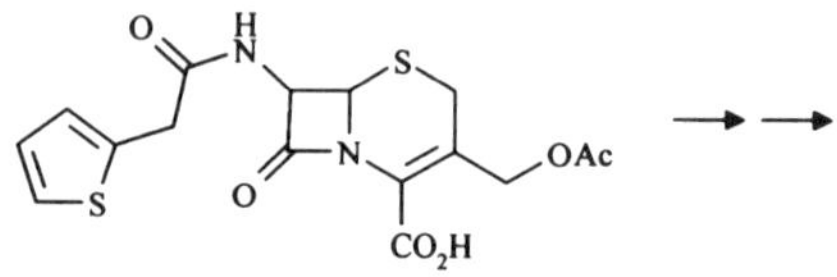

31

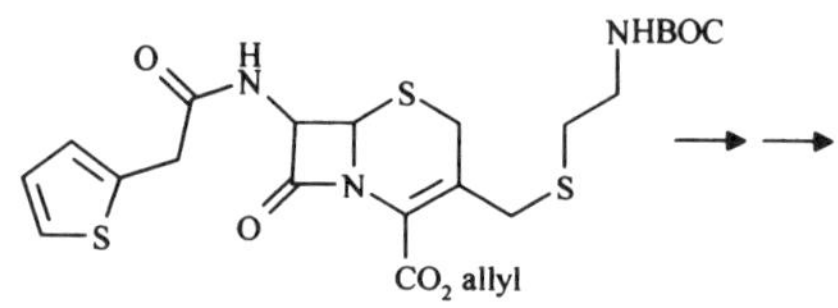

32

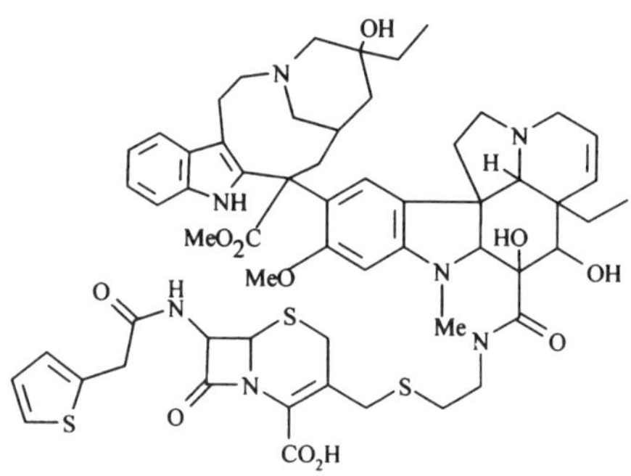

30

keflin **31** by introducing a protected aminoethanethiol group and allyl ester protection to give intermediate **32**. After acidic removal of the BOC-protecting group, the resulting amine is coupled with desacetylvinblastine azide, and the allyl ester is removed to give an isomeric mixture of products. Reverse-phase chromatography provides pure Δ-3 isomer **30**. *In vitro* experiments show that prodrug **30** is a good substrate for β-lactamase. A reproducible fivefold (mol/mol) difference in cytotoxicity between drug and prodrug was observed against LS174T colon carcinoma cells. In the presence of enzyme the prodrug is equipotent with the drug. In further studies, the same group found that cephalosporin sulfoxide analogs are better substrates for lactamase than the parent cephalosporin and this prompted the synthesis of prodrug **33**, a cephalosporin sulfoxide derivative of the

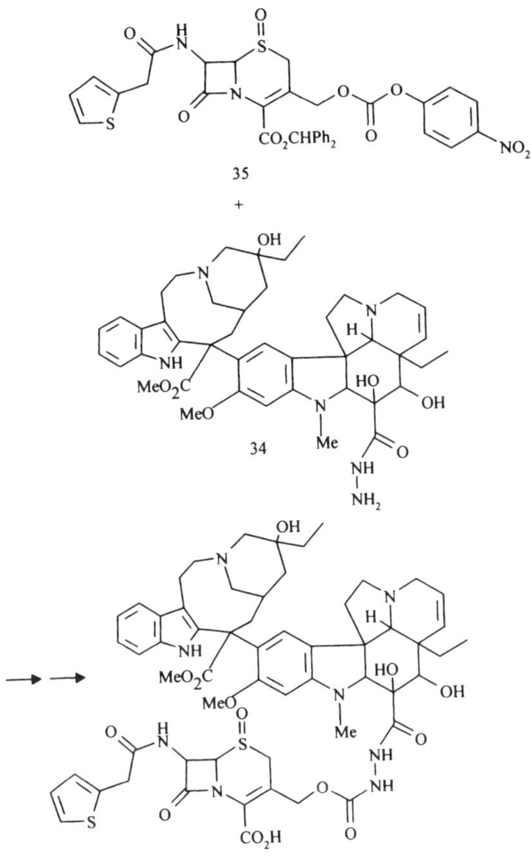

33

potent agent deseacetylvinblastine hydrazide **34** (Jungheim *et al.*, 1992). Prodrug **33** is synthesized by condensing **34** with carbonate **35** followed by removal of the benzhdryl group. The prodrug is an excellent substrate for lactamase and, as was shown by HPLC, rapidly releases **34**. The compound is less cytotoxic than **34** and caused regression of established tumors in mouse xenograft studies (Meyer *et al.*, 1993).

A similar cephalosporin nucleus was used to make a prodrug of doxorubicin (Jungheim *et al.*, 1993). Prodrug **36** is readily prepared by displacing 4-nitrophenol from **37** by the amino group of doxorubicin, followed by removal of the allyl protecting group. Prodrug **36** is a good substrate for β-lactamase and is less cytotoxic than the parent drug. Combined with an appropriate β-lactamase conjugate, prodrug **36** shows good antitumor effects in a mouse xenograft model.

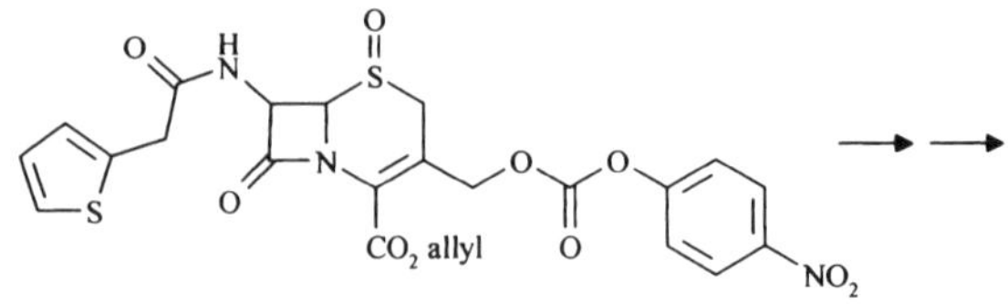

37

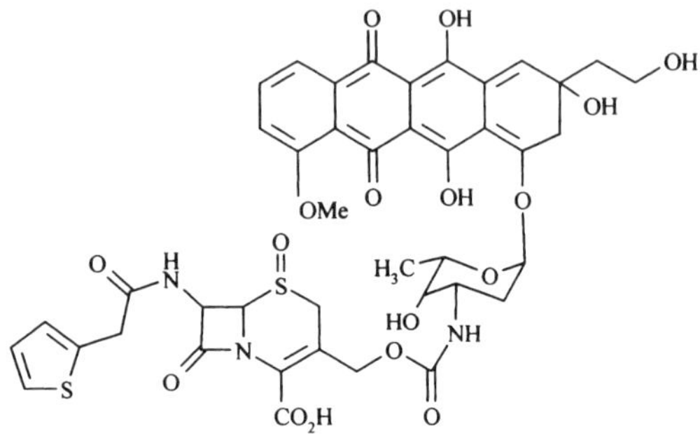

36

A doxorubicin prodrug based on a cephalosporin nucleus similar to that used in **36** was also reported (Hudyma *et al.*, 1993). Pharmacokinetic studies of a mouse model showed that the intratumoral levels of drug after treatment with conjugate and prodrug were higher than those obtained by systemic treatment of drug. This data correlated with greater therapeutic effects for the ADEPT system (Svensson *et al.*, 1995).

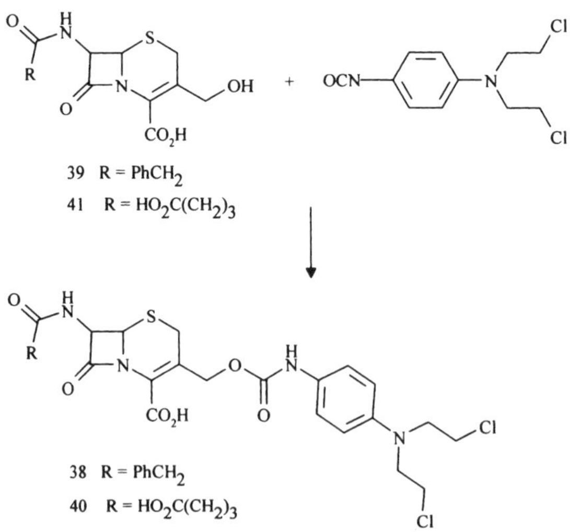

39 $R = PhCH_2$

41 $R = HO_2C(CH_2)_3$

38 $R = PhCH_2$

40 $R = HO_2C(CH_2)_3$

Senter *et al.* have synthesized phenylenediamine mustard prodrugs based upon cephalosporin (Svensson *et al.*, 1992). Prodrug **38**, prepared by condensing **39** with phenylenediamine mustard isocyanate, causes severe tail vein necrosis upon injection. This problem was overcome by substituting the side chain of **38** with a more water-soluble carboxylic acid derivative. Condensation of intermediate **41** with the mustard isocyanate gives prodrug **40**, which exhibits good antitumor effects in a mouse xenograft model (Kerr *et al.*, 1995; Vrudhula *et al.*, 1993).

A cephalosporin-based prodrug of taxol has been synthesized for use with β-lactamase (Rodrigues *et al.*, 1995). The compound shows a 10-fold reduction in cytotoxicity over taxol and releases the drug on prolonged exposure to the enzyme. A cephalosporin-carboplatinum prodrug, which is a substrate for β-lactamase, has been reported but no further biological data have been presented (Hanessian and Wang, 1993).

2.5. Prodrugs for Penicillin V/G Amidase

Kerr and co-workers have synthesized prodrugs of melphalan and doxorubicin designed to be used with penicillin V amidase (Kerr *et al.*, 1990). Prodrug **42** is prepared by activating the carboxylic acid group of 4-hydroxyphenoxy acetic acid with *N*-hydroxysuccinimide/DCCI and coupling with the amino group of doxorubicin. Prodrug **43** is prepared similarly by substituting melphalan HCl for doxorubicin. Both prodrugs are significantly less toxic than their parent drugs to H2981 lung adenocarcinoma cells and Daudi lymphoma cells. However, the cytotoxicity

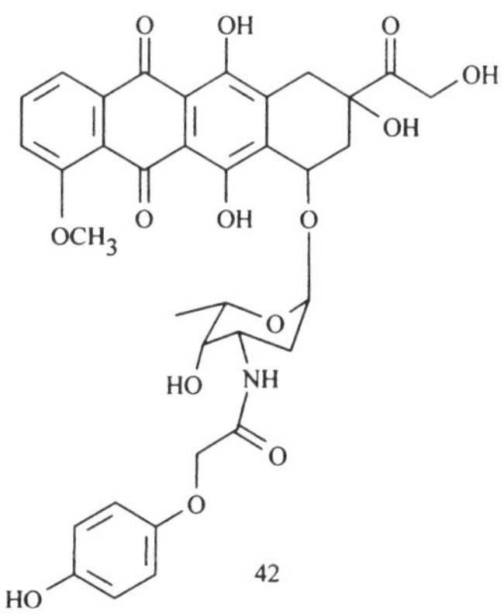

42

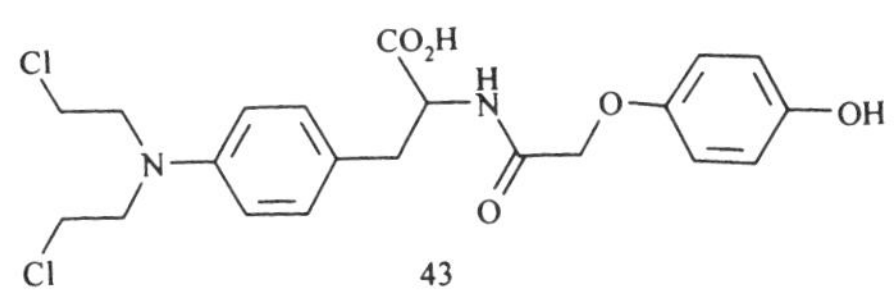

43

of **43** was not enhanced by pretreating the cells with the appropriate mAb-amidase conjugate. The authors speculated that this resulted from slow kinetics of enzyme turnover, such that a cytotoxic concentration of drug is never reached (Kerr *et al.*, 1990).

Prodrugs **44** and **45**, designed to be activated by penicillin G amidase, were synthesized by Vrudhula and co-workers by forming the phenyl acetamide of melphalan and doxorubicin, respectively (Vrudhula *et al.*, 1993). Both prodrugs are substrates for penicillin G amidase and are less cytotoxic than their parent drugs.

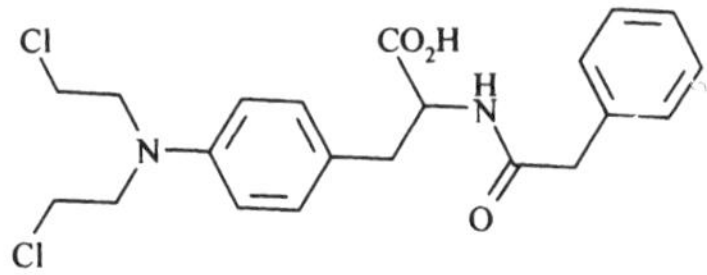

44

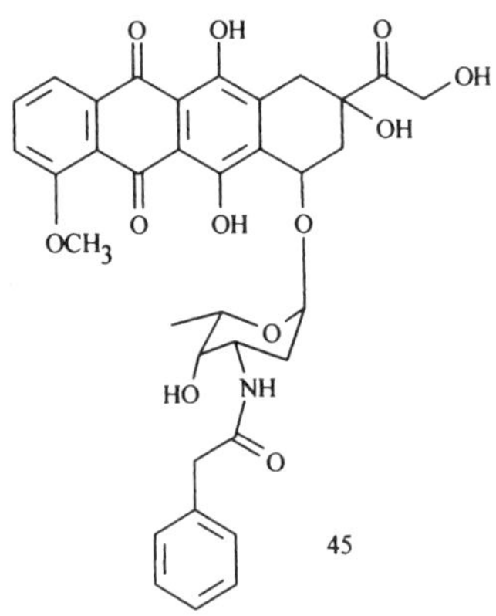

45

2.6. Prodrugs for Nitroreductase

Several groups have synthesized prodrugs for activation by a nitroreductase enzyme isolated from *E. coli.* Thus, prodrugs of phenol mustard, phenylenediamine mustard, actinomycin D, mitomycin C, and doxorubicin, respectively, have been reported (Mauger *et al.*, 1994). In each case the prodrugs were formed by reacting the parent drugs with 4-nitrobenzylchloroformate to generate 4-nitrobenzyloxycarbonyl derivatives. Reduction of the nitro group to the corresponding 4-hydroxylamine resulted in fragmentation that releases the parent drug. Prodrug **46** proved to be the most promising candidate for ADEPT and rapidly generates

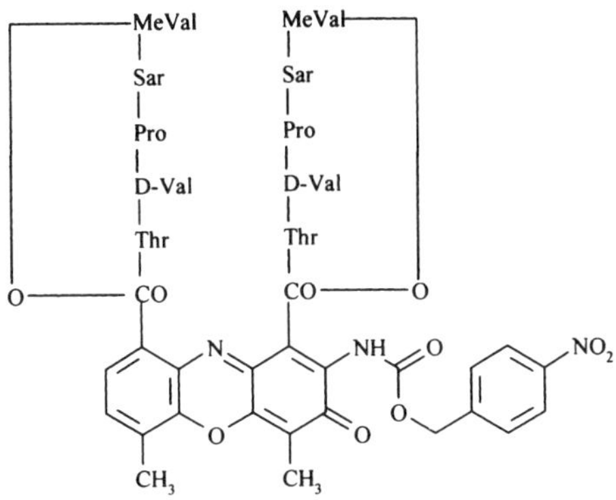

46

actinomycin D on incubation with nitroreductase in the presence of NADH cofactor. The cytotoxicity of **46** against V76 cells is considerably lower than that of the parent drug and this is mirrored by toxicity studies in mice. Prodrugs based on an enediyne and tallimustine, which were activated similarly, have also been reported (Hay *et al.*, 1995; Lee *et al.*, 1997).

The synthesis of a seco-CI prodrug designed to be activated by nitroreductase has been reported (Tercel *et al.*, 1996). Prodrug **47** is prepared from 2-chloro-5-

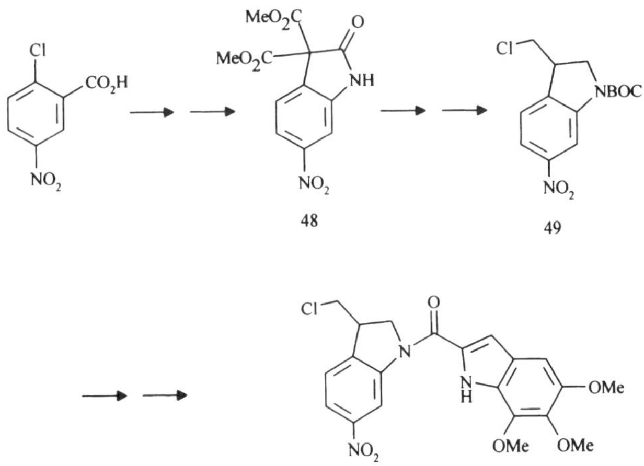

47

nitrobenzoic acid *via* the lactam **48**. Treatment of the chloromethyl derivative **49** with 5,6,7-trimethoxyindole-2-carboxylic acid gives prodrug **47**. A 400-fold increase in cytotoxicity was observed when **47** was incubated with nitroreductase and NADH in the presence of UV4 cells.

It has been proposed that CB 1954 (5-(aziridin-1-yl)-2,4-dinitrobenzamide)

(**50**) may be a suitable prodrug for nitroreductase-based ADEPT (Knox, *et al.*, 1993; Sunters *et al.*, 1991). The compound alone shows little cytotoxicity to Chinese hamster V79 cells but is activated in the presence of a nitroreductase enzyme isolated from *E. coli* to give 5-(aziridin-1-yl)-4-hydroxylamino-2-nitrobenzamide (**51**) that, upon further, nonenzymatic metabolism, is converted into a potent bifunctional cross-linking agent (Knox *et al.*, 1993). *In vitro* data have shown that a 10,000-fold difference in cytotoxicity may be obtained between **50** alone and **50** in the presence of enzyme and cofactor (Knox *et al.*, 1993). A series

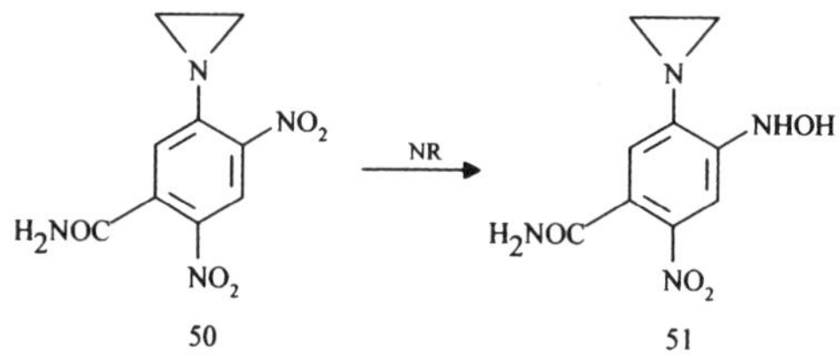

of related dinitrobenzamide mustards have also been evaluated as potential prodrugs for nitroreductase (Atwell *et al.*, 1996).

A potential disadvantage of using nitroreductase for ADEPT is the enzyme's requirement for a cofactor to supply reducing equivalents. The biogenic cofactors for reductases, NADH, and NADPH, are unsuitable for this purpose because they are unstable in serum. However, it has been suggested that a serum-stable, enzyme-selective cofactor, such as reduced nicotinate riboside may allow the development of an ADEPT system that uses a bioreductive enzyme (Knox *et al.* 1993).

2.7. Prodrugs for Cytosine Deaminase

The conversion of 5-fluorocytosine **52** to 5-fluorouracil **53** by cytosine deaminase has been examined by Senter and colleagues (Senter *et al.*, 1991). **52** is nontoxic to H2981 human lung adenocarcinoma cells but in the presence of

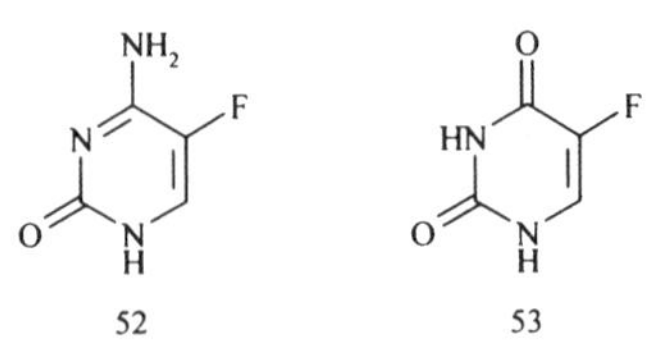

enzyme displays cytotoxicity equivalent to that of the drug **53**. This promising *in vitro* data was confirmed by studies in a mouse tumor model which showed that a significantly greater dose of **52** relative to **53** could be administered to mice which had received antibody-enzyme conjugate followed by a clearing antibody (Wallace *et al.*, 1994). In the same ADEPT model the intratumoral concentration of **53** was 17 times greater than that obtained by systemic administration of **53**, whereas blood levels were similar in both cases. Because **52** and **53** are clinically approved agents it is hoped that this exciting data may rapidly be translated to a clinical setting.

3. SUMMARY

In the past ten years a substantial number of antibodies, enzymes, and prodrugs have been evaluated as candidates for ADEPT. Although most studies have been confined to synthesizing and *in vitro* testing of prodrug/drug pairs, a number of systems have been transferred to xenograft mouse models, where great promise has been shown. To date only one ADEPT system has been clinically tested, but the results from this trial were encouraging and demonstrated the feasibility of the approach in a clinical setting.

Prodrugs based on a wide variety of cytotoxic agents have been synthesized. Those of alkylating agents, particularly aromatic nitrogen mustards, have received considerable attention and this has proved to be fully merited. It is fitting that a prodrug based on benzoic acid mustard should have been the first to clinical trial. A system based on phenol mustard is currently in clinical development, and the results of a trial are awaited with interest. Several other systems, notably a cephalosporin/vinblastine/lactamase system, a cephalosporin/doxorubicin/lactamase system, and a β-glucuronidase/doxorubicin system have shown exciting preclinical results, and it is hoped that they will soon progress to the clinic.

Prodrug design has gone a long way toward meeting the requirements of a successful ADEPT system, and the prodrugs in current use will be refined and optimized as data from further studies, particularly on prodrug pharmacokinetics, become available. The major challenge for the further development of ADEPT systems is the production of highly specific, nonimmunogenic conjugates. This would allow administering increased doses of prodrug and permit carrying out more than one cycle of therapy.

REFERENCES

Atwell, G. J., Boyd, M., Palmer, B. D., Anderson, R. F., Pullen, S. M., Wilson, W. R., and Denny, W. A., 1996, Synthesis and evaluation of 4-substituted analogues of 5-[*N*-*N*-bis(2-chloroethyl)-amino]-2-nitrobenzamide as bioreductively activated prodrugs using an *Escherichia coli* nitroreductase, *Anticancer Drug Des.* **11:**553–567.

Azoulay, M., Florent, J-C., Monneret, C., Gesson, J. P., Jacquesy, J-C., Tillequin, F., Koch, M.,

Bosslet, K., Czech, J., and Hoffman, D., 1995, Prodrugs of anthracycline antibiotics suited for tumour-specific activation, *Anticancer Drug Des.* **10:**441–450.

Bagshawe, K. D., 1987, Antibody directed enzymes revive anti-cancer prodrugs concept, *Br. J. Cancer* **56:**531–532.

Bagshawe, K. D., Springer, C. J., Searle, F., Antoniw, P., Sharma, S. K., Melton, R. G., and Sherwood, R. F., 1988, A cytotoxic agent can be generated selectively at cancer sites, *Br. J. Cancer* **58:** 700–703.

Bagshawe, K. D., Sharma, S. K., Springer, C. J., and Antoniw, P., 1995, Antibody-directed enzyme prodrug therapy: A pilot scale clinical trial, *Tumor Targeting* **1:**17–29.

Blakey, D. C., Valcaccia, B. E., East, S., Wright, A. F., Boyle, F. T., Springer, C. J., Burke, P. J., Melton, R. G., and Bagshawe, K. D., 1993, Antitumour effects of an antibody-carboxypeptidase G2 conjugate in combination with a benzoic acid mustard prodrug, *Cell Biophys.* **222:**1–8.

Blakey, D. C., Davies, D. H., Dowell, R. I., East, S. J., Burke, P. J., Sharma, S. K., Springer, C. J., Mauger, A. B., and Melton, R. G., 1995, Anti-tumour effects of an antibody-carboxypeptidase G2 conjugate in combination with phenol mustard prodrugs, *Br. J. Cancer* **72:**1083–1088.

Blakey, D. C., Burke, P. J., Davies, D. H., Dowell, R. I., East, S. J., Eckersley, K. P., Fitton, J. E., McDaid, J., Melton, R. G., Niculescu-Duvaz, I. A., Pinder, P. E., Sharma, S. K., Wright, A. F., and Springer, C. J., 1996, ZD2767, an improved system for antibody-directed enzyme prodrug therapy which results in tumour regressions in colorectal tumour xenografts, *Cancer Res.* **56:**3287–3292.

Bosslet, K., Czech, J., and Hoffmann, D., 1994, Tumour selective prodrug activation by fusion protein-mediated catalysis, *Cancer Res.* **54:**2151–2159.

Bukhari, M. A., Everett, J. L., and Ross, W. C. J., 1972, Aryl-2-halogenoalkylamines. XXVI. Glucuronic, sulphuric and phosphoric esters of p-di-2-chloroethylaminophenol, *Biochem. Pharmacol.* **21:**963–967.

Dowell, R., Springer, C. J., Davies, D. H., Hadley, E. M., Burke, P. J., Boyle, F. T., Melton, R. G., Connors, T. A., Blakey, D. C., and Mauger, A. B., 1996, New mustard prodrugs for antibody-directed enzyme prodrug therapy (ADEPT): Alternative to the amide link, *J. Med. Chem.* **39:** 1100–1105.

Gesson, J.-P., Jacquesy, J.-C., Mondon, M., Petit, P., Renoux, B., Andrianomenjanahary, S., Dufat-Trinh Van, H., Koch, M., Michel, S., Tillequin, F., Florent, J.-C., Monneret, C., Bosslet, K., Czech, J., and Hoffmann, D., 1994, Prodrugs of anthracyclines for chemotherapy *via* enzyme-monoclonal antibody conjugates, *Anticancer Drug Des.* **9:**409–423.

Haenseler, E., Esswein, A., Vitols, K. S., Montejano, Y., Mueller, B. M., Reisfeld, R. A., and Huennekens, F. M., 1992, Activation of methotrexate-α-alanine by carboxypeptidase A monoclonal antibody conjugate, *Biochemistry* **31:**891–897.

Hanessian, S., and Wang, J., 1993, Design and synthesis of a cephalosporin-carboplatinum prodrug activatable by a β-lactamase, *Can. J. Chem.* **71:**896–906.

Hay, M. P., Wilson, W. R., and Denny, W. A., 1995, A novel enediyne prodrug for antibody-directed enzyme prodrug therapy (ADEPT) using *E. coli* B nitroreductase, *Bioorg. Med. Chem. Lett.* **5:**2829–2834.

Houba, P. H. J., Leenders, R. G. G., Boven, E., Scheeren, J. W., Pinedo, H. M., and Haisma, H. J., 1996, Characterisation of novel anthracycline prodrugs activated by human β-glucuronudase for use in antibody-directed enzyme prodrug therapy, *Biochem. Pharm.* **52:**455–463.

Hudyma, T. W., Bush, K., Colson, K. L., Firestone, R. A., and King, H. D. (1993). Synthesis and release of doxorubicin from a cephalosporin based prodrug by a β-lactamase-immunoconjugate. *Bioorg. Med. Chem. Lett.* **3:**323–328.

Jungheim, L. N., Shepherd, T. A., and Meyer, D. L., 1992, Synthesis of acylhydrazido-substituted cephems. Design of cephalosporin-vinca alkaloid prodrugs: Substrates for an antibody-targeted enzyme, *J. Org. Chem.* **57:**2334–2340.

Jungheim, L. N., Shepherd, T. A., and Kling, J. K., 1993, Synthesis of a cephalosporin-doxorubicin antitumour prodrug: A substrate for an antibody-targeted enzyme, *Heterocycles* **35:**339–348.

Kerr, D. A., Senter, P. D., Burnett, W. V., Hirschberg, D. L., Hellstrom, I., and Hellstrom, K., 1990, Antibody-penicillin-V-amidase conjugates kill antigen positive cells when combined with doxorubicin phenoxyacetamide, *Cancer Immunol. Immunother.* **31**:202–205.

Kerr, D. E., Schreiber, G. J., Vrudhula, V. M., Svensson, H. P., Hellstrom, I., Hellstrom, K. E., and Senter, P. D., 1995, Regressions and cures of melanoma xenografts following treatment with monoclonal antibody β-lactamase conjugates in combination with anticancer prodrugs, *Cancer Res.* **55**:3558–3563.

Knox, R. J., Friedlos, F., and Boland, M. P., 1993, The bioactivation of CB1954 and its use as a prodrug in antibody-directed enzyme prodrug therapy, *Cancer Metastasis Rev.* **122**:195–212.

Lee, M., Simpson, J. E., Woo, S., Kaenzig, C., Anlezark, G. M., Eno-Amooquaye, E., and Burke, P. J., 1997. Synthesis of an aminopropyl analog of the experimental anticancer drug tallimustine, and activation of its 4-nitrobenzylcarbamoyl prodrug by nitroreductase and NADH, *Bioorg. Medicinal Chem. Lett.* **7**:1065–1070.

Mauger, A. B., Burke, P. J., Somani, H. H., Friedlos F., and Knox, R. J., 1994, Self-immolative prodrugs: Candidates for antibody-directed enzyme prodrug therapy in conjunction with a nitroreductase enzyme, *J. Med. Chem.* **37**:3452–3458.

Meyer, D. L., Jungheim, L. N., Mikolajczyk, S. D., Shepherd, T. A., Starling, J. J., and Ahlem, C. N., 1992, *Bioconjugate Chem.* **3**:42–47.

Meyer, D. L., Jungheim, L. N., Law, K. L., Mikolajczyk, S. D., Shepherd, T. A., Mackensen, D. G., Briggs, S. L., and Starling, J. J., 1993, Site-specific prodrug activation by antibody-β-lactamase conjugates: Regression and long-term growth inhibition of human colon carcinoma xenograft models, *Cancer Res.* **53**:3956–3963.

Rodrigues, M. L., Carter, C., Wirth, C., Mullins, S., Lee, A., and Blackburn, B. K., 1995, Synthesis and β-lactamase-mediated activation of a cephalosporin-taxol prodrug, *Chem. Biol.* **2**:223–227.

Roffler, S. R., Wang, S. M., Chern, J. W., Yeh, M. Y., and Tung, E., 1991, Anti-neoplastic glucuronide prodrug treatment of human tumour cells targeted with a monoclonal antibody-enzyme conjugate, *Biochem. Pharmacol.* **42**:2062–2065.

Schmidt, F., Florent, J-C., Monneret, C., Straub, R., Czech, J., Gerken, M., and Bosslet, K., 1997, Glucuronide prodrugs of hydroxy compounds for antibody directed enzyme prodrug therapy (ADEPT): A phenol nitrogen mustard carbamate, *Bioorg. Medicinal Chem. Lett.* **7**:1071–1076.

Senter, P. D., Schreiber, J. G., Hirschberg, D. L., Ashe, S. A., Hellstrom, K. E., and Hellstrom, I., 1989, Enhancement of the in vitro and in vivo antitumour activities of phosphorylated mitomycin C and etoposide derivatives by monoclonal antibody-alkaline phosphatase conjugates, *Cancer Res.* **49**:5789–5792.

Senter, P. D., 1990, Activation of prodrugs by antibody-enzyme conjugates: A new approach to cancer therapy, *FASEB J.* **4**:188–193.

Senter, P. D., Su, P. C. D., Katsuragi, T., Sakai, T., Cosland, W. L., Hellstrom, I., and Hellstrom, K. E., 1991, Generation of 5-fluorouracil from 5-fluorocytosine by monoclonal antibody-cytosine deaminase conjugates, *Bioconjugate Chem.* **2**:447–451.

Shepherd, T. A., Jungheim, L. N., Meyer, D. L., and Starling, J. (1991). A novel targeted delivery system using a cephalosporin-oncolytic prodrug activated by an antibody-β-lactamase conjugate for the treatment of cancer. *Bioorg. Med. Chem. Lett.*, **1**:21–26.

Springer, C. J., Antoniw, P., Bagshawe, K. D., Searle, F., Bisset, G. M. F., and Jarman, M., 1990, Novel prodrugs which are activated to cytotoxic alkylating agents by carboxypeptidase G2, *J. Med. Chem.* **33**:677–681.

Springer, C. J., Bagshawe, K. D., Sharma, S. K., Searle, F., Boden, J. A., Antoniw, P., Burke, P. J., Rogers, G. T., Sherwood, R. F., and Melton, R. G., 1991, Ablation of human choriocarcinoma xenografts in nude mice by antibody-directed enzyme prodrug therapy (ADEPT) with three novel compounds, *Eur. J. Cancer* **27**:1361–1366.

Springer, C. J., Dowell, R., Burke, P. J., Hadley, E., Davies, D. H., Blakey, D. C., Melton, R. G., and Niculescu-Duvaz, I., 1995, Optimisation of alkylating agent prodrugs derived from phenol and aniline mustards: A new clinical candidate prodrug (ZD2767) for ADEPT, *J. Med. Chem.* **38:**5051–5065.

Springer, C. J., Bavetsias, V., Jackman, A. L., Boyle, F. T., Marshall, D., Pedley, R. B., Bisset, G. M. F., 1996, Prodrugs of thymidylate synthase inhibitors: Potential for antibody directed enzyme prodrug therapy (ADEPT), *Anticancer Drug Des.* **11:**625–636.

Sunters, A., Baer, J., and Bagshawe, K. D., 1991, Cytotoxicity and activation of CB1954 in a human tumour cell line, *Biochem. Pharm.* **41:**1293–1298.

Svensson, H. P., Kadow, J. F., Vrudhula, V. M., Wallace, P. M,., and Senter, P. D., 1992, Monoclonal antibody-β-lactamase conjugates for the activation of a cephalosporin mustard prodrug, *Bioconjugate Chem.* **3:**176–181.

Svensson, H. P., Vrudhula, V. M., Emswiler, J. E., MacMaster, J. F., Cosand, W. L., Senter, P. D., and Wallace, P. M., 1995, *In vitro* and *in vivo* activities of a doxorubicin prodrug in combination with monoclonal antibody β-lactamase conjugates, *Cancer Res.* **55:**2357–2365.

Tercel, M., Denny, W. A., and Wilson, W. R., 1996, A novel nitro-substituted *seco*-CI:Application as a reductively activated ADEPT prodrug, *Bioorg. Medicinal Chem. Lett.* **6:**2741–2744.

Vitols, K. S., Haag-Zeino, B., Baer, T., Montejano, Y. D., and Huennekens, F. M., 1995, Methotrexate-α-phenylalanine: Optimisation of methotrexate prodrug for activation by carboxypeptidase A monoclonal antibody conjugate, *Cancer Res.* **55:**478–481.

Vrudhula, V. M., Svensson, H. P., Kennedy, K. A., Senter, P. D., and Wallace, P. M., 1993, Anti-tumour activities of a cephalosporin prodrug in combination with monoclonal antibody-β-lactamase conjugates, *Bioconjugate Chem.* **4:**334–340.

Vrudhula, V. M., Senter, P. D., Fischer, K. J., and Wallace, P. M., 1993, Prodrugs of doxorubicin and melphalan and their activation by a monoclonal antibody-penicillin G amidase conjugate, *J. Med. Chem.* **36:**919–923.

Wallace, P. M., and Senter, P. D., 1991, In vitro and in vivo activities of monoclonal antibody-alkaline phosphatase conjugates in combination with phenol mustard phosphate, *Bioconjugate Chem.* **2:**349–352.

Wallace, P. M., MacMaster, J. F., Smith, V. F., Kerr, D. E., Senter, P. D., and Cosand, W. L., 1994, Intratumoral generation of 5-fluorouracil mediated by an antibody-cytosine deaminase conjugate in combination with 5-fluorocytosine, *Cancer Res.* **54:**2719–2723.

Wang, S. M., Chern, J. W., Yeh, M. Y., Ng, J. C., Tung, E., and Roffler, S. R., 1992, Specific activation of glucuronide prodrugs by antibody-targeted enzyme conjugate for cancer therapy, *Cancer Res.* **52:**4484–4491.

6

Preparation and Purification of Antibody-Enzyme Conjugates for Therapeutic Applications

Roger G. Melton

1. INTRODUCTION

Techniques for preparing chemically linked antibody–enzyme conjugates have long been established for manufacturing such conjugates used in enzyme-linked immunosorbent assays (ELISAs), but the methodology used in producing conjugates for therapeutic applications differs from that normally employed to make conjugates for ELISA applications. For immunoassay usage, the primary requirement is to achieve high yields with good retention of antigen-binding and enzymic activity. For therapeutic purposes, although the criteria of yield and activity retention are still crucial, it is also important to generate a product which is as closely defined as possible and which can be manufactured reproducibly. Thus it is desirable that the therapeutic antibody-enzyme product should be a 1:1 antibody:enzyme conjugate, to minimize the conjugate size and achieve optimal extravasation and penetration of the target tumor, and to be free of large aggregates which, even if soluble, are rapidly taken up by cells of the reticuloendothelial system. Therefore the older methods of antibody–enzyme coupling, such as periodate oxidation or glutaraldehyde activation (Boorsma and Streefkerk, 1979) have been largely superseded by methods based on heterobifunctional reagents, using methodology which was initially developed for producing antibody–toxin conjugates (Thorpe and Ross, 1982), which are themselves highly specialized

Roger G. Melton • Enzacta Ltd., Salisbury SP4 0JQ, United Kingdom.

Enzyme-Prodrug Strategies for Cancer Therapy, edited by Melton and Knox. Kluwer Academic/Plenum Publishers, New York, 1999.

enzymes in the case of the ribosome-inactivating proteins, such as ricin. The use of such reagents allows careful control of the stoichiometry of the reaction to optimize the formation of the desired 1:1 conjugate. The methods used for producing antibody-glucose oxidase conjugates for possible use as anticancer agents, which may be regarded as the precursors of antibody-directed enzyme prodrug therapy (ADEPT) systems, represent a transitional phase in the development of conjugation methodology. Some workers in this field used periodate oxidation of carbohydrate groups on the antibody (Muzykantov *et al.*, 1988, 1990) or glutaraldehyde activation (Stanislawski *et al.*, 1989), and others described the use of the homobifunctional reagent diethylmalonimidate (Philpott *et al.*, 1973). The production of conjugates of defined composition was not a priority in these experiments, and the conjugates generated were typically not purified beyond centrifugation to remove insoluble aggregates (Muzykantov *et al.*, 1988, 1990). The conjugates produced were used only *in vitro*, and under these conditions the presence of high molecular weight aggregates would be of relatively low importance to their efficacy.

2. PRODUCTION OF ANTIBODY FRAGMENTS

The manufacture of the enzyme and antibody components in general is not discussed in detail here. However in many cases it is desirable to use $F(ab')_2$ or Fab fragments of antibodies, rather than the intact form because antibodies and conjugates of low molecular mass penetrate tumors better than larger ones (Fjeld *et al.*, 1991; Sutherland *et al.*, 1997; Yokota *et al.*, 1992). Fab/c fragments have been described that consist of half an intact IgG (Wines and Easterbrooksmith, 1991) but have not been applied in any ADEPT system to date. The various antibody fragments available for consideration are illustrated diagrammatically in Figure 1. The lower molecular masses of antibody fragments potentially leads to more rapid clearance of conjugate, better tissue penetration, and improved specificity of localization as a result of diminished interaction with nonspecific Fc receptors (Adams *et al.*, 1993; Hand *et al.*, 1992; Owens *et al.*, 1991; Wahl *et al.*, 1983). As yet, there is no standardized method of producing Fv fragments of antibodies by chemical digestion. However, methods for producing $F(ab')_2$ and Fab are well established, and commercial kits are readily available for this purpose, although such kits are based only on the long established enzymes pepsin and papain. Other enzymes, such as bromelain, ficin, elastase (Mariani *et al.*, 1991), and lysyl endopeptidase (Werlen *et al.*, 1994), have also been successfully used to generate $F(ab')_2$ and may offer some advantages, particularly for antibodies which are difficult to fragment successfully in high yield. For most antibodies of the IgG1 subclass, bromelain or papain work satisfactorily.

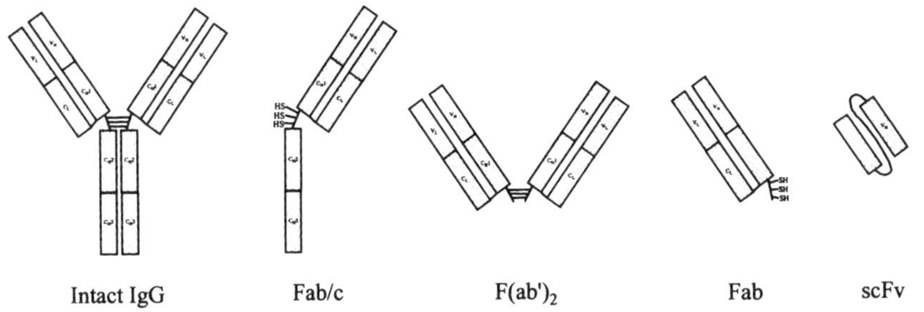

Figure 1. Intact immunoglobulin and the smaller antigen-binding fragments that can be derived from it by proteolytic digestion and/or reduction [$F(ab')_2$, Fab/c, Fab] or molecular engineering (scFv).

The early method of using pepsin digestion to produce $F(ab')_2$ fragments (Lamoyi and Nisonoff, 1983) is still employed by some workers but has been largely superseded by the use of the sulfhydryl-activated enzymes papain or bromelain in the absence of reducing agent. These have the advantage of being more efficient than pepsin and permit the use of less protease and therefore diminish the problem of removing the proteolytic enzyme from the final product. They also are much more rapid in their action. Digestion is complete in 1–4 hr, whereas pepsin digestion usually requires overnight incubation. It is normally possible to use about 0.5–2% bromelain or papain by weight, compared with 3% or more of pepsin. Some commercially available kits use pepsin or papain immobilized on an insoluble support that permits their removal by centrifugation once digestion is complete. The yields obtained using sulfhydryl-activated papain or bromelain are also higher than those achievable with pepsin (Kurkela *et al.*, 1988; Milenic *et al.*, 1989). Depending on the conditions used, bromelain and papain can be used to generate either $F(ab')_2$ or Fab. The product depends on whether incubation takes place in the presence or absence of sulfhydryl groups, although in both cases the proteolytic enzyme must be activated by preincubation with sulfhydryl. Then the activating sulfhydryl must be removed, for example, by gel filtration, if $F(ab')_2$ is required. Fab can also be generated from $F(ab')_2$ by mild reduction of $F(ab')_2$ with about 20 mM of a sulfhydryl reducing agent, such as dithiothreitol (DTT). However this process must be carefully controlled because overreduction leads to cleavage of the interchain disulfide bridge and dissociation of heavy and light chains. The pH at which reduction is carried out is also important. Low pH facilitates the reduction of interchain disulfide bonds. If heavy and light chains dissociate, it is possible to reoxidize the heavy and light chains to reform Fab by using dehydroascorbic acid. Prolonged incubation with this agent also leads to the reformation of $F(ab')_2$. However, this latter process occurs very slowly, and in practice, the reaction can be easily contained at the Fab stage

(Parham *et al.*, 1982). Once Fab has been generated, the sulfhydryl groups released by cleavage of the hinge disulfide bridge are capped by addition of *N*-ethylmaleimide or can be used to couple other proteins.

The ability of copper ions to cause specific cleavage of a recombinant human IgG_1 within its hinge region has also been described. The antibody fragments at a lysine–threonine peptide bond (Smith *et al.*, 1995), and the rate of fragmentation is related to copper concentration. Although complete cleavage occurs over 24 hr at pH 8.0 in the presence of 1mM $CuCl_2$, this does not seem to be a potentially routine method of antibody fragment production but is important information because it suggests that the presence of copper in storage solutions may be responsible for the gradual degradation of antibodies occasionally encountered.

There is some evidence that natural $F(ab')_2$ molecules are unstable *in vivo*. The degree of instability is related to the number of disulfide bridges in the hinge region (Buchegger *et al.*, 1992). The use of a homobi/trifunctional maleimide-based linking agent has been described as a method of producing stable bi- or tri-Fab molecules which have enhanced stability *in vivo* (Casey *et al.*, 1996). These could be used as the basis of antibody–enzyme conjugates if stability of the antibody component of a $F(ab')_2$-based conjugate is a problem.

Fv fragments, normally produced by recombinant technology in the single-chain Fv form, provide the basis of fusion protein constructs (Goshorn *et al.*, 1993), although Fab fragments have also been used for producing fusion proteins (Bosslet *et al.*, 1992). One group described the enzymatic digestion (using clostripain) of a mutated IgG_{2a}, in which the entire CH1 domain is deleted, to yield stable Fv fragments (Takahashi *et al.*, 1991). However, this is not a widely applicable route to the routine production of Fv fragments. Other workers have described the direct production of Fv from a series of morphine-binding monoclonal antibodies by direct digestion with papain and suggest that although it is not a generally applicable method, it may be an approach which is general for λ-chain-containing antibodies (Ornatowska and Glasel, 1991).

The majority of monoclonal antibodies used for ADEPT applications to date have been of the IgG_1 subclass and generally present few problems in fragmentation. The use of a rat IgG_{2a}, ICR12, has also been reported (Eccles *et al.*, 1994), but in this instance whole antibody was used, partly because of the known difficulties of fragmenting antibodies of this subclass. Other proteolytic enzymes, such as lysyl endopeptidase, have different specificities and can be successfully used to digest immunoglobulins of subclasses, such as IgG_{2a} or IgG_{2b}, which can resist the normally employed enzymes, such as pepsin and papain (Yamaguchi *et al.*, 1995). In some cases it is possible to produce $F(ab')_2$ fragments of antibodies, but the fragments produced are unstable *in vivo* and therefore of little practical value (Buchegger *et al.*, 1992).

3. CHEMICAL COUPLING OF ANTIBODIES TO ENZYMES

The differing modes of action of toxin– and enzyme–antibody conjugates has led to developing different coupling protocols to suit the differing conjugate stability requirements of the two approaches, but the underlying chemistry is essentially similar. In all the commonly used coupling protocols, a heterobifunctional reagent couples via modified lysine residues on one protein to sulfhydryl (thiol) groups on the second protein. The lysine residues are modified by using of a heterobifunctional reagent consisting of an *N*-hydroxysuccinimide functional group, combined with maleimide or a protected sulfhydryl group. The linkage between the proteins is one of two basic types, a disulfide bridge or a thioether bond (Fig. 2) depending on whether the group introduced is a sulfhydryl or maleimide, respectively. The sulfhydryl group in the second protein may be an endogenous free sulfhydryl if one is available (for example, a cysteine generated by reducing an inter- or intramolecular disulfide bridge), or it may be chemically introduced, again by modifying lysine residues.

3.1. Choice of Linkage

The choice of a disulfide or thioether linkage is largely governed by the stability required of the linkage. Disulfide linkages are inherently unstable in the plasma (Blakey *et al.*, 1987; Worrell *et al.*, 1986), but a certain degree of instability

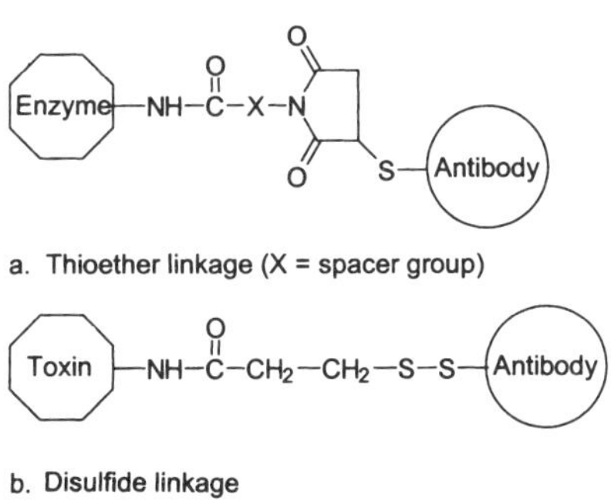

Figure 2. The two main types of protein–protein linkages which are commonly used for producing chemically linked antibody–enzyme (a) or antibody–toxin (b) conjugates. The positions of the proteins relative to the linker group may be reversed, depending on the sensitivity of the proteins to modification.

is required for antibody–toxin conjugates to allow intracellular release of the free toxin, following internalization of the antibody-enzyme toxin, and nonreducible conjugates have reduced potency (Worrell *et al.*, 1986). For ADEPT applications, however, such instability is definitely undesirable *in vivo* because a period of some days may be required to elapse between administering an antibody-enzyme conjugate and prodrug to allow unbound conjugate to clear from normal tissues (Bagshawe *et al.*, 1988). In the case of a conjugate of CPG_2 with ICR12, an intact IgG_{2a} directed against c-*erb*B2, a period of 14 days was necessary between administration of conjugate and prodrug, when used in a mouse xenograft model, to allow sufficient clearance of nonlocalized conjugate from the blood to give workable tumor to blood ratios. Yet the antibody–enzyme conjugate remained active at the end of this period, and a potent antitumor effect was achieved on administration of prodrug (Eccles *et al.*, 1994). Although accelerated clearance systems have been developed to allow administering prodrug while tumor uptake remains high (Kerr *et al.*, 1993; Sharma *et al.*, 1990), a minimum of 24 hr is usually allowed between administration of conjugate and clearing agent to achieve optimal tumor localization of conjugate. The inherent instability of disulfide-linked conjugates can be controlled to some extent by using sterically hindered disulfide cross-linking agents, such as *N*-succinimidyloxycarbonyl-α-(2-pyridyldithio)-toluene (SMPT) (Thorpe *et al.*, 1987) or *N*-succinimidyl-3-(2-pyridyldithio)-butyrate (SPDB) (Worrell *et al.*, 1986). However, these reagents still do not possess stability comparable with those of thioether-linked conjugates. With very few exceptions, therefore, the preferred linkage for conjugates used in ADEPT systems has been the thioether bond. So far, the use of disulfide-linked conjugates is confined to *in vitro* studies (Wang *et al.*, 1992). There have been no reports that conjugates for use in ADEPT were produced by using hindered disulfide linkers.

3.2. Insertion of Maleimide Groups into Proteins

A number of maleimide-introducing reagents are available for generating thioether linkages. Their chemistry is essentially common to all of the systems described. *N*-Succinimidyl-4-(*N*-maleimidomethyl-cyclohexane-1-carboxylate (SMCC) or its more soluble sulphated form (Sulfo-SMCC) are the most commonly used reagents (Esswein *et al.*, 1991; Haisma *et al.*, 1992a,b; Kerr *et al.*, 1991; Meyer *et al.*, 1992; Shepherd *et al.*, 1991b; Wang *et al.*, 1992), although *N*-succinimidyl-4-(*p*-maleimidophenyl)-butyric acid (SMPB) (Melton *et al.*, 1993a) and *N*-maleimidobenzoyl-*N*-hydroxysuccinimide (MBS) (Searle *et al.*, 1986) have also been used. These compounds differ primarily in the carbon spacer length between the *N*-hydroxysuccinimide (NHS) and maleimide functional groups. SMPB has the greatest spacer length, and MBS has the shortest. Extended length coupling agents based on SMCC, where the basic nine-carbon spacer of SMCC is extended by aminocaproic acid homologation to give 16-, 23- and 30-carbon spacers, have recently been described and are claimed to improve antibody binding by lowering steric hindrance of the complementarity-determining region in conjugates constructed with these reagents (Bieniarz *et al.*, 1996) The structures of all the heterobifunctional reagents described are illustrated in Fig. 3, which shows the increase in spacer length. An early compound, the *N*-hydroxysuccinimide ester of iodoacetic acid (Rector *et al.*, 1978) gives a very short spacer and has been used by many workers. In all of these compounds, the NHS ester moiety reacts with the primary amines of lysine residues at slightly alkaline pH (ca. 7.5) to form an amide bond, linking the maleimide group to the protein and releasing *N*-hydroxysuccinimide (Fig. 4a), which can be removed, together with unreacted reagent, by dialysis or gel filtration. Proprietary prepacked columns, such as the PD10 columns containing Sephadex G25 marketed by Pharmacia, are ideal for this purpose. After removing excess reagent, the maleimide-activated protein is mixed with the second, thiolated protein. Then the maleimide group reacts with the sulfhydryl groups in the second protein under slightly acidic to neutral (pH 6.5–7.5) conditions to form a thioether bond (Fig. 4b). Both reactions are rapid and are essentially complete within a few minutes at room temperature under the condi-

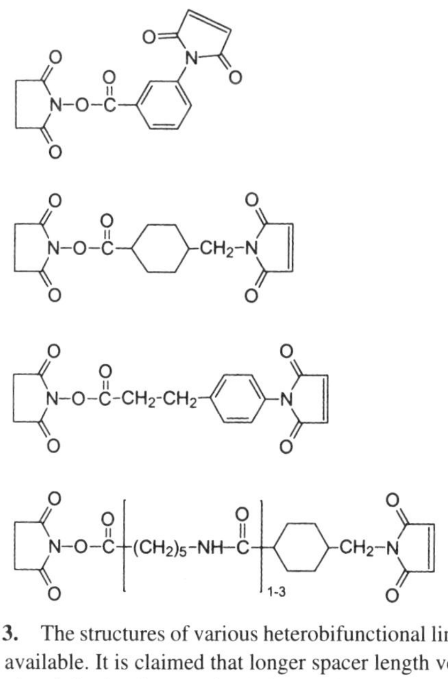

	Spacer length
MBS	8 atoms
SMCC	9 atoms
SMPB	10 atoms
	16 - 30 atoms

Figure 3. The structures of various heterobifunctional linking agents, illustrating the range of spacer lengths available. It is claimed that longer spacer length versions of SMCC lead to better retention of biological activity by the proteins being conjugated.

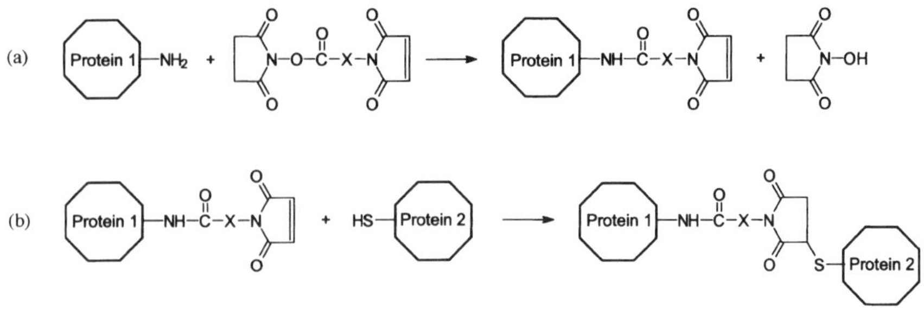

Figure 4. The generalized reaction scheme for coupling proteins using *N*-hydroxysuccinimide ester/maleimide heterobifunctional agents. × represents spacer groups of differing chain lengths, as illustrated in Fig. 3. Thiol groups may be inserted using a variety of methods, as illustrated in Fig. 5, or endogenous thiols may be used if present.

tions outlined. It is clearly necessary that the protein to have the maleimide group attached should be free of endogenous sulfhydryl groups in the form of cysteine residues, otherwise homopolymerization may occur. If both proteins contain such groups, they may be blocked on one, before reaction with the heterobifunctional reagent, by the adding of *N*-ethylmaleimide which can also be used to cap unreacted sulfhydryl residues, once conjugation is complete, to prevent subsequent aggregation when further manipulation of the conjugate takes place for example, concentration for loading onto columns. Treatment with a slight molar excess of N-ethylmaleimide, followed by a slightly larger molar excess of mercaptoethanol, caps all unreacted sulfhydryl and maleimide groups. However, enzymes that have endogenous free sulfhydryl residues may be susceptible to loss of activity on blocking with *N*-ethylmaleimide (Vingerhoeds *et al.*, 1996).

3.3. Thiolation of Proteins

The method of inserting a sulfhydryl group into the second component protein is less standardized. In common with antibody–toxin conjugation methodology, *N*-succinimidyl-3-(2-pyridyldithio)propionate (SPDP) has often been used. The reaction proceeds rapidly at neutral or slightly alkaline pH and is essentially complete within 15 minutes. SPDP inserts a 2-pyridyldisulfide-protected sulfhydryl which must be reduced with a suitable reducing agent, such as dithiothreitol, to generate the free sulfhydryl required (Fig. 5a). Thus the process is somewhat complicated by the need for an additional step which is necessary to remove excess reducing agent before coupling because otherwise this would provide a competing source of sulfhydryl groups. The release of pyridine

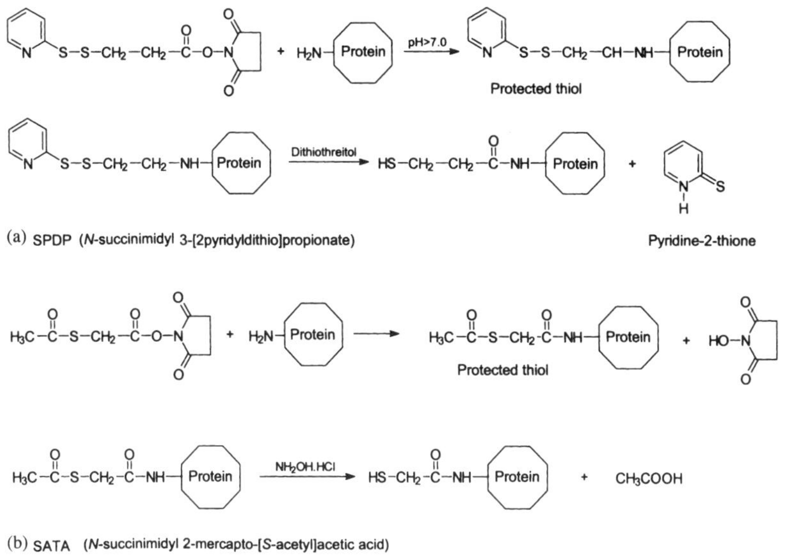

Figure 5. Reaction mechanisms for two *N*-hydroxysuccinimide-based reagents used for thiolating proteins and deprotection of the inserted thiols.

-2-thione residues in this process, used as a measure of the number of sulfhydryl groups inserted, provides a useful and rapid in-process test (Carlsson *et al.*, 1978). The reduction is normally carried out at pH 4.5 to prevent the reduction of internal disulfide bridges within the protein, but this can prove problematical because not all proteins tolerate such low pH well. Antibodies do, so it is advisable to try sulfhydrylating the antibody first when developing new coupling methods. In the absence of internal disulfide bridges, however, the reduction step can be carried out at neutral pH. *N*-Succinimidyl-2-mercapto-[*S*-acetyl]acetic acid (SATA) (Duncan *et al.*, 1983) also inserts a protected sulfhydryl and is a closely related reagent to SPDP in terms of its reaction conditions and properties, but in this case the protecting S-acetyl group may be cleaved off by hydroxylammonium hydrochloride at neutral pH (Fig. 5b), and excess cleaving reagent does not have to be removed prior to the coupling reaction. In common with SPDP, the protected sulfhydryl generated by using SATA has the advantage that the thiolated protein can be stored in the protected form before use. A disadvantage of SATA, however, is that the deprotection reaction cannot be used to measure the degree of modification achieved analogously to SPDP. Therefore separate assays to determine sulfhydryl incorporation are necessary.

The common feature of SPDP and SATA is the *N*-hydroxysuccinimide function by which the protected sulfhydryl group is attached to lysine residues. But in the case of some enzymes, modification of lysine residues by NHS reagents results in extensive loss of enzyme activity (Haisma *et al.*, 1992a) and reversal of the system, that is, introducing the maleimide group on to the enzyme does not help because the maleimide-insertion reaction also uses NHS esters. In such instances 2-iminothiolase hydrochloride (Traut's reagent, 2-IT) provides a useful alternative to both SPDP and SATA and thiolates without loss of enzymic activity (Haisma *et al.*, 1992a). As shown in Fig. 6, it has the further advantage that it does not alter the charge properties of the modified protein because it converts the normally positively charged lysine residue to another positively charged group. In contrast, modification with NHS-based reagents converts positively charged lysines to neutral groups, which may be the cause of loss of bioactivity if the charged residue so modified is involved in the catalytic or binding site or is involved with stabilizing protein folding. Unlike the NHS-based reagents, however, 2-IT reacts relatively slowly and progressively with time. Therefore in optimizing coupling conditions, it is necessary to determine the rate of thiolation over a period of time and to maintain and control this incubation period.

Occasionally there are endogenous free sulfhydryl residues on one of the proteins, which may be used to eliminate the need for a thiolation step. One such example is the sulfhydryl groups exposed by cleavage of the hinge region of the $F(ab')_2$ fragment of immunoglobulin to yield Fab fragments (Meyer *et al.*, 1992, 1993). Conversely, free sulfhydryl residues may be present which, although they are not available for coupling purposes because of steric hindrance, may nevertheless form internal disulfide bridges if the protein is thiolated, leading to loss of functionality (Haisma *et al.*, 1992a).

3.4. Coupling Conditions

A detailed study of the coupling chemistry initially developed for conjugating carboxypeptidase G_2 to the $F(ab')_2$ fragment of A5B7, an anticarcinoembryonic antigen monoclonal antibody, was carried out to optimize the coupling conditions for use on a large scale (Melton *et al.*, 1993a). It was found that insertion of 1–3 active groups (both sulfhydryl and inserted maleimide) per protein molecule gives the best yields of 1:1 conjugate with minimal formation of high molecular weight aggregates. It should be noted, however, that although

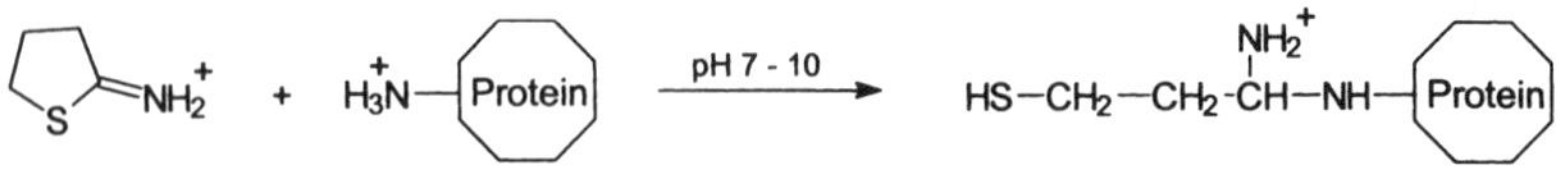

Figure 6. The mechanisms for thiolating proteins using 2-iminothiolane.

these conditions provide a general approximate starting point, detailed study of the precise requirements for any given system is necessary. In general, however, it is advisable to use the minimum level of substitution compatible with an acceptable yield of product. High substitution levels cause degradation of enzymic activity and antigen-binding properties and also increase the risk of forming high molecular weight aggregates. An important, third, interrelated variable is the concentration of the components on mixing. If high substitution levels are needed to obtain a good yield of conjugate, it is possible to control unwanted aggregation by reducing the concentration of the components at the mixing step. Conversely, it may be possible to use low substitution levels and high protein concentrations to preserve bioactivity. This latter approach is preferable for large-scale production of conjugate because it minimizes the volumes of materials handled.

The availability of lysine or sulfhydryl residues on the native protein for coupling may be readily determined by conventional analytical biochemistry (Habeeb, 1966; Mueller *et al.*, 1988; Sedlak and Lindsay, 1968), as may be the substituent levels (Melton *et al.*, 1993a). Determining optimal coupling conditions in this empirical manner is relatively costly in terms of materials but is essential to attain if reproducibility of batches of conjugate for *in vitro* and, more particularly, *in vivo* studies is to be achieved. The overall yields of purified conjugate obtained by chemical linkage are not high. Typical figures range from about 35–40% yield as 1:1 conjugate in the crude incubation mix (as estimated by integrating A_{280} traces from analytical gel filtration columns) but 15% or less as purified conjugate is the norm.

3.5. Site-Specific Coupling of Proteins

The coupling methods described previously have some disadvantages. There is little or no control over the location of the substitution site, so that linkage may occur via any of the available lysine residues. Specificity of location can be ensured by using endogenous sulfhydryl residues because these are relatively rare and their position is fixed in the amino acid sequence of the protein. But this specificity applies only to one of the two proteins to be conjugated. As described before, modification of amino acids in critical areas of the protein may result in deleterious effects on the protein's function, for example, loss of enzymic activity or antigen-binding function. Until recently, very little effort has been devoted to devising methods for protecting the active sites of enzymes or antibodies. Where intact antibodies are employed, a study of more traditional coupling strategies based on periodate oxidation of glycosyl residues might be worthy of consideration because the majority of immunoglobulin glycosylation is associated with the Fc domain. Therefore this could be used to achieve domain-specific coupling away from the variable regions of the antibody which constitute its binding site (Murayama *et al.*, 1978; O'Shannessy and Quarles, 1987) and indeed such an approach has recently been described for constructing antibody–hapten conjugates, albeit with the added sophistication of using an engineered unique glycosylation site (Leung *et al.*, 1995). Other workers have described coupling peptides to antibodies that have been desialylated to expose galactose residues which are then periodate oxidized, but such an approach is less precise (Brumeanu *et al.*, 1995). Quite high levels of activation are achieved by periodate oxidation of endogenous carbohydrate, as measured by the incorporation of a tritium label on reduction of the Schiff bases with tritiated sodium cyanoborohydride (Lenz *et al.*, 1989). However the use of high levels of sodium periodate can lead to loss of binding avidity and immunoreactivity (Abraham *et al.*, 1991), as can reduction with sodium borohydride (Peng *et al.*, 1987). Other workers have used the carbohydrate-directed linker *S*-(2-thiopyridyl)-L-cysteine hydrazide and compared this with the conventional disulfide bridge-forming reagent SPDP for producing antibody (IgM)-cobra venom factor conjugates (Zara *et al.*, 1995). The carbohydrate-directed modification allowed significantly higher levels of derivatization without impairment of antigen-binding activity, although the cobra venom factor activity was retained better when coupled via SPDP. This work suggests that conjugation via sugar residues is a potentially useful approach where such residues are available.

An alternative approach is to bind the protein to be modified to an affinity matrix or to carry out the reaction in the presence of substrate or a reversible inhibitor to protect the active site residues while they are undergoing modification. Such an approach has been described for modifying yeast invertase (Pillai and Bachhawat, 1977). The enzyme was bound to a column of concanavalin A-Sepharose and activated with glutaraldehyde. Uricase was allowed to react with the invertase, and the conjugate was eluted from the column. This method gave conjugates of fairly well-defined molecular weight and presumably provided protection for the active site of invertase, if not uricase. Similar systems may have potential for conjugating enzymes and antibodies which are very sensitive to modification, although affinity systems based on proteinaceous ligands would present alternative reactive sites for coupling reagents and the use of nonprotein affinity ligands might be preferable where these are available.

An alternative approach to protein conjugation which does produce conjugates with well-defined linkage sites has been described by Fisch *et al.* (1992) and Rose *et al.* (1991). This technique, based on reverse proteolysis and illustrated in Fig. 7, works under mild conditions. A carbohydrazide residue is specifically attached to the carboxyl terminus of one protein using a proteolytic enzyme under conditions which force it to act in reverse, as a ligase. Then this carbohydrazide group is coupled to an oxidized aldehyde or ketone group on the second protein. If the second protein has a suitable amino acid (serine or threonine) as its

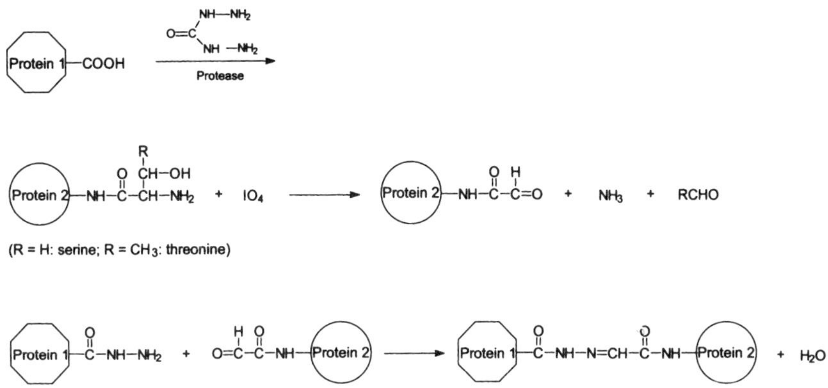

Figure 7. The mechanism of coupling of proteins by site-specific reverse proteolysis. Carbohydrazide is attached to the C-terminal carboxylic acid of one protein and coupled to an oxidized N-terminal threonine or serine on the second.

N-terminus, this can be specifically oxidized to provide a second defined linkage site. The technique has been successfully used to conjugate both carboxypeptidase G_2 (Werlen *et al.*, 1994) and β-lactamase (Mikolajczyk *et al.*, 1994) and, in the absence of the desired serine or threonine N-terminus, an N-terminal threonine was generated on carboxypeptidase G_2 by site-directed mutagenesis. This conjugation method offers significant benefits compared with less specific chemical conjugation techniques. The method could become the approach of first choice in the future, although it is likely that it will largely superseded by the development of antibody-enzyme fusion proteins and may well represent the ultimate chemical conjugation technique. The yields obtainable are at least comparable with those achieved by more conventional means (Werlen *et al.*, 1994). The problem of purifying the desired conjugate from the uncoupled components remains, although this technique minimizes the formation of higher molecular mass aggregates.

4. ALTERNATIVES TO CHEMICAL COUPLING

The avoidance of chemical conjugation methods for producing antibody–enzyme conjugates is clearly desirable if alternative methods are available, given the relatively low yields of material attained and the problems of conjugate heterogeneity and loss of bioactivity. The simplicity of chemical conjugation methods makes their continued use for initial studies probable, but alternative approaches to the conjugation of proteins are gradually being developed. Although some are covered elsewhere in this volume, a brief outline of the available approaches is appropriate here.

4.1. Bispecific Antibodies

Sahin *et al.*, (1990) described the generation of a bispecific antibody, one arm of which binds to the CD30 antigen and the other to alkaline phosphatase, providing a capture mechanism for this enzyme. In this case the two fragments were co-expressed in a mammalian cell line. When the fragments are randomly recombined some 50% of the total antibody produced are present as the bispecific form, which can be isolated by two affinity purification steps, one for each binding function. The affinity matrices are selected so that only the bifunctional antibody binds to the column at both stages of the purification process. For such an enzyme capture strategy to work, it is important that binding to the target epitope in the enzyme does not inhibit the active site of the enzyme. De Sutter and Fiers (1994) have taken the process a step further and avoided the latter problem by producing a mouse-human chimeric antibody in which one antigen-binding arm has been replaced by bacterial β-lactamase enzyme.

4.2. Fusion Proteins

A number of workers have begun producing fusion proteins for ADEPT applications. The first to be described was that of Bosslet *et al.* (1992) who fused placental β-glucuronidase to the Fab fragment of a CDR-grafted humanized antibody to carcinoembryonic antigen. Similarly, Goshorn *et al.* (1993) fused β-lactamase from *Bacillus cereus* with a single chain Fv form of the L6 antibody, which binds to a tumor-associated glycoprotein on a range of tumor cell types. Both of these enzymes are monomeric proteins that circumvent the potential problems of subunit assembly which may occur with dimeric proteins, such as carboxypeptidase G_2 and were encountered by Michael *et al.* (1996). Fusion constructs of CPG_2 with antibody can potentially use the CPG_2 subunit interaction to generate a "pseudo $F(ab')_2$" construct, but some cleavage of fusion protein monomers occurs, giving rise to a series of possible combinations, illustrated in Figure 8.

A major problem identified in pilot clinical trials with chemically linked A5B7 $F(ab')_2$-carboxypeptidase G_2 conjugate is that of the immunogenicity of the conjugate. Human antibody responses to both the antibody and enzyme components of the administered antibody-enzyme conjugate developed within 10 days of conjugate administration (Bagshawe *et al.*, 1991; Sharma *et al.*, 1991) necessitating

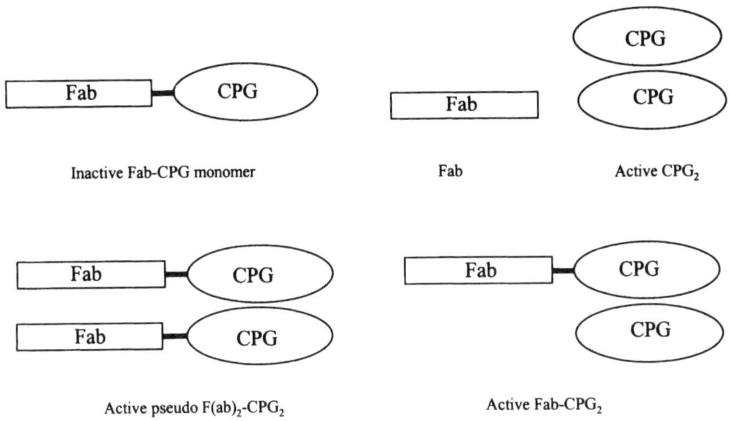

Figure 8. Possible molecular species arising from partial cleavage of a Fab-CPG monomer fusion protein showing the effect on enzymatic activity.

the use of immunosuppressive drugs if repeat cycles of therapy are to be possible in humans (Sharma *et al.*, 1996). Fusion proteins could possibly overcome this problem, if constructed using a human enzyme and human, or humanized, antibody components. A construct described by Bosslet *et al.*, (1992) comes closest to this ideal in its use of a humanized murine antibody fragment and human placental glucuronidase, but there have been no data reported yet regarding its immunogenicity. This can be tested only in a clinical trial. The ultimate constructs may be derived from human bispecific catalytic antibodies, although anti-idiotypic responses may be encountered even with such molecules. Although catalytic antibodies catalyze a wide range of reactions with specificity equal to or greater than the corresponding enzyme, in general the rates of reaction are not comparable with those of enzymes. The technology of catalytic antibodies is in its infancy at present, but developments in the field are rapid, and it can be confidently predicted that catalytic antibodies will eventually play an important role in ADEPT systems of the future.

5. PURIFICATION OF CONJUGATES

Once the antibody and enzyme have been coupled, the desired conjugate must be separated from the unlinked components because free antibody competes with the conjugate for antigen-binding sites and free enzyme may persist in the circulation and cause nonspecific activation of prodrug in plasma. In practice, the rate of clearance of free enzyme is far more rapid than that of conjugate or free antibody, and it is therefore probably more important to aim for complete removal of free antibody.

The most common method of purification used for conjugates is size exclusion chromatography on low-pressure Sephadex-type matrices or high-pressure counterparts, although its success is governed by the difference in molecular weight between the conjugate and its components. Thus, a 1:1 conjugate constructed of $F(ab')_2$ with a low molecular weight enzyme will be poorly separated from the antibody component, although well separated from the free enzyme. The disadvantage of this type of chromatography is the relatively low sample capacity (1–2% of total column volume for good separation) and the low flow rates of gel filtration columns based on the older Sephadex gels. Modern gel filtration matrices, such as Sephacryl and Superdex, have alleviated the problem of flow rates considerably and improved resolution. Their use is practically universal, although sample capacities remain limited by volume and viscosity considerations, and therefore production of larger amounts of conjugates presents problems in terms of the size of the columns needed. A purification protocol based on two gel filtration steps, using Sephacryl S300 and Superdex G200, has been successfully used to produce A5B7 $F(ab')_2$-carboxypeptidase G_2 conjugates on milligram and multigram scales for pilot clinical trials (R. G. Melton, unpublished data), and the conditions were successfully adapted to manufacture and purify a murine anti-Ly 2.1-carboxypeptidase G_2 conjugate, in this case using intact antibody (Melton *et al.*, 1993b).

Ion exchange chromatography would normally be the method of choice for purifying conjugates. The capacities of ion exchange matrices are high compared with those of size exclusion gels, whereas elution conditions are milder than those commonly used for immunoaffinity columns. The technique is, however, relatively little used to purifying of antibody-enzyme conjugates, although a number of workers have described the use of anion (Haisma *et al.*, 1992a; Haenseler *et al.*, 1992; Kerr *et al.*, 1991; Wallace and Senter, 1991) or cation exchange (Svensson *et al.*, 1992) matrices for separating uncoupled antibody from conjugate. The reason for this probably lies in the fact that, as discussed earlier, NHS-based reagents modify the charge properties and convert the initially homogeneous charge of a protein to become heterogeneous on modification, resulting in elution profiles which are ill defined under salt or pH gradient elution conditions compared with that of the native protein. Under these circumstances the use of 2-iminothiolane, which does not modify the charge of the group it modifies, may be helpful in assisting the development of ion exchange purification techniques.

Hydrophobic interaction columns have relatively high capacities and may be worth considering for purifying antibody enzyme conjugates, but their use has not been described to date.

Protein A-Sepharose immunoaffinity chromatography has been used to purify an antibody-β-glucuronidase conjugate (Wang *et al.*, 1992) and can be used to remove free enzyme, but it does not separate free antibody from conjugate. Its use is largely restricted to conjugates produced from intact antibody because binding of antibody to protein A occurs primarily via the Fc domain. A more recently developed immunoglobulin-binding protein, protein L from *Peptostreptococcus magnus*, binds via the Fv domain and may offer significant advantages for purifying conjugates constructed from antibody fragments (Nilson *et al.*, 1993). It may be of particular use for purifying antibody–enzyme fusion proteins, where there is no background of unreacted component antibody. A possible disadvantage of immunoaffinity chromatography systems is that relatively harsh conditions, for example, low pH, may be required to elute bound material, and although antibodies tolerate such conditions, enzymes may not. An example of this is carboxypeptidase G_2, which is irreversibly inactivated at low pH (R. G. Melton, unpublished data), but each case must be judged on its individual merits. The purification of an anti-CD30/alkaline phosphatase bispecific antibody has been achieved using immobilized calf alkaline phosphatase, with elution by a pH gradient. In this instance, the single binding moiety, the bispecific antibody, is recovered by elution at pH 5.5, whereas the bivalent form requires a pH of 3.5 for elution, presumably because of the higher avidity of the bivalent binding (Sahin *et al.*, 1990). Similarly, an immunoaffinity column using rabbit polyclonal anti-β-lactamase has been used to purify a murine::human chimeric antibody in which one antigen-binding arm is replaced by β-lactamase (Desutter and Fiers, 1994). Elution in this case required the use of 0.1M glycine at pH 3.0.

Conventional affinity chromatography, using the affinity of the enzyme for its substrate or substrate analogs, has rarely been used to purify antibody–enzyme conjugates, but β-lactamase fusion proteins have been purified on phenylboronic acid affinity columns with elution at pH 7.0, by a salt gradient (Rodrigues *et al.*, 1995), and chemically linked ub-lactamase conjugates have been similarly purified on boronic acid affinity columns (Vrudhula *et al.*, 1993).

Although affinity purification techniques are powerful, they have certain disadvantages. The cost of immunoaffinity matrices based on antibodies is high, and their capacity is limited compared with ion exchange matrices. The large-scale purification of conjugates for clinical trial purposes is not feasible based on such technology and is complicated by the need to quantify the rate of leaching of the immobilized ligand from the column for quality assurance purposes. The use of the immunoglobulin-binding proteins is more appropriate for large-scale production purposes, if an affinity-based system is required, because such matrices are commonly used to produce monoclonal antibodies, and the loss rate of immobilized protein is well studied for such materials (Bessos *et al.*, 1991).

The specificity of enzyme-substrate interactions gives rise to the possibility of developing methods of conjugate purification based on such interactions, using nonsubstrate analogs. Such compounds may be generated in the course of prodrug optimization screens and if they act as reversible inhibitors, may form the basis of an affinity matrix from which bound conjugate could be eluted by using a competitive substrate. The latter low molecular weight material could be readily recovered by conventional desalting or buffer exchange techniques.

6. POSTPURIFICATION MODIFICATIONS

The problems associated with immunogenicity of enzymes and antibodies from nonhuman sources have already been alluded to. Although the use of human enzymes is a possible way around this problem, such enzymes are naturally endogenous and bring with them the risk of nonspecific activation of prodrug. Bacterial enzymes also have the advantage that their properties are often superior in terms of their kinetics and the range of enzymes available. Recently, attempts have been made to carry out postconjugation modifications to a Fab′-β-lactamase conjugate to reduce its immunogenicity (Mikolajczyk *et al.*, 1996). It has long been recognized that attaching various biopolymers to proteins can enhance their plasma persistence (Melton *et al.*, 1987; Sherwood *et al.*, 1977) and decrease their immunogenicity (Fagnani *et al.*, 1990; Wileman *et al.*, 1986), but most workers have produced large, amorphous aggregates because the proteins being modified have a high content of available ε-amino acids in the form of lysines, and the modifying polymers are also highly activated, leading to multiple attachment sites and extensive intermolecular cross-links. By pretreating the Fab′-β-lactamase conjugate with the reversible amine-blocking agents methyl acetimidate or *N*-succinimidyl (methylsulphonyl)ethyl carbonate, Mikolajczyk *et al.*, (1996) controlled the degree of modification of the protein by periodate oxidised dextran, so that between 0.5 and 9.2 dextran molecules could be attached to the protein conjugate. The use of a dextran of relatively low molecular mass (6000 daltons) also helped to ensure that the molecular mass of the modified conjugate remained relatively low. The dextran modification was successful in reducing the immune response of animals to β-lactamase, but there was some loss of the antibody component's immunoreactivity, manifested by lower tumor uptake of the modified conjugate compared with non-dextran-linked control. These experiments used periodate oxidized dextran, in which unreacted aldehyde groups were reduced by sodium cyanoborohydride. However, this chemistry may be unsuitable for some enzymes and antibodies, as described earlier. There are alternative polymers, such as polyethylene glycol, for which different coupling chemistries are available, however (Greenwald *et al.*, 1996; Kodera *et al.*, 1992; Schiavon *et al.*, 1991).

7. FUTURE DEVELOPMENTS

It seems likely that the current widespread use of chemically linked antibody–enzyme conjugates in ADEPT will eventually fall into relative disuse as more sophisticated approaches, such as fusion proteins, become more practical. But the chemically linked conjugate is likely to play an important role in early preclinical studies, given the relative simplicity and rapidity of production.

As antibody-enzyme fusion proteins become more widely used, it is likely that purification protocols based on the use of histidine tags that permit using immobilized metal affinity chromatography (IMAC) (Hutchens and Yip, 1995) or alternative systems, using cleavable affinity tags which can be removed after purification of the protein, will become important routes for conjugate purification. Such systems, based on glutathione-S-transferase (Smith and Johnson, 1988) or maltose binding protein (Maina *et al.*, 1988), use a site-specific protease, such as thrombin or Factor Xa, to cleave off the tag, and the plasmid vector includes a suitable recognition site sequence immediately upstream of the fusion protein cloning site. There are some potential problems with these systems, notably, ensuring complete consistency of tag removal. For this reason, histidine tags, which are not normally removed and have no adverse effects *in vivo*, at least in the mouse (Casey *et al.*, 1995), may be the preferred option. However, there is yet relatively limited information on the *in vivo* interactions of the histidine tag. There has also been one report that the use of a chloramphenicol resistance selection marker impaired purification of a histidine-tagged fusion protein from recombinant *E. coli* (Oswald and Rinas, 1996). A second protein, identified as chloramphenicol acetyltransferase, co-eluted from the IMAC column with the desired fusion protein and is known to have a relatively high (5.5%mole) histidine content compared with normal *E. coli* proteins. Therefore it seems advisable to avoid using chloramphenicol-resistant *E. coli* strains to express histidine-tagged proteins.

REFERENCES

Abraham, R., Moller, D., Gabel, D., Senter, P., Hellstrom, I., and Hellstrom, K. E., 1991, The influence of periodate oxidation on monoclonal antibody avidity and immunoreactivity, *J. Immunol. Meth.* **144:**77–86.

Adams, G. P., Mccartney, J. E., Tai, M. S., Oppermann, H., Huston, J. S., Stafford, W. F., Bookman, M. A., Fand, I., Houston, L. L., and Weiner, L. M., 1993, Highly specific *in vivo* tumor targeting by monovalent and divalent forms of 741F8 anti-c-erbB-2 single-chain Fv, *Cancer Res.* **53:**4026–4034.

Bagshawe, K. D., Sharma, S. K., Springer, C. J., Antoniw, P., Boden, J. A., Rogers, G. T., Burke, P. J., Melton, R. G., and Sherwood, R. F., 1991, Antibody Directed Enzyme Prodrug Therapy (ADEPT)—Clinical Report. *Dis. Marker* **9:**233–238.

Bagshawe, K. D., Springer, C. J., Searle, F., Antoniw, P., Sharma, S. K., Melton, R. G., and Sherwood, R. F., 1988, A cytotoxic agent can be generated selectively at cancer sites, *Br. J. Cancer* **58:**700–703.

Bessos, H., Appleyard, C., Micklen, L. R., and Pepper, D. S., 1991, Monoclonal antibody leakage from gels: Effect of support, activation and eluant composition. *Prep. Chromatogr.* **1:**207–220.

Bieniarz, C., Husain, M., Barnes, G., King, C. A., and Welch, C. J., 1996, Extended length heterobifunctional coupling agents for protein conjugations. *Bioconjugate Chem.* **7:**88–95.

Blakey, D. C., Watson, G. J., Knowles, P. P., and Thorpe, P. E., 1987, Effect of chemical deglycosylation of ricin A chain on the *in vivo* fate and cytotoxic activity of an immunotoxin composed of ricin A chain and anti-Thy 1.1 antibody, *Cancer Res.* **47:**947–952.

Boorsma, D. M., and Streefkerk, J. G., 1979, Periodate or glutaraldehyde for preparing peroxidase conjugates? *J. Immunol. Meth.* **30:**245–255.

Bosslet, K., Czech, J., Lorenz, P., Sedlacek, H. H., Schuermann, M., and Seemann, G., 1992, Molecular and functional characterisation of a fusion protein suited for tumour specific prodrug activation, *Br. J. Cancer* **65:** 234–238.

Brumeanu, T. D., Dehazya, P., Wolf, I., and Bona, C. A., 1995, Enzymatically mediated, glycosidic conjugation of immunoglobulins with viral epitopes, *J. Immunol. Meth.* **183:**185–197.

Buchegger, F., Pelegrin, A., Hardman, N., Heusser, C., Lukas, J., Dolci, W., and Mach, J. P., 1992, Different behaviour of mouse-human chimeric antibody $F(ab')_2$ fragments of IgG_1, IgG_2 and IgG_4 sub-class *in vivo*, *Int. J. Cancer* **50:**416–422.

Carlsson, J., Drevin, H., and Axen, R., 1978, Protein thiolation and reversible protein–protein conjugation. *N*-succinimidyl 3-(2-pyridyldithio)propionate, a new heterobifunctional reagent, *Biochem. J.* **173:**723–737.

Casey, J. L., Keep, P. A., Chester, K. A., Robson, L., Hawkins, R. E., and Begent, R. H. J., 1995, Purification of bacterially expressed single chain Fv antibodies for clinical applications using metal chelate chromatography, *J. Immunol. Meth.* **179:**105–116.

Casey, J. L., King, D. J., Chaplin, L. C., Haines, A. M. R., Pedley, R. B., Mountain, A., Yarranton, G. T., and Begent, R. H. J., 1996, Preparation, characterisation and tumour targeting of cross-linked divalent and trivalent anti-tumour Fab′ fragments, *Br. J. Cancer* **74:**1397–1405.

Desutter, K., and Fiers, W., 1994, A bifunctional murine-human chimeric antibody with one antigen-binding arm replaced by bacterial beta-lactamase, *Mol. Immunol.* **31:**261–267.

Duncan, R. J. S., Weston, P. D., and Wrigglesworth, R., 1983, A new reagent which may be used to introduce sulphydryl groups into proteins and its use in the preparation of conjugates for immunoassay, *Anal. Biochem.* **132:**68–73.

Eccles, S. A., Court, W. J., Box, G. A., Dean, C. J., Melton, R. G., and Springer, C. J., 1994, Regression of established breast carcinoma xenografts with antibody-directed enzyme prodrug therapy against c-erbB2 p185, *Cancer Res.* **54:**5171–5177.

Esswein, A., Hanseler, E., Montejano, Y., Vitols, K. S., and Huennekens, F. M., 1991, Construction and therapeutic potential of a carboxypeptidase A/monoclonal antibody conjugate, *Adv. Enzyme Regul.* **31:**3–12.

Fagnani, R., Hagan, M. S., and Bartholomew, R., 1990, Reduction of immunogenicity by covalent modification of murine and rabbit immunoglobulins with oxidised dextrans of low molecular weight, *Cancer Res.* **50:**3638–3645.

Fisch, I., Kunzi, G., Rose, K., and Offord, R. E., 1992, Site-specific modification of a fragment of a chimeric monoclonal antibody using reverse proteolysis, *Bioconjugate Chem.* **3:**147–153.

Fjeld, J. G., Michaelsen, T. E., Benestad, H. B., and Nustad, K., 1991, The effect of the biodistribution differences between IgG, $F(ab')_2$ and Fab′ on their immunotargeting potential for human tumor cells in immunocompetent mice, *Antibody Immunoconjugate and Radiopharmaceuticals* **4:** 443–451.

Goshorn, S. C., Svensson, H. P., Kerr, D. E., Somerville, J. E., Senter, P. D., and Fell, H. P., 1993, Genetic construction, expression, and characterization of a single chain anti-carcinoma antibody fused to beta- lactamase, *Cancer Res.* **53:**2123–2127.

Greenwald, R. B., Pendri, A., Martinez, A., Gilbert, C., and Bradley, P., 1996, PEG thiazolidine-2-thione, a novel reagent for facile protein modification: Conjugation of bovine hemoglobin, *Bioconjugate. Chem.* **7:**638–641.

Habeeb, A. F. S. A., 1966, Determination of free amino groups in proteins by trinitrobenzenesulfonic acid, *Anal. Biochem.* **14:**328–336.

Haenseler, E., Esswein, A., Vitols, K. S., Montejano, Y., Mueller, B. M., Reisfeld, R. A., and Huennekens, F. M., 1992, Activation of methotrexate-alpha-alanine by carboxypeptidase-A monoclonal antibody conjugate, *Biochemistry* **31:**891–897.

Haisma, H. J., Boven, E., Vanmuijen, M., Dejong, J., Vandervijgh, W. J. F., and Pinedo, H. M., 1992a. A monoclonal antibody-beta-glucuronidase conjugate as activator of the prodrug epirubicin-glucuronide for specific treatment of cancer, *Br. J. Cancer* **66:**474–478.

Haisma, H. J., Boven, E., Vanmuijen, M., Devries, R., and Pinedo, H. M., 1992b, Analysis of a conjugate between anti-carcinoembryonic antigen monoclonal antibody and alkaline phosphatase for specific activation of the prodrug etoposide phosphate, *Cancer Immunol. Immunother.* **34:**343–348.

Hand, P. H., Calvo, B., Milenic, D., Yokota, T., Finch, M., Snoy, P., Garmestani, K., Gansow, O., Schlom, J., and Kashmiri, S. V. S., 1992, Comparative biological properties of a recombinant chimeric anticarcinoma mAb and a recombinant aglycosylated variant, *Cancer Immunol. Immunother.* **35:**165–174.

Hutchens, T. W., and Yip, T.-T., 1995, Synthetic metal-binding protein surface domains for metal ion-dependent interaction chromatography. II. Immobilisation of synthetic metal-binding peptides from metal ion transport proteins as model bioactive protein surface domains, *J. Chromatogr.* **604:**133–141.

Kerr, D. E., Garrigues, U. S., Wallace, P. M., Hellstrom, K. E., Hellstrom, I., and Senter, P. D., 1993, Application of monoclonal antibodies against cytosine deaminase for the *in vivo* clearance of a cytosine deaminase immunoconjugate, *Bioconjugate Chem.* **4:**353–357.

Kerr, D. E., Senter, P. D., Burnett, W. V., Hirschberg, D. L., Hellstrom, I., and Hellstrom, K. E., 1991, Antibody-penicillin-V-amidase conjugates kill antigen-positive tumor cells when combined with doxorubicin phenoxyacetamide, *Cancer Immunol. Immunother.* **31:**202–206.

Kodera, Y., Tanaka, H., Matsushima, A., and Inada, Y., 1992, Chemical modification of L-asparaginase with a comb-shaped copolymer of polyethylene glycol derivative and maleic anhydride, *Biochem. Biophys. Res. Commun.* **184:**144–148.

Kurkela, R., Vuolas, L., and Vihko, P., 1988, Preparation of $F(ab')_2$ fragments from monoclonal mouse IgG_1 suitable for use in radioimaging, *J. Immunol. Meth.* **110:**229–236.

Lamoyi, E., and Nisonoff, A., 1983, Preparation of $F(ab')_2$ fragments from mouse IgG of various subclasses, *J. Immunol. Meth.* **56:**235–243.

Lenz, A. G., Costabel, U., Shaltiel, S., and Levine, R. L., 1989, Determination of aldehyde groups in oxidatively modified proteins by reduction with tritiated sodium borohydride, *Anal. Biochem.* **177:**419–425.

Leung, S. O., Losman, M. J., Govindan, S. V., Griffiths, G. L., Goldenberg, D. M., and Hansen, H. J., 1995, Engineering a unique glycosylation site for site-specific conjugation of haptens to antibody fragments. *J. Immunol.* **154:**5919–5926.

Maina, C. V., Riggs, P. D., Grandea, A. G. I., Slatko, B. E., Moran, L. S., Tagliamonte, J. A., McReynolds, L. A., and Guan, C. D., 1988, An *Escherichia coli* vector to express and purify foreign proteins by fusion to and separation from maltose-binding protein, *Gene* **74:**365–373.

Mariani, M., Camagna, M., Tarditi, L., and Seccamani, E., 1991, A new enzymatic method to obtain high-yield $F(ab')_2$ suitable for clinical use from mouse IgG_1, *Mol. Immunol.* **28:**69–77.

Melton, R. G., Boyle, J. M. B., Rogers, G. T., Burke, P., Bagshawe, K. D., and Sherwood, R. F., 1993a, Optimisation of small-scale coupling of A5B7-monoclonal antibody to carboxypeptidase G_2, *J. Immunol. Meth.* **158:**49–56.

Melton, R. G., Rowland, J. A., Pietersz, G. A., Sherwood, R. F., and McKenzie, I. F. C., 1993b, Tumour necrosis factor increases tumour uptake of co-administered antibody-carboxypeptidase G_2 conjugate, *Eur. J. Cancer* **29A:**1177–1183.

Melton, R. G., Wiblin, C. N., Foster, R. L., and Sherwood, R. F., 1987, Covalent linkage of carboxypeptidase G_2 to soluble dextrans—I. Properties of conjugate and effects on plasma persistence in mice, *Biochem. Pharmacol.* **36:**105–112.

Meyer, D. L., Jungheim, L. N., Law, K. L., Mikolajczyk, S. D., Shepherd, T. A., Mackensen, D. G., Briggs, S. L., and Starling, J. J., 1993, Site-specific prodrug activation by antibody-beta-lactamase conjugates—regression and long-term growth inhibition of human colon carcinoma xenograft models, *Cancer Res.* **53:**3956–3963.

Meyer, D. L., Jungheim, L. N., Mikolajczyk, S. D., Shepherd, T. A., Starling, J. J., and Ahlem, C. N., 1992, Preparation and characterization of a beta-lactamase-Fab′ conjugate for the site-specific activation of oncolytic agents, *Bioconjugate Chem.* **3:**42–48.

Michael, N. P., Chester, K. A., Melton, R. G., Robson, L., Nicholas, W., Boden, J. A., Pedley, R. B., Begent, R. H. J., Sherwood, R. F., and Minton, N. P., 1996, *In vitro* and *in vivo* characterisation of a recombinant carboxypeptidase G_2::anti-CEA scFv fusion protein, *Immunotechnology* **2:**47–57.

Mikolajczyk, S. D., Meyer, D. L., Fagnani, R., Hagan, M. S., Law, K. L., and Starling, J. J., 1996, Dextran modification of a Fab′-beta-lactamase conjugate modulated by variable pretreatment of Fab′ with amine-blocking reagents, *Bioconjugate Chem.* **7:**150–158.

Mikolajczyk, S. D., Meyer, D. L., Starling, J. J., Law, K. L., Rose, K., Dufour, B., and Offord, R. E., 1994, High field, site-specific coupling of N-terminally modified beta- lactamase to a proteolytically derived single-sulfhydryl murine Fab′, *Bioconjugate Chem.* **5:**636–646.

Milenic, D. E., Esteban, J. M., and Colcher, D., 1989, Comparison of methods for the generation of immunoreactive fragments of a monoclonal antibody (B72.3) reactive with human carcinomas, *J. Immunol. Meth.* **120:**71–83.

Mueller, B. M., Wrasidlo, W. A., and Reisfeld, R. A., 1988, Determination of the number of ϵ-amino groups available for conjugation of effector molecules to monoclonal antibodies, *Hybridoma* **7:** 453–456.

Murayama, A., Shimada, K., and Yamamoto, T., 1978, Modification of immunoglobulin G using specific reactivity of sugar moiety, *Immunochemistry* **15:**523–528.

Muzykantov, M. R., Sakharov, D. V., Sinitsyn, V. V., Domogatsky, S. P., Goncharov, N. V., and Danilov. S. M., 1988, Specific killing of human endothelial cells by antibody-conjugated glucose oxidase, *Anal. Biochem.* **169:**383–389.

Muzykantov, V. R., Trubetskaya, O. V., Puchnina, E. A., Sakharov, D. V., and Domogatsky, S. P., 1990, Cytotoxicity of glucose oxidase conjugated with antibodies to target cells: Killing efficiency depends on the conjugate internalization, *Biochim. Biophys. Acta* **1053:**27–31.

Nilson, B. H. K., Lögdberg, L., Kastern, W., Björck, L., and Äkerström, B., 1993, Purification of antibodies using protein L-binding framework structures in the light chain variable domain, *J. Immunol. Meth.* **164:**33–40.

O'Shannessy, D. J., and Quarles, R. H., 1987, Labeling of the oligosaccharide moieties of immunoglobulins, *J. Immunol. Meth.* **99:**153–161.

Ornatowska, M., and Glasel, J. A., 1991, Direct production of Fv-fragments from a family of monoclonal IgGs by papain digestion, *Mol. Immunol.* **28:** 383–391.

Oswald, T., and Rinas, U., 1996, Chloramphenicol resistance interferes with purification of histidine-tagged fusion proteins from recombinant *Escherichia coli*, *Anal. Biochem.* **236:**, 357–358.

Owens, R. J., King, D. J., Howat, D., Lisle, H., Mountain, A., Harvey, A., Bergin, S., Turner, A., Pedley, B., Boden, J., Begent, R., and Yarranton, G. T., 1991, Tumor binding properties of B72.3 Fv fragments, *Antibody Immunoconjugates and Radiopharmaceuticals* **4:**459–467.

Parham, P., Androlewicz, M. J., Brodsky, F. M., Holmes, N. J., and Ways, J. P., 1982, Monoclonal antibodies: Purification, fragmentation and application to structural and functional studies of class I MHC antigens. *J. Immunol. Meth.* **53:**133–173.

Peng, L., Carlton, G. J., and Burnett, J. W., 1987, Effect of borohydride reduction on antibodies, *Appl. Biochem. Biotechnol.* **14:**91–99.

Philpott, G. W., Shearer, W. T., Bower, R. J., and Parker, C. W., 1973, Selective cytotoxicity of hapten-substituted cells with an antibody-enzyme conjugate, *J. Immunol.* **111:**921–929.

Pillai, S., and Bachhawat, B. K., 1977, Protein–protein conjugation on a lectin matrix, *Biochem. Biophys. Res. Commun.* **75:**240–245.

Rector, E. S., Schwenk, R. J., Tse, K. S., and Sehon, A. H., 1978, A method for the preparation of protein–protein conjugates of predetermined composition, *J. Immunol. Meth.* **24:**321–336.

Rodrigues, M. L., Presta, L. G., Kotts, C. E., Wirth, C., Mordenti, J., Osaka, G., Wong, W. L. T., Nuijens, A., Blackburn, B., and Carter, P., 1995, Development of a humanized disulfide-stabilized anti-beta p185(HER2) Fv-beta-lactamase fusion protein for activation of a cephalosporin doxorubicin prodrug, *Cancer Res.* **55:**63–70.

Rose, K., Vilaseca, L. A., Werlen, R., Meunier, A., Fisch, I., Jones, R. L., and Offord, R. E., 1991, Preparation of well defined protein conjugates using enzyme-assisted reverse proteolysis, *Bioconjugate Chem.* **2:**154–159.

Sahin, U., Hartmann, F., Senter, P., Pohl, C., Engert, A., Diehl, V., and Pfreundschuh, M., 1990, Specific activation of the prodrug mitomycin phosphate by a bispecific anti-CD30/anti-alkaline phosphatase monoclonal antibody, *Cancer Res.* **50:**6944–6948.

Schiavon, O., Caliceti, P., Sartore, L., and Veronese, F. M., 1991, Surface modification of enzymes for therapeutic use—monomethoxypoly (ethylene glycol) derivatization of ribonuclease, *Farmaco* **46:**967–978.

Searle, F., Bier, C., Buckley, R. G., Newman, S., Pedley, R. B., Bagshawe, K. D., Melton, R. G., Alwan, S. M., and Sherwood, R. F., 1986, The potential of carboxypeptidase G_2-antibody conjugates as anti-tumour agents. I. Preparation of antihuman chorionic gonadotrophin-carboxypeptidase G_2 and cytotoxicity of the conjugate against JAR choriocarcinoma cells in vitro, *Br. J. Cancer* **53:**377–384.

Sedlak, J., and Lindsay, R. H., 1968, Estimation of total, protein bound and non-protein sulphydryl groups in tissue with Ellman's reagent, *Anal. Biochem.* **25:**192–205.

Sharma, S. K., Bagshawe, K. D., Burke, P. J., Boden, R. W., and Rogers, G. T., 1990, Inactivation and clearance of an anti-CEA carboxypeptidase G_2 conjugate in blood after localisation in a xenograft model, *Br. J. Cancer* **61:**659–662.

Sharma, S. K., Bagshawe, K. D., and Melton, R. G., 1991, Immunogenicity of antibody-enzyme conjugates in antibody directed enzyme prodrug therapy, *Antibody Immunoconjugates and Radiopharmaceuticals* **4:**226.

Sharma, S. K., Bagshawe, K. D., Melton, R. G., and Begent, R. H. J., 1996, Effect of cyclosporine on immunogenicity of a bacterial enzyme carboxypeptidase G_2 in ADEPT, *Transplant. Proc.* **28:** 3154–3155.

Shepherd, T. A., Jungheim, L. N., Meyer, D. L., and Starling, J. J., 1991, A novel targeted delivery system utilizing a cephalosporin-oncolytic prodrug activated by an antibody beta-lactamase conjugate for the treatment of cancer, *Bioorg. Medicinal Chem. Lett.* **1:**21–26.

Sherwood, R. F., Baird, J. K., Atkinson, T., Wiblin, C. N., Rutter, D. A., and Ellwood, D. C., 1977, Enhanced plasma persistence of therapeutic enzymes by coupling to soluble dextran, *Biochem. J.* **164:**461–464.

Smith, D. B., and Johnson, K. S., 1988, Single-step purification of polypeptides expressed in *Escherichia coli* as fusions with glutathione-*S*-transferase, *Gene* **67:**31–40.

Smith, M. A., Easton, M., Everett, P., Lewis, G., Payne, M., Riverosmoreno, V., and Allen, G., 1996, Specific cleavage of immunoglobulin G by copper ions, *Int. J. Pept. Protein Res.* **48:**48–55.

Stanislawski, M., Rousseau, V., Goavec, M., and Ito, H., 1989, Immunotoxins containing glucose oxidase and lactoperoxidase with tumoricidal properties: *In vitro* killing effectiveness in a mouse plasmacytoma cell model, *Cancer Res.* **49:**5497–5504.

Sutherland, R., Buchegger, F., Schreyer, M., Vacca, A., and Mach, J.-P., 1997, Penetration and binding of radiolabelled anti-carcinoembryonic antigen monoclonal antibodies and their antigen binding fragments in human colon cancer multicellular tumor spheroids, *Cancer Res.* **47:**1627–1633.

Svensson, H. P., Kadow, J. F., Vrudhula, V. M., Wallace, P. M., and Senter, P. D., 1992, Monoclonal antibody beta-lactamase conjugates for the activation of a cephalosporin mustard prodrug, *Bioconjugate Chem.* **3:**176–181.

Takahashi, H., Igarishi, T., Shimada, I., and Arata, Y., 1991, Preparation of the Fv fragment from a short-chain mouse IgG_{2a} anti-dansyl monoclonal antibody and use of selectively deuterated Fv analogues for two-dimensional H-1 NMR analyses of the antigen-antibody interactions, *Biochemistry* **30**:2840–2847.

Thorpe, P. E., and Ross, W. C. J., 1982, The preparation and cytotoxic properties of antibody-toxin conjugates, *Immunol. Rev.* **62**:119–158.

Thorpe, P. E., Wallace, P. M., Knowles, P. P., Relf, M. G., Brown, A. N. F., Watson, G. J., Knyba, R. E., Wawrzynczak, E. J., and Blakey, D. C., 1987, New coupling agents for the synthesis of immunotoxins containing hindered disulphides with improved stability *in vivo*, *Cancer Res.* **47**:5924–5931.

Vingerhoeds, M. H., Haisma, H. J., Belliot, S. O., Smit, R. H. P., Crommelin, D. J. A., and Storm, G., 1996, Immunoliposomes as enzyme-carriers (immuno-enzymosomes) for antibody-directed enzyme prodrug therapy (ADEPT): Optimization of prodrug activating capacity, *Pharm. Res.* **13**:604–610.

Vrudhula, V. M., Svensson, H. P., Kennedy, K. A., Senter, P. D., and Wallace, P. M., 1993, Antitumor activities of a cephalosporin prodrug in combination with monoclonal antibody-beta-lactamase conjugates, *Bioconjugate Chem.* **4**:334–340.

Wahl, R. L., Parker, C. W., and Philpott, G. W., 1983, Improved radioimaging and tumor localisation with monoclonal $F(ab')_2$, *J. Nucl. Med.* **24**:316–325.

Wallace, P. M., and Senter, P. D., 1991, *In vitro* and *in vivo* activities of monoclonal antibody-alkaline phosphatase conjugates in combination with phenol mustard phosphate, *Bioconjugate Chem.* **2**: 349–352.

Wang, S. M., Chern, J. W., Yeh, M. Y., Ng, Y. C., Tung, E., and Roffler, S. R., 1992, Specific activation of glucuronide prodrugs by antibody-targeted enzymes for cancer therapy, *Cancer Res.* **52**:4484–4491.

Werlen, R. C., Lankinen, M., Rose, K., Blakey, D., Shuttleworth, H., Melton, R., and Offord, R. E., 1994, Site-specific conjugation of an enzyme and an antibody fragment. *Bioconjugate Chem.* **5**: 411–417.

Wileman, T. E., Roster, R. L., and Elliott, P. N. C., 1986, Soluble dextran-asparaginase conjugates show increased circulatory persistence and lowered antigen reactivity, *J. Pharm. Pharmacol.* **38**: 264–271.

Wines, B. D., and Easterbrooksmith, S. B., 1991, The Fab/c fragment of IgG produced by cleavage at cyanocysteine residues, *Mol. Immunol.* **28**:855–863.

Worrell, N. R., Cumber, A. J., Parnell, G. D., Mirza, A., Forrester, J. A., and Ross, W. C. J., 1986, Effect of linkage variation on pharmacokinetics of ricin A chain-antibody conjugates in normal rats, *Anticancer Drug Des.* **1**:179–188.

Yamaguchi, Y., Kim, H., Kato, K., Masuda, K., Shimada, I., and Arata, Y., 1995, Proteolytic fragmentation with high specificity of mouse immunoglobulin G—Mapping of proteolytic cleavage sites in the hinge region, *J. Immunol. Meth.* **181**:259–267.

Yokota, T., Milenic, D. E., Whitlow, M., and Schlom, J., 1992, Rapid tumor penetration of a single-chain Fv and comparison with other immunoglobulin forms, *Cancer Res.* **52**:3402–3408.

Zara, J., Pomato, N., McCabe, R. P., Bredehorst, R., and Vogel, C.-W., 1995, Cobra venom factor immunoconjugates: Effects of carbohydrate directed versus amino group directed conjugation, *Bioconjugate Chem.* **6**:367–372.

7

Phage Technology for Producing Antibody-Enzyme Fusion Proteins

K. A. Chester, Roger G. Melton, and R. E. Hawkins

1. INTRODUCTION

For ADEPT approaches to be successful, site-specific delivery of the enzyme to the tumor is clearly essential. This is most readily achieved with antibodies, of which an extensive range exists covering a wide variety of antigens. Most of the early trials of ADEPT systems, from preclinical studies through to clinical trials, used conjugates whose attachment of antibody to enzyme was performed by chemical conjugation. Now these molecules can be joined at a molecular level by linking the genes. The resulting expressed product, known as a fusion protein, has both antibody and enzymic activities. For effective therapy with ADEPT, the antibody–enzyme conjugate must localize in the tumor and remain there after it clears from normal tissues. Then a nontoxic prodrug can be administered, and it is activated by the localized enzyme to produce a cytotoxic drug at the tumor site. A perceived advantage of ADEPT systems is that they potentially include an amplification step (each enzyme molecule produces many molecules of the active drug within the tumor) and a bystander effect (destruction of neighboring cells and those binding the antibody fragment) because the final cytotoxic molecule is

---

K. A. Chester • Department of Clinical Oncology, Royal Free Hospital School of Medicine, London NW3 2PF, United Kingdom. *R. E. Hawkins* • Department of Oncology, MRC Centre, Cambridge CB2 2HQ, United Kingdom. *Roger G. Melton* • Enzacta Ltd., Salisbury SP4 0JQ, United Kingdom.

Enzyme-Prodrug Strategies for Cancer Therapy, edited by Melton and Knox. Kluwer Academic/Plenum Publishers, New York, 1999.

produced at the cell surface and is small, allowing diffusion throughout the tumor. The potential disadvantages of the approach are that the antibody–enzyme conjugate must be stable for long periods in the tumoral environment and it must be nonimmunogenic to allow repeated therapy. The use of fusion proteins should result in very stable antibody enzyme conjugates, and there is the potential to construct them from either human proteins or proteins engineered to minimize immunogenicity.

2. ANTIBODY-TARGETED CANCER THERAPY

Tumor imaging has been successfully achieved with radiolabeled monoclonal antibodies and their derivatives in the clinic, but major therapeutic successes with antibodies, so far, have been confined to radioimmunotherapy of lymphoma (Press *et al.*, 1993) and therapy of colorectal cancer in an adjuvant setting (Reitmuller *et al.*, 1994). Now it is well established that in general only a small fraction of antibody reaches the tumor (typically only 0.001–0.0001% of the injected dose/g in humans) (Begent, 1985) and that it normally does not penetrate far from the tumor vasculature. The main reasons for this disappointing performance are discussed in some detail in Chapter 3 of this volume, but are primarily caused by poor tumor penetration, poor target specificity, and lower than optimal affinity, resulting in antibodies that bind sluggishly or remain bound only for a short time. The combined effects of all these factors lead to modest tumor to normal tissue ratios. Furthermore, for repeated treatment, the problem of immunogenicity of the therapeutic antibody and the attached effector protein arises. Phage technology and antibody engineering will allow optimization of antibodies to overcome some if not all of these obstacles.

3. POTENTIAL OF ENGINEERED ANTIBODIES FOR THERAPY

The range of antibody fragments which can be engineered is shown in Fig. 1 in comparison with their natural antibody counterparts. The ideal molecule for therapy would probably be small to give optimal penetration and have high affinity and specificity for its antigen. It should also be seen as "self" to avoid immune responses during repeated therapy. To overcome tumor antigen heterogeneity, the effector protein component of the targeting molecule should also have the potential to generate a bystander effect because it is not possible to reach every tumor cell with current antibody-targeting agents. Finally, given the relatively low proportion of injected antibody localizing in the tumor, the preferred effector protein should provide an amplification effect but should be nontoxic in its own

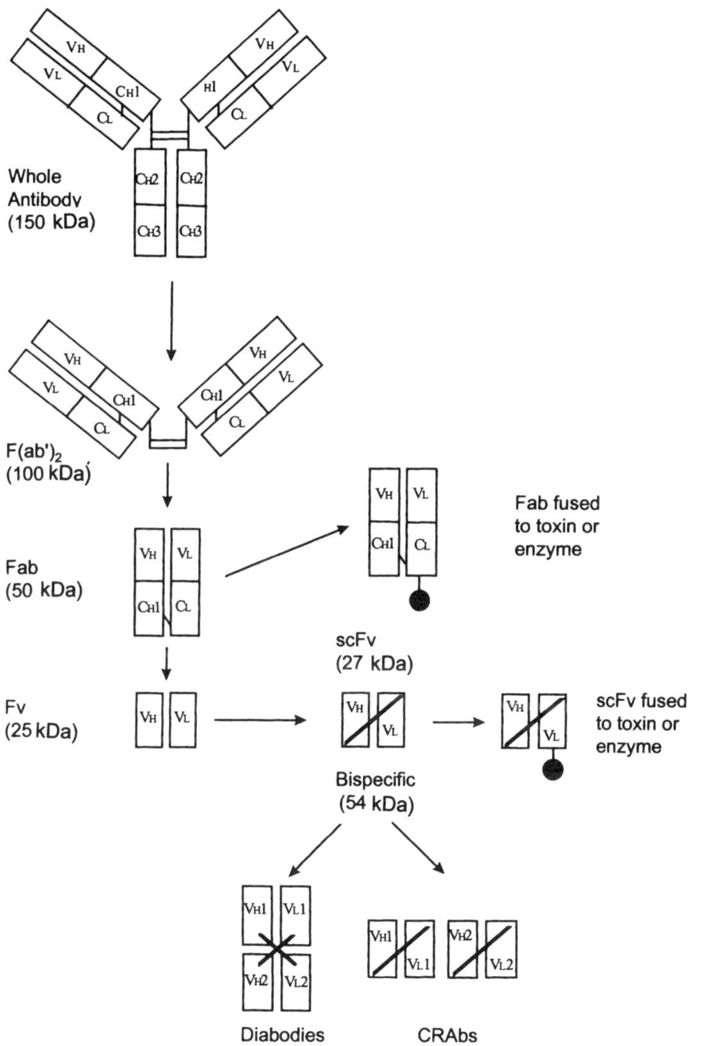

Figure 1. The various formats of engineered antibody-derived fragments compared with native immunoglobulin. Molecular size is important for tumor penetration and in determining the circulating half-life. The Fab and scFv can also be produced as bispecific antibodies. Diabodies and CRAbs may be expressed as bispecific fragments without further manipulation. These fragments can be produced in eukaryotic cells or bacteria.

right. Most of these criteria have been met by the various ADEPT systems described to date, but most require improvement in some areas. The extent to which the genetic manipulation of antibodies allows us to improve on existing systems is discussed following.

4. SINGLE-CHAIN FV ANTIBODIES

A single-chain Fv or scFv antibody (see Fig. 1) is a heavy-chain variable region (VH) tethered by a flexible 15 amino acid linker, for example, (gly gly gly gly ser)$_3$ (Huston *et al.*, 1993), to a light-chain variable region (VL). ScFvs have a molecular mass of 27 kDa and they are the smallest molecules which retain full (although monovalent) antibody-binding activity. They are being used extensively as the building blocks of therapeutic molecules because they can be readily cloned and expressed in bacteria (Chiswell and McCafferty, 1992) and can be made as genetic fusions with a variety of effector proteins. Examples of such proteins include cytokines (Hoogenboom *et al.*, 1991; Savage *et al.*, 1993), toxins, such as *Pseudomonas* exotoxin (Brinkmann *et al.*, 1993; Kreitman *et al.*, 1994), and immunostimulatory agents, such as superantigens (Dohlsten *et al.*, 1994). In addition, scFvs can be displayed as functional antibody fragments on the surface of bacteriophage (McCafferty *et al.*, 1990), a technique which allows selecting potentially clinically useful antibodies from vast combinatorial libraries, as described later. One advantage of using scFvs obtained by the phage system is that it permits selecting antibodies packaged together with cloned DNA encoding them. This facilitates attaching enzymes for ADEPT and the many other available novel effector mechanisms.

4.1. Tumor Penetration and Biodistribution

Poor penetration of antibody into a solid tumor mass remains one of the greatest barriers to effective therapy, and this is likely to be worsened by adding an effector protein component because the resultant fusion protein will be larger than the antibody alone. There is ample evidence that the penetration of tumors by antibodies depends on their size (Rowlinson-Busza *et al.*, 1996; Sutherland *et al.*, 1997). For this reason it is desirable to use antibody fragments, rather than whole IgG. ScFvs permeate more rapidly and deeper into tumors than whole IgG (Yokata *et al.*, 1992). However, consistent with their small size, scFvs are rapidly cleared through the kidneys, resulting in a short serum half-life (Milenic *et al.*, 1991). Kidney uptake of scFvs is commonly observed: the extent differs with different scFvs (Verhaar *et al.*, 1995), but the controlling factors are not clear. Therefore scFvs are unlikely to be useful as single entities for therapy, but as fusion proteins with enzymes they provide attractive molecules for ADEPT. These larger molecules should be above the glomerular filtration threshold and therefore should avoid the kidney uptake encountered with unmodified scFvs.

There may be other important considerations to be taken into account with the use of scFv's. For example, unmodified scFvs are univalent, so they must rely on intrinsic affinity rather than avidity to remain bound in the tumor. This suggests that larger multivalent molecules may be advantageous, but this inevitably leads to an increase in size, and so other approaches to improving penetration may be needed. One possible approach to improving tumor penetration of antibody conjugates in general is pretreating with antibodies conjugated to vasoactive agents, such as histamine or interleukin-2 (IL-2), to increase local vascular permeability (Khawali *et al.*, 1994). The use of antibody engineering to produce a scFv-IL-2 fusion (Savage *et al.*, 1993) may be particularly useful in this respect.

4.2. Avidity of Binding

The avidity of the targeting molecule is important. Thus natural antibodies of the IgG isotype are bivalent, whereas IgMs are pentavalent. Both use this multivalency to improve binding. The affinity of binding results from univalent interactions but binding of two or more heads results in a much more stable interaction, known as the avidity effect. This can result in large increases in functional affinity (up to 1000-fold for an IgG compared to an Fab) but depends critically on the density of the target antigen and the spatial separation of the epitopes relative to the spacing of the antibody heads (Holland and Stewart, 1991; Kaufman and Jain, 1992; Cheong *et al.*, 1990). Whole antibodies use this feature, and by careful design it may be possible to make molecules that are small and thus penetrate well but are avid and so bind strongly. Bivalent scFv molecules can be made in a number of ways, for example, by using tags of amphipathic helices that are covalently associated or in dynamic equilibrium with the monomer (Pack and Plückthun, 1991), or creating diabodies by shortening the single-chain linker of scFv's to force dimerization (Holliger *et al.*, 1993; McGuinness *et al.*, 1996). Chelating recombinant antibodies (CRAbs) (Neri *et al.*, 1995), which consist of two scFvs specific for different epitopes on the same molecule, are particularly interesting because these give increased affinity and specificity for their target antigen. The structure of diabodies and CRAbs relative to native immunoglobin is shown in Fig. 1.

For ADEPT, avidity has been increased by cross-linking the antibody fragments as described later from work of Bosslet *et al.* (1992, 1994). Theoretically, avidity is also be increased by using the dimeric structure of the enzyme, such as

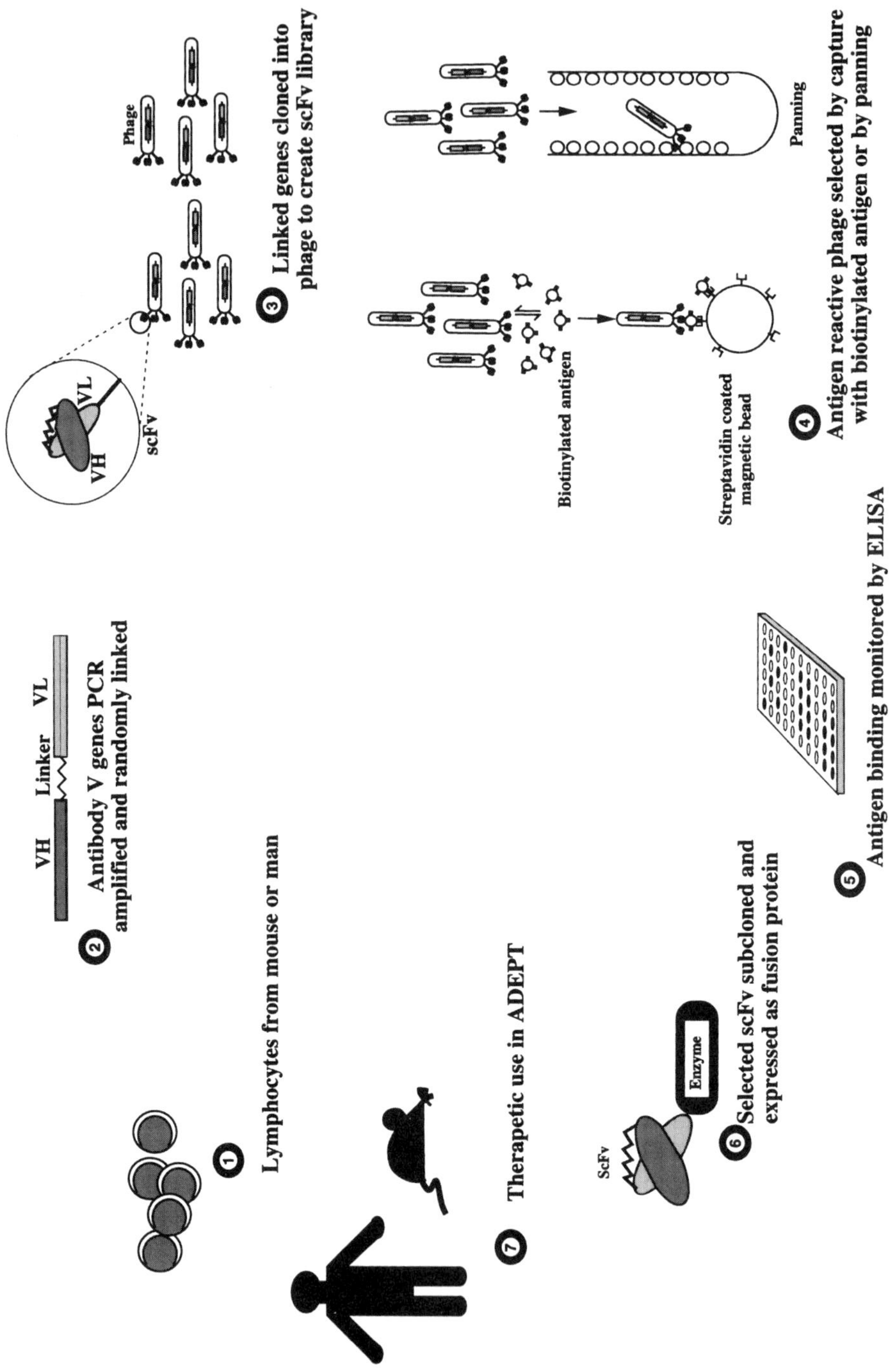
1 Lymphocytes from mouse or man
VH
Linker
VL
2 Antibody V genes PCR amplified and randomly linked
VH
VL
scFv
Phage
3 Linked genes cloned into phage to create scFv library
Biotinylated antigen
Streptavidin coated magnetic bead
Panning
4 Antigen reactive phage selected by capture with biotinylated antigen or by panning
5 Antigen binding monitored by ELISA
ScFv
Enzyme
6 Selected scFv subcloned and expressed as fusion protein
7 Therapetic use in ADEPT

carboxypeptidase G_2, to effectively mimic the hinge region of an antibody and form a bivalent pseudo IgG molecule.

5. THE PHAGE SYSTEM

The first reports describing the expression of antibody fragments on the surface of bacteriophage appeared in 1990 (McCafferty *et al.*, 1990). Since then rapid development of this approach to making antibodies has occurred (Hoogenboom, 1997). The underlying technology behind the use of phage to make antibodies is the polymerase chain reaction (PCR). Using suitable primers, PCR amplifies rearranged murine (Orlandi *et al.*, 1989; Ward *et al.*, 1989; Zhou *et al.*, 1994) or human (Marks *et al.*, 1991) V-genes. Once amplified, these gene libraries can be joined together as an antibody fragment and cloned into phage expression vectors. The display of antibodies on the surface of phage means that the phage can be treated exactly as if it were an antibody and thus can be selected for specificity and desired binding characteristics (Clackson *et al.*, 1991; Hawkins *et al.*, 1992). The antibody gene is selected along with the phage antibody that it encodes. A diagram outlining the stages in creating a scFv library, isolating an antigen specific scFv from this library, and creating an antibody-enzyme fusion protein is shown in Fig. 2. After the initial selection, various methods can be used to improve the antibodies originally selected. These include mutagenesis, either random (Hawkins *et al.*, 1992) or rational (Reichmann and Weill, 1993; Roberts *et al.*, 1987), and 'chain shuffling' (Marks *et al.*, 1992), where a single heavy (or light) chain can be combined with many light (or heavy) chains to produce combinations of higher affinity. These approaches have been successful with Fabs and scFvs. In one instance an increase in affinity of over 400-fold was achieved by combining optimal mutations from hypervariable regions (Wei-Ping *et al.*, 1995).

←

Figure 2. Stages involved in making an antibody–enzyme fusion with an scFv from a filamentous phage library. **1.** Lymphocytes from human or immunized mouse are used as a source of mRNA-encoding antibody V regions. **2.** cDNA is produced and V regions are amplified by PCR, using redundant primers in antibody frameworks 1 and 4. The V regions are randomly combined by PCR using overhanging linking primers encoding a flexible peptide [typically $(\text{gly gly gly gly ser})_3$] to tether VH to VL. This is followed by PCR with external primers encoding restriction sites to clone into phage vectors. **3.** Linked scFvs are cloned into phage to give a diverse combinatorial library, and each phage displays an individual scFv. **4.** Phages expressing scFvs reactive with antigen are selected by "panning" on antigen-coated tubes or by reaction with biotinylated antigen and subsequent capture on streptavidin-coated magnetic beads. The latter has the advantage that higher affinity binders can be selected by reducing the concentration of biotinylated antigen. Selection may also be performed on cells. **5.** The success of selection is monitored by ELISA using individual phages or soluble scFvs derived from them. **6.** The genes of selected scFvs are fused with that of the enzyme of choice and subloned into a suitable expression vector. **7.** The fused scFv-enzyme gene is expressed, and the fusion protein is purified and tested for therapeutic efficacy in animal models and subsequently in humans.

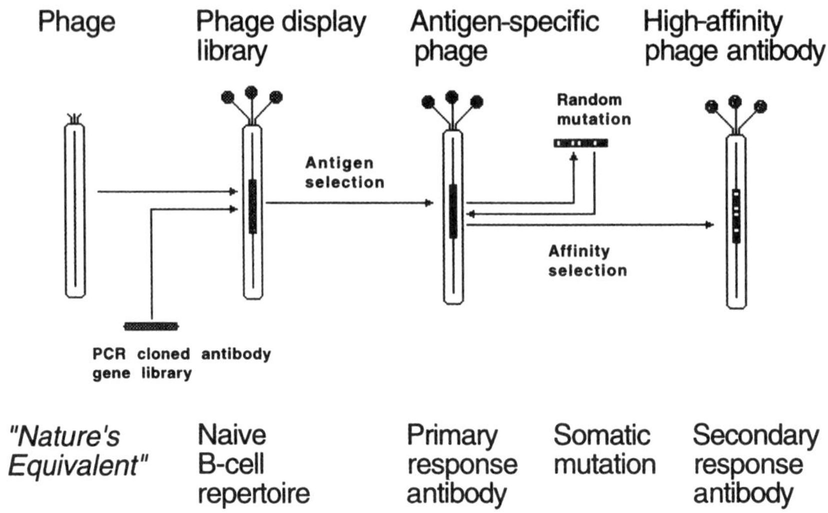

Figure 3. The phage system for improving antibody affinity. The initial repertoire of scFv fragments is equivalent to the naive B-cell repertoire. Amongst these, low-affinity antibodies can be found to any antigen. Subsequently affinity can be improved by using methods of mutagenesis and phage selection. The use of very large libraries, naive libraries, or libraries from immunized sources allows isolating directly antibodies equivalent to secondary immune responses. Then mutagenesis and selection could be used to produce even higher affinity forms.

Whatever method or combination of methods is used, the system is essentially an *in vitro* mimic of affinity maturation, as illustrated in Figure 3, and depends on the phage system to select higher affinity variants. Selection using biotinylated antibody and streptavidin-coated magnetic beads for capture of variants (Hawkins *et al.*, 1992), as shown in Fig. 2, is particularly powerful and can be used at any stage of phage antibody production.

Where the phage system has been used to make antibodies from immunized mice and where these have been compared to conventional monoclonal antibodies, they have exhibited very favorable characteristics (higher affinity and better specificity). They also behave well in imaging studies both in mice and humans (Begent *et al.*, 1996; Chester *et al.*, 1994).

5.1. Immunogenicity

Murine antibodies are immunogenic in humans, resulting in a human anti-mouse (HAMA) response and progressively shorter half-life of injected antibody with repeated dosage (Meredith *et al.*, 1993). This results in lack of tumor localization and can also result in toxic side effects, such as serum sickness or anaphylaxis. Furthermore, according to all precedents except for Burkitt's lymphoma, repeated therapy sustained over months is essential for substantially effective cancer therapy. The need to eradicate or minimize immunogenicity of antibody-targeted therapeutics is evident, and for the engineered antibody component of the molecule there are a number of approaches to be taken. Production of humanized (Reichmann *et al.*, 1988) or "resurfaced" (Roguska *et al.*, 1994) murine antibodies can allow immunogenicity to be reduced or avoided, but rapid methods of making human antibodies directly are desirable.

In principle, the phage antibody system can be combined with immunization using mice transgenic for the human immunoglobulin locus (Lonberg *et al.*, 1994) or mice engineered with human heavy- and light-chain yeast artificial chromosomes (YACs) (Green *et al.*, 1994; Mendez *et al.*, 1997). However, it is also possible to bypass the need for immunization completely and develop a system which mimics and could even improve on the immune system, entirely *in vitro*. By using repertoires from nonimmunized humans (Marks *et al.*, 1991), it is possible to make human antibodies to many different antigens and affinity mature them, just as in the natural immune system. As the technology advances and the libraries become larger, so the quality and number of the antibodies produced improves (Griffiths *et al.*, 1994; Vaughan *et al.*, 1996). Now, it seems likely that the process of making useful human antibodies will eventually be feasible from a "single-pot" library (Winter *et al.*, 1994). However, the problem of an anti-idiotype response still remains (Viale *et al.*, 1987). It seems likely that some antibodies will be better than others in this respect, and the problem of the immunogenicity of the enzyme component of a conjugate used for ADEPT remains. The bacterial enzymes used in the majority of ADEPT systems to date are highly immunogenic (Sharma *et al.*, 1993).

A possible solution to this problem would be to use bispecific antibodies, one binding arm of which reacts with the tumor and the other acts as an enzymically catalytic center (Pollack *et al.*, 1986; Tramontano *et al.*, 1986). One problem has been that catalytic antibodies (abzymes) have very low activity compared with enzymes. However it is possible that a low turnover of prodrug could be advantageous for ADEPT, if the prodrug and abzyme can be designed to prolong active drug delivery at the tumor site. Recently an abzyme generated to a transition-state analog achieved *in vitro* cell killing using a phenylcorbamoyl L-glutamate mustard prodrug (Wentworth *et al.*, 1996) and this cell-kill assay could form the basis for directly screening phage antibody libraries for suitable enzymic activity. The use of phage technology to produce a diversity of scFvs to transition-state analogs certainly offers an attractive approach because both parts of the therapeutic molecule can be made from human sources. Techniques already described could be used to link and screen the antigen-binding/abzyme arms (McGuinness *et al.*,

1996), and it is possible that antibody-directed abzyme prodrug therapy, or ADAPT, will eventually provide a solution to immunogenicity, although here again there is the specter that an anti-idiotypic response will arise.

In view of the advances previously described, it can be anticipated that immunogenicity of the antibody arm of an ADEPT therapeutic will probably not be the main obstacle to avoiding the patient's immune response. Unless, as discussed later, a human enzyme is used, the enzyme is likely to be the main immunogenic component. It is possible that phage may be used to express libraries displaying wild-type enzyme epitopes and their mutants (Soumillion *et al.*, 1994) and that these libraries may be used to characterize and select non-immunogenic variants.

5.2. Affinity

The interaction of an antibody with its antigen is usually described by its affinity, which is a composite of the kinetic "on rate" and "off rate." It seems reasonable to suppose that the fastest "on rate" and the slowest "off rate" would be desirable for an antibody targeted therapeutic, and indeed with radiolabelled antibodies there is evidence that the use of higher affinity leads to improved antitumor activity and improved survival in animal models (Schlom *et al.*, 1992; Velders *et al.*, 1996). Furthermore, greater tumor uptake has been shown with a higher affinity scFv compared with a lower affinity scFv to carcinoembryonic antigen (CEA) (Verhaar *et al.*, 1995). Another study with affinity-matured scFvs against c-erbB-2 showed that a 12-fold enhancement of targeting is achieved by a 30-fold improvement in affinity (Adams *et al.*, 1996). This latter report is particularly indicative of the importance of affinity because, unlike previous studies, the higher affinity mutants were derived from the lower affinity parent scFv and therefore would be expected to be essentially the same antibody and react with the same epitope. Although as yet there is little experimental data to define optimal affinity, the data so far encourages obtaining the highest possible affinity antibodies for use in ADEPT. Under such circumstances, the use of filamentous bacteriophage technology is invaluable for creating high-affinity antibodies as described earlier.

5.3. Diversity

Because polyclonal antisera are more effective than monoclonal antibodies in treating recurrent Hodgkin's disease with radiolabeled antiferritin antibodies

(Vriesendorp *et al.*, 1991), it may be that cocktails of monoclonal antibodies will prove more successful than single antibodies. Indeed, it is well established that the expression of tumor antigens is not heterogenous throughout the tumor and in the face of this multiple antibodies to multiple antigens will be necessary for effective therapy. Many different antibodies can be obtained from the large human antibody libraries (Griffiths *et al.*, 1994; Vaughan *et al.*, 1996), and this may bring further benefits for therapy.

6. TARGET ANTIGENS

A potential advantage of ADEPT systems is that it is not essential that the target antigen be attached to the cell surface because prodrug can be converted to active drug in the tumor interstitial space and diffuse into the tumor cells. Targets that are secreted or cleaved from tumor cells or produced in tumoral stroma may be usable. Similarly, target antigens which are heterogeneously expressed, as often occurs with tumor antigens, are also potentially usable. Nevertheless, the specificity of the antibody arm for the target tumor antigen is important. Even weakly cross-reactive tissue may be problematic (Pai *et al.*, 1991), especially if it is more accessible, because the amount of antibody that binds is very dependent on the amount that reaches the target (Kennel *et al.*, 1991).

The power of phage technology for generating the well defined and highly specific antibodies desired can best be exploited if the target antigen is available in purified form. The availability of milligram quantities of antigen enables antigen-driven affinity selection, as described earlier, and purified antigen is essential to quantitate binding. This has already been successful in selecting high-affinity, phage-derived antibodies to CEA (Chester *et al.*, 1994). Should the purified antigen be available only in microgram amounts, phage selections may be performed by panning (Griffiths *et al.*, 1994; Marks *et al.*, 1991), or alternatively, now it is possible to select tumor-reactive phage directly on cells (Chi and Garen, 1996; De Kruif *et al.*, 1996). The task of antibody selection is undoubtedly easier when large amounts of purified material are available, however, and this should be considered when selecting a target antigen for an ADEPT application. Indeed, in practical terms this consideration may be more important than absolute specificity for tumor because, as discussed in Chapter 3 of this volume, the choice of truly tumor-specific targets is limited. Although such targets exist, for example, the idiotype in lymphomas. the relative abundance of tumor-associated antigens compared to other tissues (Urban and Schreiber, 1992) may give adequate selectivity for ADEPT, although the normal tissue reactivity should be studied. In some cases, the target antigen expressed in normal tissue may be relatively inaccessible to antibody, as for example, with CEA which is expressed mainly on the lumenal surface of normal intestinal glands and has been used for ADEPT in patients and animal models (Bagshawe *et al.*, 1995; Bosslet *et al.*, 1994; Sharma *et al.*, 1991).

Another target tumor-associated antigen which has been investigated is $p185^{HER2}$, which is up-regulated in breast cancer. This antigen is also expressed on a variety of normal epithelial cells, but *in vitro* experiments have shown that the relative abundance of $p185^{HER2}$ on a breast tumor cell line gives sufficient differential from normal tissues to allow specific targeting in ADEPT (Rodrigues *et al.*, 1995), and *in vivo* experiments using chemically conjugated antibody-enzyme conjugates have shown that highly effective therapy is achieved (Eccles *et al.*, 1994). Where tumor-associated antigens are also nonspecifically expressed in normal tissues, their pattern of distribution should be considered when choosing a suitable toxic agent. For example, the enzyme β-lactamase, genetically fused to anti- $p185^{HER2}$, has been used to generate the drug doxorubicin, for which heart tissue is a site of chronic and dose-limiting toxicity and bone marrow is a site of acute toxicity (Rodrigues *et al.*, 1995). Because there is no detectable expression of $p185^{HER2}$ in heart or bone marrow, the previous ADEPT combination seems particularly suitable for this antigen.

Tumor vasculature is an attractive target for some antibody therapies because it is readily accessible and essential for tumor growth. Moreover, experimental models have demonstrated the potential efficacy of targeted immunotoxins for killing tumor endothelial cells (Burrows and Thorpe, 1993; Thorpe and Burrows, 1995). Tumor vasculature may provide a good target for ADEPT if suitable prodrugs can be designed with very short half lives, so that active drug decays to a harmless form before it leaks back into normal tissues via the blood.

7. ENGINEERED ANTIBODY-ENZYME FUSION PROTEINS IN ADEPT

It is obvious that the technology described in this chapter can be extended to produce antibody-enzyme fusion proteins. Potentially, such molecules can be engineered to optimize their performance in ADEPT applications and overcome known problems with conjugates prepared by chemical linkage of native antibodies, such as immunogenicity and tumor penetration. Another advantage is that the complexity, expense, low yields, and difficulty of obtaining a reproducible product, inherent in using chemical conjugation methods, are also avoided. If high enough yields can be obtained, the fusion protein may be produced from prokaryotic cells which would represent a further advantage, because bacterial production of clinical products is rapid and largely avoids the potential for contamination with pathogenic viruses or their DNA inherent with mammalian cells.

Fusion proteins based on the scFv fragment and small monomeric enzymes can be made, and these smaller reagents have the advantage of potentially better ability to penetrate tumor deposits. The first scFv-enzyme fusion protein used to activate an anticancer prodrug was this type and consisted of an engineered scFv obtained from the L6 monoclonal antitumor antibody fused to the bacterial enzyme β-lactamase to give a therapeutic molecule of 52 kDa (Goshorn *et al.*, 1993). The fusion protein, purified by affinity chromatography, yielded about 2.5 μg of fusion protein/ml of initial bacterial culture supernatant. The purified material bound to tumor cells at least as well as the chemically prepared F(ab′) and retained β-lactamase activity comparable with that of the native enzyme. The latter point was demonstrated by experiments in which tumor cells coated with the fusion protein were as sensitive to a cephalosporin mustard prodrug, but because β-lactamase is a bacterial enzyme and the L6 antibody variable regions are of murine origin, the problem of immunogenicity remains with this construct.

A similar fusion protein, also produced in bacteria, combines the enzyme β-lactamase with a humanized disulfide stabilized Fv reactive with the $p185^{HER2}$ antigen (Rodrigues *et al.*, 1995). This construct used a humanized antibody which, when tested alone has not stimulated an antibody response in clinical trials, although immunogenicity of the fusion protein in humans is still untested, and the enzyme remains of bacterial origin. β-lactamase has also been expressed in eukaryotic cells where it has been used to replace the Fab arm of a whole murine–human chimeric antibody directed against the tumor marker, human placental alkaline phosphatase. The enzymic activity was retained in this system and there was no steric hindrance of the antibody (De Sutter and Fiers, 1994). Such an approach may also be applicable to other enzymes and would presumably be similarly effective with a fully humanized antibody although again the immunogenicity of the enzyme remains.

An obvious approach to addressing the issue of immunogenicity would be to use a human enzyme and a humanized antibody. This approach was taken by Bosslet *et al.* (1992), who constructed a fusion protein of a humanized anti-CEA Fab with the human lysosomal enzyme β-glucuronidase linked to the C-terminus. By incorporating the hinge region of human IgG3 between the Fab and the β-glucuronidase, they showed that it is possible to obtain a bivalent species from eukaryotic cells whose association constants are the same as those of the humanized monoclonal antibody. As discussed previously, one of the drawbacks of scFvs is their monovalent binding which can reduce functional affinity avidity, so this work is of interest in offering a potential solution to the avidity obstacle. The relatively large size of the molecule (>250 kDa), however, may be a limiting factor for tumor penetration, and there is always the danger inherent with human enzymes that active drug may be released by endogenous enzyme in nontumor tissues. Nevertheless, preclinical *in vivo* studies with this fusion protein have shown high specificity ratios (>100:1) between tumor and normal tissues and therapeutic effects superior to those of conventional chemotherapy. Tumor concentrations of doxorubicin were 4- to 12-fold higher in tumors of mice treated with fusion protein and prodrug than in those treated with the maximum tolerated dose of doxorubicin (Bosslet *et al.*, 1994). The fusion protein was stable in human plasma and there was no detectable toxicity at the therapeutic dose. Reports of clinical trials with this fusion protein are awaited with interest.

If the problem of immunogenicity can be overcome, a fusion protein of anti-CEA carboxypeptidase G_2 would be attractive because results of the first ADEPT clinical trial have already shown encouraging responses in patients with colorectal cancer receiving anti-CEA chemically conjugated to carboxypeptidase G_2 to activate a benzoic acid mustard prodrug (Bagshawe *et al.*, 1995). A recent report has described a bacterially produced fusion protein of this enzyme with the anti-CEA scFv MFE-23 (Michael *et al.*, 1996). MFE-23 (Chester *et al.*, 1994) was produced from immunized mice using phage technology, and it has a 10-fold higher affinity for CEA than A5B7, the monoclonal antibody used in the ADEPT clinical trial mentioned before. Because carboxypeptidase G_2 is a bacterial enzyme and MFE-23 is of murine origin, genetic manipulation of both components to reduce immunogenicity is essential before sustained therapy can be achieved. As discussed earlier, the strategy for this is established for the antibody arm but it remains to be seen if this can be extended to the enzyme.

8. PRACTICAL CONSIDERATIONS

ScFv's are a most attractive tumor recognition unit and a basic feature of bispecific molecules. However, scFv's can have some technical problems in stability, aggregation, and low yield. Improved tumor penetration would be lost if scFv's or products derived from them form multimers after purification. In our experience this can be overcome by choosing an antibody which has been created and selected for high expression as an scFv. In one case we have shown that a murine scFv produced by filamentous phage technology has distinct production advantages over scFv produced from a murine hybridoma (Verhaar *et al.*, 1995).

ScFv's from phage libraries made from immunized mouse spleens can have high affinity and stability and give good yields, but they may also be immunogenic in patients and could lose some of their desirable properties if humanized. This may be overcome by initially selecting scFv from a fully human library, but affinity is likely to be low unless the library is vast and construction of such large libraries is not to be undertaken lightly because they require great technical skill (Griffiths *et al.*, 1994). Alternative solutions, such as the use of human immunoglobulin GACs (Green *et al.*, 1994; Mendez *et al.*, 1997) or mice transgenic for the human immunoglobulin locus (Lonberg *et al.*, 1994) have potential for the future but are not yet proven methods. For many workers, however, the choice of a murine or human library is not the issue. They are committed to an established antibody specificity which has already shown success in clinical application as a hybridoma. In such cases, adaptation of the V regions to scFv format is required before other engineering considerations, and the cloned product may give technical problems (Verhaar *et al.*, 1995) that require subsequent manipulation. Although there is, as yet, no single foolproof approach to obtaining the antibody of choice, there are many possible routes, and the great potential of these cloned products is rapidly driving research to indicate successful routes to the clinic.

9. CONCLUSIONS

Antibody phage technology and antibody engineering produce novel antibodies with reduced immunogenicity, improved affinity, altered size, and increased avidity. These engineered antibodies can be genetically linked to enzymes and further tailored to give optimal performance in ADEPT systems. The choice of an appropriate antibody fragment is only one of the first of many hurdles in the challenge of developing an engineered product for ADEPT. The choice of an appropriate enzyme and prodrug to match the biodistribution and biological activity of a particular antibody moiety are equally important in what is a multicomponent system. Equally challenging is the development work required to solve the technical problem of large-scale production of the engineered fusion protein to clinical standards (Begent *et al.*, 1991; Horn *et al.*, 1997; Kipriyanov, Moldenhauer, and Little, 1997; Kipriyanov *et al.*, 1997; Nieba *et al.*, 1997).

ACKNOWLEDGMENTS. We thank Prof. R. H. J. Begent for help in preparation. Kerry Chester is supported by the Ronald Raven Chair in Clinical Oncology Trust. Robert Hawkins is a Cancer Research Campaign Senior Clinical Research Fellow.

REFERENCES

Adams, G. P., Shier, R., Mcartney, J. E., McCall, A. M., Wolf, E. J., Tai, M-S., Stafford, W. F., Houston, L. L., Huston, J. S., Marks, J. D., and Weiner, L. M., 1996, Influence of affinity, avidity and size on the tumour retention of anti-c-erbB-2 single chain Fv molecules. *Poster 300, Keystone Symposia*: Exploring and exploiting antibody and Ig superfamily combining sites (D1) Taos, New Mexico Feb. 22–28, 1996, *Immunotechnology*, in press.

Bagshawe, K. D., Sharma, S. K., Springer, C. J., and Antoniw, P., 1995, Antibody directed prodrug therapy: Pilot scale clinical trial, *Tumour Targeting* **1**:1–13.

Begent, R. H. J., 1985, Recent advances in tumour imaging: Use of radiolabelled monoclonal antibodies, *Biochim. Biophys. Acta* **780**:151–166.

Begent, R. H. J., Chester, K. A., Connors, T. A., Crowther, D., Fox, B., Griffiths, E., Hince, T. A., Ledermann, J. A., McVie, J. G., Minor, P., Secher, D. S., Schwartsmann, G., Thorpe, R., Wilbin, C., and Zwierzina, H., 1991, Cancer Research Campaign operation manual for control recommendations for products derived from recombinant DNA technology prepared for investigational administration to patients with cancer in Phase I trials, *Eur. J. Cancer* **29A:**1907–1910.

Begent, R. H. J., Verhaar, M. J., Chester, K. A., Casey, J. L., Green, A. J., Napier, M. P., Hopestone, L. D., Cushen, N., Keep, P. A., Johnson, C. J., Hawkins, R. E., Hilson, A. J. W., and Robson, L., 1996, Clinical evidence of efficient tumor targeting based on single-chain Fv antibody selected from a combinatorial library, *Nature Med.* **2:**979–984.

Bosslet, K., Czech, J., and Hoffmann, D., 1994, Tumour-selective prodrug activation by fusion protein-mediated catalysis, *Cancer Res.* **54:**2151–2159.

Bosslet, K., Czech, J., Lorenz, P., Sedlacek, H. H., Sheuermann, M., and Seemann, G., 1992, Molecular and functional characterisation of a fusion protein suited for tumour specific prodrug activation, *Br. J. Cancer* **65:**234–238.

Brinkmann, U., Lee, B. K., and Pastan, I., 1993, Recombinant immunotoxins containing the VH or VL domain of monoclonal antibody B3 fused to *Pseudomonas* exotoxin, *J. Immunol.* **150:**2774–2782.

Burrows, F. J., and Thorpe, P. E., 1993, Eradication of large solid tumors in mice with an immunotoxin directed against tumor vasculature, *Proc. Natl. Acad. Sci. USA* **90:**8996–9000.

Cheong, H. S., Chang, J. S., Park, J. M., and Byun, S. M., 1990, Affinity enhancement of bispecific antibody against two different epitopes in the same antigen, *Biochem. Biophys. Res. Commun.* **173:**795–800.

Chester, K. A., Begent, R. H. J., Robson, L., Keep, P., Pedley, R. B., Boden, J. A., Boxer, G., Green, A., Winter, G., Cochet, O., and Hawkins, R. E., 1994, Phage libraries for generation of clinically useful antibodies, *Lancet* **343:**455–456.

Chi, X., and Garen, A., 1996, Anti-melanoma antibodies from melanoma patients immunised with genetically modified autologous tumour cells: Selection of specific antibodies from single-chain Fv fusion phage libraries, *Proc. Natl. Acad. Sci. USA* **92:**6537–6541.

Chiswell, D. J., and McCafferty, J., 1992, Phage antibodies: Will new 'coliclonal' antibodies replace monoclonal antibodies? *TIBTECH* **10:**80–84.

Clackson, T., Hoogenboom, H. R., Griffiths, A. D., and Winter, G., 1991, Making antibody fragments using phage display libraries, *Nature* **352:**624–628.

De Kruif, J., Terstappen, L., Boel, E., and Logtenberg, T., 1996, Rapid selection of cell subpopulation-specific human monoclonal antibodies from a synthetic phage antibody library, *Proc. Natl. Acad. Sci. USA* **92:**3938–3942.

De Sutter, K., and Fiers, W., 1994, A bifunctional murine::human chimeric antibody with one antigen-binding arm replaced by bacterial β-lactamase, *Mol. Immunol.* **31:**261–267.

Dohlsten, M., Abrahmsen, L., Bjork, P., Lando, P. A., Hedlund, G., Forsberg, G., Brodin, T., Gascoigne, N. R. J., Forberg, C., Lind, P., and Kalland, T., 1994, Monoclonal antibody-superantigen fusion proteins: Tumor-specific agents for T-cell-based tumor therapy. *Proc. Natl. Acad. Sci. USA* **91:**8945–8949.

Eccles, S. A., Court, W. J., Box, G. A., Dean, C. J., Melton, R. G., and Springer, C. J., 1994, Regression of established breast carcinoma xenografts with antibody-directed enzyme prodrug therapy against c-erbB2p185, *Cancer Res.* **54:**5171–5177.

Goshorn, S. C., Svensson, H. P., Kerr, D. E., Somerville, J. E., Senter, P. D., and Fell, H. P., 1993, Genetic construction, expression and characterisation of a single chain anti-carcinoma antibody fused to beta-lactamase, *Cancer Res.* **53:**2123–2127.

Green, L. L., Hardy, M. C., Maynard-Currie, C. E., Tsuda, H., Louie, D. M., Mendez, M. J., Abderrahim, H., Noguchi, M., Smith, D. H., Zeng, Y., David, N. E., Sasai, H., Garza, D., Brenner, D. G., Hales, J. F., McGuinnes, R. P., Capon, D. J., Klapholz, S., and Jakobovits, A., 1994,

Antigen-specific human monoclonal antibodies from mice engineered with human Ig heavy and light chain YACs, *Nature Genet.* **7**:13–21.

Griffiths, A. D., Williams, S. C., Hartley, O., Tomlinson, I. M., Waterhouse, P., Crosby, W. L., Kontermann, R. E., Jones, P. T., Low, N. M., Allison, T. J., Prospero, T. D., Hoogenboom, H. R., Nissim, A., Cox, J. P. L., Harrison, J. L., Zacculo, M., Gherardi, E., and Winter, G., 1994, Isolation of high affinity human antibodies directly from large synthetic repertoires, *EMBO J.* **13**:3245–3260.

Hawkins, R. E., Russell, S. J., and Winter, G., 1992, Selection of phage antibodies by binding affinity: mimicking affinity maturation, conjugate. *J. Mol. Biol.* **226**:889 –896.

Holland, G. P., and Steward, M. W., 1991, The influence of epitope density on the estimation of the affinity of antibody for complex antigens, *J. Immunol. Methods* **138**:245–255.

Holliger, P., Prospero, T., and Winter, G., 1993, Diabodies: Small bivalent and bispecific antibody fragments, *Proc. Natl. Acad. Sci. USA* **90**:6444–6448.

Hoogenboom, H. R., Designing and optimizing library selection strategies for high-affinity antibodies, 1997, *TIBTECH* **15**:62–70.

Horn, U., Strittmatter, W., Krebber, A., Knupfer, U., Kujau M., Wenderoth, R., Matzku, S., Pluckthun, A., and Reisenberg, D., 1996, High volumetric yields of functional dimeric miniantibodies in *Escherichia coli*, using an optimised expression vector and high cell-density fermentation under non-limited growth conditions. *Appl. Microbiol. Biotechnol.* **46**:524–532.

Huston, J. S., McCartney, J., Tai, M-S., Mottola-Hartshorn, C., Jin, D., Warren, F., Keck, P., and Oppermann, H., 1993, Medical applications of single chain antibodies, *Int. Rev. Immunol.* **10**:195–217.

Kaufman, E. N., and Jain, R. K., 1992, Effect of bivalent interaction upon apparent antibody affinity: Experimental confirmation of theory using fluorescence photobleaching and implications for antibody binding assays, *Cancer Res.* **52**:4157–4167.

Kennel, S. J., 1991, Effects of target antigen competition on distribution of monoclonal antibody to solid tumors, *Cancer Res.* **52**:1284–1290.

Khawali, L. A., Miller, G. K., Epstein, A. L., 1994, Effect of seven new vasoactive immunoconjugates on the enhancement of monoclonal antibody uptake in tumours, *Cancer*, Supplement **73**:824–831.

Kipriyanov, S. M., Moldenhauer, G., and Little, M., 1997, High level production of soluble single chain antibodies in small scale *Escherichia coli* cultures, *J. Immunol. Methods* **200**:69–77.

Kipriyanov, S. M., Moldenhauer, G., Martin, A. C. R., Kupriyanova, O. A., and Little, M., 1997, Two amino acid mutations in an anti-human CD3 single chain Fv antibody fragment that affect the yield on bacterial secretion but not the affinity. *Protein Eng.* **10**:445–453.

Kreitman, R. J., Bailon, P., Chaudhary, V. K., Fitzgerald, D. J. P., and Pastan, I., 1994, Recombinant immunotoxins containing anti-Tac(Fv) and derivatives of *Pseudomonas* exotoxin produce complete regression in mace of an interleukin-2 receptor-expressing human carcinoma, *Blood* **83**:426–434.

Lonberg, N., Taylor, L. D., Harding, F. A., Trounstine, M., Higgins, K. M., Schramm, S. R., Kuo, C-C., Mashayekh, R., Wymore, K., McCabe, J. G., O'Regan, D. M., O'Donnell, S. L., Lapachet, E. S. G., Benoechea, T., Fishweld, D. M., Carmack, C. E., Kay, R. M., and Huszar, D., 1994, Antigen-specific human antibodies from mice comprising four distinct genetic modifications, *Nature* **368**:856–859.

Marks, J. D., Griffiths, A. D., Malmqvist, M., Clackson, T. P., Bye, J. M., and Winter, G., 1992, By-passing immunisation: Improving the affinity of a human antibody by chain shuffling, *Bio/Technology* **10**:779–783.

Marks, J. D., Hoogenboom, H. R., Bonnert, T. P., MacCafferty, J., Griffiths, A. D., and Winter, G., 1991, By-passing immunization: Human antibodies from V-gene libraries displayed on bacteriophage, *J. Mol. Biol.* **222**:581–597.

McCafferty, J., Griffiths, A. D., Winter, G., and Chiswell, D. J., 1990, Phage antibodies: Filamentous phage displaying antibody variable domains, *Nature* **348**:552–554.

McGuinness, B. T., Walter, G., Fitzgerald, K., Schuler, P., Mahoney, W., Duncan, A. R., and Hoogenboom, H. R., 1996, Phage diabody repertoires for selection of large numbers of bispecific antibody fragments, *Nat. Biotechnol.* **14:**1149–1154.

Mendez, M. J., Green, L. L., Corvalan, J. R. F., Jia, X-C., Maynard-Currie, C. E., Yang, X-D., Gallo, M. L., Louie, D. M., Lee, D. V., Erickson, K. E., Luna, J., Roy, C. M-N., Abderrahim, H., Kirschenbaum, F., Noguchi, M., Smith, D. H., Fukushima, A., Hales, J. F., Finer, M. H., Davis, C. G., Zsebo, K. M., and Jakobovits, A., 1997, Functional transplant of megabase human immunoglobulin loci recapitulates human antibody response in mice, *Nature Genet.* **15:**146–156.

Meredith, R. F., Khazaeli, M. B., Plott, W. E., Liu, T., Russell, C. D., Wheeler, R. H., and Lobuglio, A. F., 1993, Effect of human immune response on repeat courses of ^{131}I-chimeric B72.3 antibody therapy, *Antibody Immunoconjugate and Radiopharmaceuticals* **6:**39–46.

Michael, N. P., Chester, K. A., Melton, R. G., Robson, L., Nicholas, W., Boden, J. A., Pedley, R. B., Begent, R. H. J., Sherwood, R. F., and Minton, N. P., 1996, *In vitro* and *in vivo* characterisation of a recombinant carboxypeptidase G2::anti-CEA scFv fusion protein, *Immunotechnology* **2:**47–57.

Milenic, D. E., Yokota, T., Filpula, D. R., Finkelman, M. A. J., Dodd, S. W., Wood, J. F., Whitlo, M., Snoy, P., and Schlom, J., 1991, Construction, binding properties, metabolism, and tumor targeting of a single-chain Fv derived from the pancarcinoma monoclonal antibody CC49, *Cancer Res.* **51:**6363–6371.

Neri, D., Mono, M., Prospero, T., and Winter, G., 1995, High affinity antigen binding by chelating recombinant antibodies (CRAbs), *J. Mol. Biol.* **246:**367–373.

Neiba, L., Honegger, A., Krebber, C., and Pluckthun, A., 1997, Disrupting the hydrophobic patches at the antibody variable/constant domain interface: Improved *in vivo* folding and physical characterisation of an engineered scFv fragment, *Protein Eng.* **10:**435–444.

Orlandi, R, Güssow, D. H., Jones, P. T., and Winter, G., 1989, Cloning immunoglobulin variable domains for expression by the polymerase chain reaction, *Proc. Natl. Acad. Sci. USA* **86:**3833–3837.

Pack, P., and Plückthun, A., 1992, Miniantibodies: Use of amphipathic helices to produce functional, flexibly linked dimeric Fv fragments with high avidity in *Escherichia coli*, *Biochemistry* **31:**1579–1584.

Pai, L. H., Bookman, M. A., Ozols, R. F., Young, R. C., Smith, J. W., Longo, D. L., Gould, B., Frankel, A., McClay, E. F., Howell, S., Reed, E., Willingham, M. C., Fitzgerald, D. J., and Pastan, I., 1991, Clinical evaluation of intraperitoneal *Pseudomonas* exotoxin immunoconjugate OVB3-PE in patients with ovarian cancer, *J. Clin. Oncol.* **9:**2095–2103.

Pollack, S. J., Jacobs, J. W., and Schultz, P. G., 1986, Selective chemical catalysis by an antibody, *Science* **234:**1570–1573.

Press, O. W., Eary, J. F., Appelbaum, F. R., Martin, P. J., Badger, C. C., Nelp, W. B., Glenn, S., Butchko, G., Fisher, D., Porter, B., Matthews, D. C., Fisher, K. D., and Bernstein, I. D., 1993, Radiolabeled-antibody therapy of B-cell lymphoma with autologous bone marrow support, *N. Engl. J. Med.* **329:**1219–1224.

Riechmann, L., Clark, M., Waldmann, H., and Winter, G., 1988, Reshaping human antibodies for therapy, *Nature* **332:**323–327.

Reichmann, L., and Weill, M., 1993, Phage display and selection of a site directed randomised single chain antibody Fv fragment for its affinity improvement, *Biochemistry* **32:**8848–8855.

Reitmuller, G., Schneider-Gadicke, E., Schlimok, G., Schmiegel, W., Raab, R., Hoffken, K., Gruber, R., Pichlmaier, H., Hirche, H., Pichlmayr, R., Buggisch, P., Witte, J., and the German Cancer Aid 17-1A Study Group, 1994, Randomised trial of monoclonal antibody for adjuvant therapy of resected Duke's C colorectal carcinoma, *Lancet* **343:**1177–1183.

Roberts, S., Cheethen, J. C., and Rees, A. R., 1987, Generation of an antibody with enhanced affinity and specificity for its antigen by protein engineering, *Nature* **328:**731–734.

Rodrigues, M. L., Presta, L. G., Kotts, C. E., Wirth, C., Mordenti, J., Osaka, G., Wong, W. L. T.,

Nuijens, A., Blackburn, B., and Carter, P., 1995, Development of a humanised disulphide stabilised anti-p185^{ER2} Fv-β-lactamase fusion protein for activation of a cephalosporin doxorubicin prodrug, *Cancer Res.* **55:**63–70.

Ruguska, M. A., Pedersen, J. T., Keddy, C. A., Henry, A. H., Searle, S. J., Lambert, J. M., Goldmacher, V. S., Blattler, W. A., Rees, A. R., and Guild, B. C., 1994, Humanisation of murine monoclonal antibodies through variable domain resurfacing, *Proc. Natl. Acad. Sci. USA.* **91:**969–973.

Rowlinson-Busza, G., Deonarain, M. P., and Epenetos, A. A., 1996, Comparison of intact monoclonal antibody, its $F(ab')_2$ and Fab fragments and recombinant single chain Fv in a human tumor xenograft model, *Tumor Targeting* **2:**37–48.

Savage, P., So, A., Spooner, R. A., and Epenetos, A. A., 1993, A recombinant single chain antibody interleukin-2 fusion protein, *Br. J. Cancer* **67:**304–310.

Schlom, J., Eggensperger, D., Colcher, D., Molinolo, A., Houchens, D., Miller, L. S., Hinkle, G., and Siler, K., 1992, Therapeutic advantage of high-affinity anticarcinoma radioimmunoconjugates, *Cancer Res.* **52:**1067–1072.

Sharma, S. K., Bagshawe, K. D., Melton, R. G., and Sherwood, R. F., 1993, Human immune response to monoclonal antibody-enzyme conjugates in ADEPT, *Cell Biophys.* **21:**109–120.

Sharma, S. K., Bagshawe, K. D., Springer, C. J., Antoniw, P., Boden, J. A., Rogers, G. T., Burke, P. J., Melton, R. G., and Sherwood, R. F., 1991, Antibody directed enzyme prodrug therapy (ADEPT)—a three phase system. *Dis. Markers* **9:**225–231.

Soumillion, P., Jespers, L., Bouchet, M., Marchand-Brynaert, J., Winter, G.,, and Fastrez, J., 1994, Selection of beta-lactamase on filamentous bacteriophage by catalytic activity, *J. Mol. Biol.* **237:** 415–422.

Sutherland, R., Buchegger, F., Schreyer, M., Vacca, A., and Mach, J., 1997, Penetration and binding of radiolabelled anti-carcinoembryonic antigen monoclonal antibodies and their antigen binding fragments in human colon cancer multicellular tumor spheroids, *Cancer Res.* **47:**1627–1633.

Thorpe, P. E., and Burrows, F. J., 1995, Antibody-directed targeting of the vasculature of solid tumors, *Breast Cancer Res. Treat.* **36:**237–251.

Tramontano, A., Janda, K. D., and Lerner, R. A., 1986, Chemical reactivity at an antibody binding site elicited by mechanistic design of a synthetic antigen, *Proc. Natl. Acad. Sci. USA* **83:**6736–6740.

Tramontano, A., Janda, K. D., and Lerner, R. A., 1985, Catalytic antibodies, *Science* **234:**1566–1570.

Urban, J. L., and Schreiber, H., 1992, Tumor antigens, *Annu. Rev. Immunol.* **10:**617–644.

Vaughan, T. J., Williams, A. J., Prithard, K., Osbourne, J. K., Pope, A. R., Earnshaw, J. C., McCafferty, J., Hodits, R. A., Wilton, J., and Johnson, K. S., 1996, Human antibodies with subnanomolar affinities isolated from a large non-immunised phage display library, *Nat. Biotechnol.* **14:**309–314.

Velders, M. P., Vanrhijn, C. M., Cornelissen, I. M. H. A., Vanmuijen, G. N. P., Briaire, I. H., Dohlsten, M., Fleuren, G. J., Warnaar, S. O., and Litvinov, S. V., 1995, The role of monoclonal antibody affinity in tumor immunotherapy evaluated in *in vivo* models for minimal residual disease, *J. Immunother.* **19:**245–256.

Verhaar, M. J., Chester, K. A., Keep, P. A., Robson, L., Pedley, R. B., Boden, J. A., Hawkins, R. E., and Begent, R. H. J., 1995, A single chain FV derived from a filamentous phage library has distinct tumour targeting advantages over one derived from a hybridoma, *Int. J. Cancer* **61:**497–501.

Viale, G., Grassi, F., Pelagi, M., Alzani, R., Menard, S., Miotti, S., Buffa, R., Gini, A., and Siccardi, A. G., 1987, Anti-human tumor antibodies induced in mice and rabbits by "internal image" anti-idiotypic monoclonal immunoglobulins, *J. Immunol.* **139:**4250–4255.

Vriesendorp, H. M., Herpst, J. M., Germack, M. A., Klein, J. L., Leichner, P. J., Loudenslager, D. M., and Order, S. E., 1991, Phase I–II studies of yttrium-labelled anti-ferritin treatment for end-stage Hodgkin's disease, including radiation therapy oncology group 87-01, *J. Clin. Oncol.* **9:**918–928.

Ward, E. S., Güssow, D., Griffiths, A. D., Jones, P. T., and Winter, G., 1989, Binding activities of a repertoire of single immunoglobulin variable domains secreted from *Escherichia coli*, *Nature* **344:**544–546.

Wei-Ping, Y., Green, K., Pinz-Sweeny, S., Briones, A. T., Burton, D. R., and Barbas, C. F. III, 1995, CDR walking mutagenesis for the affinity maturation of a potent human anti-HIV antibody into the picomolar range, *J. Mol. Biol.* **254:**392–403.

Wentworth, P., Datta, A., Blakey, D., Boyle, D., Partridge, L. J., and Blackburn, M., 1995, Toward antibody-directed "abzyme" prodrug therapy, ADAPT: Carbamate prodrug activation by a catalytic antibody and its *in vitro* application to human tumour cell killing, *Proc. Natl. Acad. Sci. USA* **93:**799–803.

Winter, G., Griffiths, A. D., Hawkins, R. E., and Hoogenboom, H. R., 1994, Making antibodies by phage display technology, *Annu. Rev. Immunol.* **12:**433–455.

Yokata, T., Milenic, D. E., Whitlow, M., and Schlom, J., 1992, Rapid tumor penetration of a single-chain Fv and comparison with other immunoglobulin forms, *Cancer Res.* **52:**3402–3408.

Zhou, H., Fisher, R. J., and Papas, T. S., 1994, Optimisation of primer sequences for mouse scFv repertoire display library construction, *Nucleic Acids Res.* **22:**888–889.

8

Early Clinical Studies with ADEPT

K. D. Bagshawe and M. Napier

1. INTRODUCTION

Although much experience had been gained with the agents available for ADEPT studies in xenografted mice (Bagshawe *et al.*, 1988), their translation into the clinic was in many ways a step into the unknown. It seemed possible, even likely, that new and unforeseen problems would emerge. To introduce a series of agents that would interact with each other *in vivo* was in itself a cause for some apprehension. One of the main objectives of the first study was to identify new problems, to obtain guidance on the dose levels of the various components likely to be effective, to test the potential of an immunosuppressive agent to allow more than one cycle of therapy to be given, and if possible, to demonstrate efficacy (Bagshawe *et al.*, 1995)

It had been known from an early stage in the development of ADEPT that the only prodrug available for the study (Springer *et al.*, 1990) was not ideal because the drug generated from it was not a highly active cytotoxic agent and had a long biological half-life which would allow it to escape from the tumor and access hematopoietic tissue. There was also uncertainty about the effectiveness of the second antibody clearance system (Sharma *et al.*, 1994). Would it cause immune complex problems in the human? Would it reduce enzyme levels at tumor sites and reduce the turnover of prodrug to ineffective levels? There was also little information on which to base the dosage of antibody-enzyme conjugate.

K. D. Bagshawe • Charing Cross & Westminster Medical School, London W6 8RP, United Kingdom. *M. Napier* • The Meyerstein Institute of Oncology, The Middlesex Hospital, London W1N 8AA, United Kingdom.

Enzyme-Prodrug Strategies for Cancer Therapy, edited by Melton and Knox. Kluwer Academic/Plenum Publishers, New York, 1999.

2. INITIAL CLINICAL STUDIES

2.1. Study Design

Because the prodrug 4-[2 chloroethyl)2 mesyloxyethyl)amino] benzoyl-L-glutamic acid (CMDA) had not previously been given to patients, it was necessary to carry out a dose escalation study with prodrug alone.

Because the main study was to be carried out in patients with colorectal adenocarcinoma, patients with advanced colorectal cancer were entered into this dose escalation study on a fully informed voluntary basis. They had all received 5-fluorouracil and folinic acid previously and had either not responded or had relapsed. Seven patients received single or multiple doses of CMDA up to a total dosage of 2.4 g/m^2. The prodrug was given by intravenous bolus in dimethyl sulfoxide (DMSO) because it was poorly soluble in aqueous solutions. No therapeutic effects were observed. Toxicity was limited to mild nausea and vomiting at the highest dose level, and the DMSO vehicle may have contributed to this. There was no myelotoxicity. Five of the seven patients went on to receive the full therapeutic protocol. The other two were judged too ill to receive further treatment. The relatively limited supply of the prodrug inhibited further dose escalation, although in retrospect this would have been desirable.

In addition to the CMDA prodrug, the other agents available for study included the $F(ab)_2$ fragment of the anticarcinoembryonic antigen monoclonal antibody A5B7 chemically conjugated to the bacterial enzyme carboxypeptidase G_2 (CPG_2) (Melton *et al.*, 1993). The second antibody SB43 inactivates CPG_2 both *in vitro* and *in vivo*. This second antibody was galactosylated (SB43-gal) because it had been shown in mouse studies that the galactosylated antibody still binds CPG_2 and clears rapidly from the blood via hepatic galactose receptors without reducing tumor enzyme levels.

The patients all had histologically proven adenocarcinoma of the lower gastrointestinal tract. If CEA levels in blood were not increased, then immunocytochemical evidence of CEA expression was required. All had progressive disease after failing conventional cytotoxic therapy.

The anti-CEA-CPG_2 conjugate was given at a dose level of 20,000 enzyme units/m^2 (250 mg/m^2) by 2 hr IV infusion. The plasma levels of enzyme were similar to those found optimal in the nude mouse for the same prodrug. The second antibody was given 24 hr later by slow intravenous infusion, initially over 3 hr. Some of the conjugate was labeled with ^{131}I, and immunoscintigraphy confirmed tumor localization. The first group of six patients received the CMDA prodrug by bolus injections in DMSO starting 72 hr after the conjugate. The total dose of drug ranged from 1.4 to 3.9 g/m^2.

2.2. Results

The three patients who received less than 2.0 g/m^2 of prodrug had minimal toxicity and no objective response. Of the three patients who received 2.7–3.9 g/m^2 of prodrug, two had partial responses, and one a mixed response. These patients had nausea and vomiting, and one of them had grade 4 myelotoxicity.

The second group consisted of five patients in whom infusion of prodrug was attempted. Technical problems occurred at this time. Loss of the gamma camera facility at this time removed our capability of monitoring localization of the conjugate, and this, unfortunately, coincided with a defective batch of conjugate. The batch had proved satisfactory with respect to antigen binding and catalytic activity but had not undergone *in vivo* testing for its ability to localize in CEA-producing tumors. Subsequent *in vivo* testing in mice confirmed poor localization of the conjugate.

So the important lesson was learned that *in vivo* testing of each batch of conjugate is essential. Another technical problem in this group resulted from the incompatibility of the DMSO solutions with several makes of plastic syringe. The one patient in the group who received an effective dose of prodrug (4.0 g/m^2) by infusion over five days had a partial response with $> 50\%$ reduction in all her pulmonary metastases at three months and stable disease for 22 months.

A third group consisted of six patients. It was clear by this stage that a single cycle of treatment with the CMDA prodrug was not likely to effect long lasting remissions in patients with very large tumor masses. It had also been shown, as expected, that the murine antibody and bacterial enzyme were potent immunogens and host antibody responses to both components of the conjugate were present in patients' blood by day 9 or 10, and titers rose rapidly thereafter. It was known from previous studies with murine antibodies that the host antibody response could be delayed by cyclosporin. Therefore it was decided to introduce cyclosporin into the protocol with the objective of giving more than one cycle of treatment

Two patients referred at this time had massive hepatic metastases. One patient's liver at autopsy weighed > 7 kg. These patients received cyclosporin by mouth starting 48 hr before receiving the conjugate. The blood levels of cyclosporin fluctuated greatly, and before they were able to receive a second cycle of treatment, hepatorenal failure occurred, and both patients died.

Subsequent patients received cyclosporin by intravenous infusion, and blood levels of cyclosporin remained mainly in the concentration range 150–350 μg/l. In contrast to all other patients in the study, those who received cyclosporin did not have detectable antibodies in their blood directed at mouse immunoglobulin and carboxypeptidase G_2 until 16–20 days after first receiving the conjugate. One patient tested a year later had maintained a high level of IgG to CPG_2 and mouse IgG.

Two patients who received cyclosporin received two cycles of therapy and two received three cycles. Two patients in this group were treated with reduced amounts of antibody–enzyme conjugate, and neither had significant myelotoxicity. One of them survived three years after treatment.

Overall, there were eight evaluable patients who received what was judged to be adequate dosage with conjugate and prodrug. Four of these had partial responses assessed radiologically, and one had a mixed response. The latter patient had metastases in porta hepatis lymph nodes at his original laparotomy, and one metastasis was noted on immunohistochemical examination to be CEA negative. He had multiple hepatic metastases at the time he received ADEPT, and all but one of these regressed during treatment. The results of this preliminary study have been reported in detail (Bagshawe *et al.*, 1995).

It was clear by the end of this first study that further studies were required where measurements of enzyme concentrations and pharmacokinetic studies could be undertaken routinely with HPLC.

3. SECOND CLINICAL STUDY

3.1. Study Design

The pharmacokinetic data available from the initial study supported the hypothesis that ADEPT has the potential to overcome some of the principal limitations of conventional antibody-targeted therapy of cancer. Although samples were collected systematically, the techniques of measurement were still being developed, and the regimen was changed during this early dose-finding study. There was a need for complete pharmacokinetic data in a group of patients receiving the same regimen. The experience of toxicity and efficacy was used to design a regimen for the second clinical trial which was undertaken at the Royal Free Hospital, London. The fundamental ADEPT approach and agents of the first trial were adhered to, that is, antibody-enzyme-conjugate (A5B7 $F(Ab')_2$-CPG_2), clearing second antibody (SB43-gal) and prodrug (CMDA). Bolus administration of the prodrug was used to allow detailed assessment of its pharmacokinetic characteristics. The complete treatment regimen is given in Table I.

Ten patients (six male, four female, median age 46 years) with advanced colorectal cancer were treated. Patient entry criteria were similar to the first clinical trial, particularly, raised serum CEA levels or immunocytochemical evidence of tumor CEA production. Plasma was collected systematically for evaluation of enzyme, prodrug, and active drug concentrations, as measured by spectrophotometric assay (McCulloch *et al.*, 1971) and the new HPLC technique (Martin *et al.*, 1997). Tumor biopsies were obtained under ultrasound guidance in five

Table I. Treatment Schedule

Day −3	Insertion of triple lumen Hickman catheter into subclavian vein
Day −2	Cyclosporin A (CyA) 5 mg/kg over 12 hr by continuous IV infusion in 5% dextrose; continue for 7 days; to be followed by oral administration: 15 mg/kg/day in divided doses for 7 days; dose to be adjusted to keep CyA levels in the range 150–350 ng/ml
Day −1	Potassium iodide 50 mg qds po for 10 days
Day 0	10,000 enzyme units/m^2 (125 mg/m^2) of A5B7 F(ab′)2 antibody-CPG2 conjugate in 500 ml of 0.9% saline over 2 hr; 10 mg of conjugate labeled with 370 MBq of ^{131}I given as an IV bolus at the end of conjugate infusion
Day +1	Infusion of SB43-gal antibody to CPG2; 7.5 mg/m^2 in 500 ml 0.9% saline over 6 hr; 15 mg/m^2 in 500 ml 0.9% saline over 18 hr; continued to day +4 at 8 mg/m^2 in 1L 0.9% saline over 24 hr
Day +2	CMDA prodrug 200 mg/m^2 infused IV over 4 min in fast running 1.26% $NaHCO_3$ (if CPG2 enzyme levels measured in plasma < 0.2 enzyme units/ml)
Day +3 and +4	Repeat Day +2

patients for direct measurement of enzyme levels by HPLC (Stribbling *et al.*, 1997). The biopsies were intentionally timed to be taken immediately after the first prodrug administration. It was also hoped to submit these biopsies for prodrug and active drug concentration analysis, but the small size of the biopsies did not allow this. Future studies need to address this fundamental point. As with the first group of patients treated at the Charing Cross Hospital, serial gamma camera imaging allowed estimations of tumor and normal tissue enzyme concentrations at various times, and these were correlated with plasma and tumor biopsy results. There was scope within the protocol to allow a second cycle of therapy if HAMA (human antimouse antibody) was not detected after the first cycle. Two patients went on to receive two treatments.

3.2. Results

Antibody-enzyme conjugate (A5CP) cleared rapidly from plasma after administering the clearing antibody such that the concentration of enzyme was less than the limits of detection by HPLC 48 hr after A5CP administration (Fig. 1). Quantitative SPECT gamma camera imaging (Green *et al.*, 1990) showed ^{131}I-A5CP localization in tumor which reached median peak levels of 6.8% of the injected radioactivity kg^{-1} 4 hr after administration. Enzyme activity measured indirectly in the five tumor biopsies taken immediately after the start of prodrug administration showed that the median tumor enzyme concentration was 0.47 (range 0.32–0.62) enzyme units g^{-1} which correlated well with the figure of 0.34 (range 0.19–0.63) calculated from the amount of radioactivity measured by

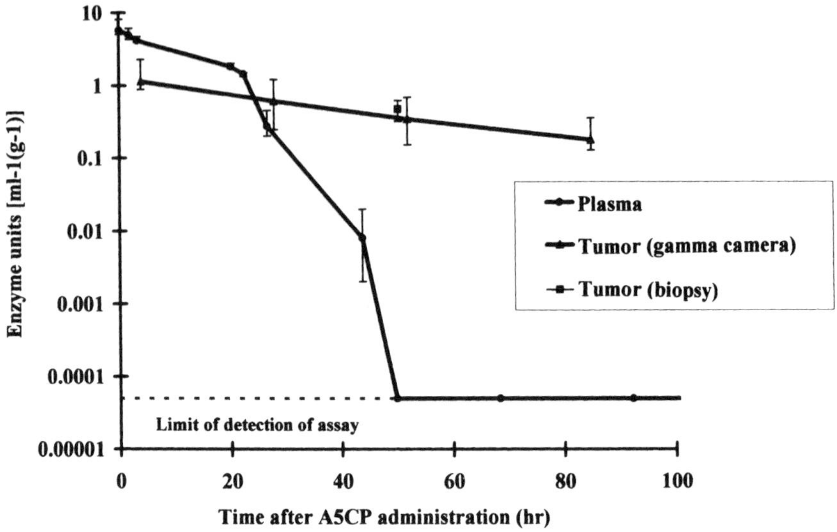

Figure 1. Enzyme levels in plasma (●) and tumor (△) following infusion of antibody-enzyme conjugate. Plasma levels are measured by methotrexate-reduction spectophotometric assay and HPLC. The dashed line indicates the lower limits of detection of this assay. Levels in tumor were estimated from SPECT gamma camera imaging. Each point is the median of 10 patients. Range includes 60% of data points. (The enzyme level in tumor (□) obtained from tumor biopsy at 50.3 hr is the median of five patients and range includes 60% of data points.)

gamma camera imaging (Fig. 1). This showed that enzyme in tumor was inaccessible to inactivation by SB43-gal and that gamma camera estimates of enzyme concentration in tumor were valid. Normal liver was obtained in biopsies of two patients and no enzyme activity was detectable in them by HPLC.

From this data tumor to plasma ratios of enzyme were >10000:1 (based on levels in plasma below the limits of detection) at the time when prodrug administration was started. The validation of gamma camera estimates of enzyme concentration given by the biopsy studies supports the view that effective enzyme levels (based on data from the first clinical trial) were found in tumor at least up to 85 hr. Tumor to liver ratios in the two patients biopsied also exceeded 10000:1.

Plasma prodrug assays suggested that the prodrug cleared from plasma with a biological half-life of 16 (range 5–27) min. Drug was detected in all cases within three minutes of the end of prodrug adminstration. The half-life recorded for drug was longer than for the prodrug at 46 (range 7–85) min. There was substantial variation in concentrations of prodrug and drug and their clearance characteristics among patients and between days in the same patient. No credible reason for this has been found. This data suggested that prodrug was activated by enzyme, and the available information on enzyme distribution suggest that this was principally occurring in the tumor with diffusion of drug into the circulation.

Neutropaenia and thrombocytopaenia were the principal toxicities with this therapy and occurred with median times to nadir of 35 and 25 days, respectively. Four patients required platelet transfusion. One patient had a febrile neutropaenic episode requiring hospitalization and antibiotic therapy. Nine patients had severe nausea or vomiting which, like the first trial, was attributable to the combination of DMSO and cyclosporin administered. Human antimouse antibody and human anti-CPG_2 antibody were found in all patients after two weeks and prevented further therapy.

As in the Charing Cross trials, there was evidence of antitumor activity. One patient with recurrent rectal cancer in his liver and stoma had a partial response (with complete resolution of two small liver metastases) lasting four months. Stabilization of disease after previous tumor progression was seen in six other patients who had median progression-free survival of four months (range 2–16), and one of these had a fall in circulating serum CA19/9 levels sustained for four months. Only three of the ten patients treated had outright progressive disease within two months of ADEPT.

These small trials show that conditions needed for effective ADEPT were achieved in patients and substantial tumor responses were achieved. The second study used a smaller conjugate dose than the first clinical trial and also used the reduced dose of clearing antibody used in the later stage of the first trial. In the first trial plasma enzyme levels were analyzed by spectrophotometric analysis alone (McCulloch *et al.*, 1971). It was clear that plasma enzyme levels below the limits of assay by this technique could also activate prodrug. The new assay procedures used in the second trial allowed detection to a far lower enzyme concentration. This analysis has shown that even the lower dose of clearing antibody was extremely efficient in clearing the circulating conjugate.

In this study we have come very close to eliminating active enzyme from the nontumor compartments. The HPLC/spectroscopy studies of tumor and liver biopsies and of plasma only demonstrated active enzyme in tumor. This makes it unlikely that significant enzyme remains in normal tissues at the time of prodrug administration. However, drug was found in plasma at all times after prodrug administration, and evidence was also found of plasma-derived toxicity (i.e., myelosuppression). Therefore the drug in plasma was probably generated in tumor and was allowed to leak back into plasma by virtue of a long biological half-life. It is also interesting that the two patients with the highest plasma drug levels had large burdens of tumor. Treating smaller tumors may be advantageous because the potential for releasing drug into circulation would be reduced in proportion to the targeted tumor volume, and drug concentration in tumor would be similar. Prodrugs that generate drugs with much shorter half-lives have also been made and should also help to minimize this problem.

The decision to use bolus administrations of CMDA allowed detailed pharmacokinetic analysis. The characteristics of the CPG2 enzyme in prodrug activation are known (Springer *et al.*, 1991), and in this study it was shown that the optimal level of prodrug concentration for conversion was maintained in the plasma for approximately 2 hr after each bolus administration. The IC_{50} of the drug was also never exceeded in plasma at any time in any of the study patients. Despite this, drug in circulation appeared to cause myelosuppression. Given that no active enzyme was found in plasma at any of the times when prodrug was given, then it is possible that drug in circulation can be explained by "leak back" from tumor. It is also probable that higher concentrations of drug were present in tumor than were measured in blood, and it is possible that a change in prodrug regimen, for instance, to an infusion over three or more days would prevent the rise of plasma drug concentration to a toxic level. The relatively long half-life of drug in plasma is consistent with it causing the dose-limiting myelosuppression. Clinical studies of ADEPT are planned with the same antibody and enzyme but with a shorter half-life drug which has shown effective therapy for colorectal cancer in animal models (Blakey *et al.*, 1996).

4. CONCLUSIONS

Despite the optimism engendered by these early trials, a number of problems have been highlighted. The immunogenicity of the components does not allow for repeated therapies which has been the cornerstone of the successes of modern oncology. The immunogenicity of CPG2 is based on its bacterial origin, but it has the important advantage over most mammalian enzymes that it has no human equivalent which could activate prodrug endogenously. Administration of cyclosporin antibody delays the production of a human antibody response to the A5B7 antibody and CPG2 but still allows only two to three therapies. Other immunosuppressive agents or drugs which induce tolerance may further delay the antibody response (Deierhoi *et al.*, 1993; Dhingra *et al.*, 1995). Humanization of the antibody may reduce immunogenicity, and it is possible that less immunogenic enzymes can be identified.

These studies show how measurement of the parameters required for the function of ADEPT gives insight into its ability for selective delivery of cancer therapy. They also show how knowledge of the effect and interactions of the components can lead to redesign of clinical protocols aiming to overcome them. The results of these early clinical trials of ADEPT suggest that it is a significant new treatment for colorectal cancer and can also be developed for other cancers many of which can be targeted now with antibodies.

REFERENCES

Bagshawe, K. D., Sharma, S. K., Springer, C. J., and Antoniw, P., 1995, Antibody directed enzyme prodrug therapy: a pilot-scale clinical trial, *Tumor Targeting* **1:**17–29.

Bagshawe, K., Springer, C. J., Searle, F., Antoniw, P., Sharma, S. K., Melton, R. G., and Sherwood, R. F., 1988, A cytotoxic agent can be generated selectively at cancer sites, *Br. J. Cancer* **58:**700–703.

Blakey, D. C., Burke, P. J., Davies, D. H., Dowell, R. I., East, S. J., Eckersley, K. P., Fitton, J. E., Mcdaid, J., Melton, R. G., Niculescu-duvaz, I. A., Pinder, P. E., Sharma, S. K., Wright, A. F., and Springer, C. J., 1996, ZD2767, an improved system for antibody-directed enzyme prodrug therapy that results in tumor regressions in colorectal tumor xenografts, *Cancer Res.* **56:**3287–3892.

Blakey, D. C., Valcaccia, B. E., East, S., Wright, A. F., Boyle, F. H., Springer, C. J., Burke, P. B., Melton, R. G., and Bagshawe, K. D., 1993, Antitumour effects of an antibody-carboxypeptidase G_2 conjugate in combination with a benzoic acid mustard prodrug, *Cell Biophys.* **22:**1–8.

Deierhoi, M. H., Kauffman, R. S., Hudson, S. L., Barber, W. H., Curtis, J. J., Julian, B. A., Gaston, R. S., Laskow, D. A., and Diethelm, A. G. (1993) Experience with mycophenolate mofetil (RS61443) in renal transplantation at a single center, *Ann. Surg.* **217:**476–482.

Dhingra, K., Fritsche, H., Murray, J. L., Lobuglio, A. F., Khazaeli, M B., Kelley, S, Tepper, M. A., Grasala, D., Buzdar, A., and Valero, V., 1995, Phase I clinical and pharmacological study of suppression of human antimouse antibody response to monoclonal antibody L6 by deoxyspergualin, *Cancer Res.* **55:**3060–3067.

Green, A. J., Dewhurst, S. E., Begent, R. H. J., and Riggs, S. J., 1990, Accurate quantitation of 131I distribution by gamma camera imaging, *Eur. J. Nucl. Med.* **16:**361–365.

Martin, J., Stribbling, S. M., Poon, G. K., Begent, R. H. J., Napier, M., Sharma, S. K., and Springer, C. J., 1997, Antibody-directed enzyme prodrug therapy: Pharmacokinetics and plasma levels of prodrug and drug in phase I clinical trial, *Cancer Chemother. Pharmacol.* **40:**189–201.

McCulloch, J. L., Chabner, B. A., and Bertino, J. R., 1971, Purification and properties of carboxypeptidase Gl, *J. Biol. Chem.* **246:**7207–7213.

Melton, R. G., Boyle, J. M. B., Rogers, G. T., Burke, P., Bagshawe, K. D., and Sherwood, R. F., 1993, Optimisation of small-scale coupling of A5B7-monoclonal antibody to carboxypeptidase G_2, *J. Immunol. Methods* **158:**49–56.

Sharma, S. K., Bagshawe, K. D., Burke, P. J., Boden, J. A., Rogers, G. T., Springer, C. J., Melton, R. G., and Sherwood, R. F., 1994, Galactosylated antibodies and antibody-enzyme conjugates in antibody-directed enzyme prodrug therapy, *Cancer* **73:**1114–1120.

Springer, C. J., Antoniw, P., Bagshawe, K. D., Searle, F., Bisset, G. M. F., and Jarman, M., 1990, Novel prodrugs which are activated to cytotoxic alkylating agents by carboxypeptidase G2, *J. Med. Chem.* **33:**677–681.

Springer, C. J., Antoniw, P., Bagshawe, K. D., and Wilman, D. E. V., 1991, Comparison of half-lives and cytotoxicity of *N*-mesyloxyethylbenzoyl compounds, products of prodrugs in antibody-directed enzyme prodrug therapy (ADEPT), *Anticancer Drug Des.* **6:**467–479.

Stribbling, S. M., Martin, J., Pedley, R. B., Boden, J. A., Sharma, S. K., and Springer, C. J., 1997, Biodistribution of an antibody-enzyme conjugate for antibody-directed enzyme prodrug therapy in nude mice bearing a human colon adenocarcinoma xenograft, *Cancer Chemother. Pharmacol.* **40:**277–284.

9

Gene-Directed Enzyme Prodrug Therapy (GDEPT) of Cancer

Richard J. Knox

1. INTRODUCTION

A new gene therapy approach for targeting cancer cells and making them sensitive to prodrugs has been proposed for human gene therapy trials (Hart and Vile, 1995; Oldfield *et al.*, 1993). Prodrug gene therapy, commonly referred to as VDEPT (virus-directed enzyme prodrug therapy) or gene-directed enzyme prodrug therapy (GDEPT) is based on the premise that a large therapeutic benefit can be gained by transferring a drug susceptibility gene into tumor cells. The gene encodes an enzyme that catalyzes the activation of a prodrug to its cytotoxic form (Fig. 1) and has been termed a "suicide gene" (Consalvo *et al.*, 1995; Freemen *et al.*, 1996; Mullen, 1994; Mullen *et al.*, 1994). However, in contrast to cytotoxic gene therapy approaches that involve expression of a toxic product [for example, diphtheria toxin (Cook *et al.*, 1994; Maxwell *et al.*, 1991, 1992; Robinson and Maxwell, 1995)], the enzyme itself is not toxic. Thus, in GDEPT, cytotoxicity results only after administering the prodrug. There is also a bystander effect because the active drug can migrate into nontransduced cells.

GDEPT is obviously related to antibody-directed enzyme prodrug therapy (ADEPT) in that both use an nonendogenous enzyme to activate a prodrug. However, there are fundamental differences between the two techniques and this means that the enzyme/prodrug combinations used for ADEPT may not be suitable for GDEPT. In GDEPT the prodrug activating enzyme is expressed inside the

---

Richard J. Knox • Department of Medical Oncology, Charing Cross Hospital, London W6 8RP, United Kingdom.

Enzyme-Prodrug Strategies for Cancer Therapy, edited by Melton and Knox. Kluwer Academic/Plenum Publishers, New York, 1999.

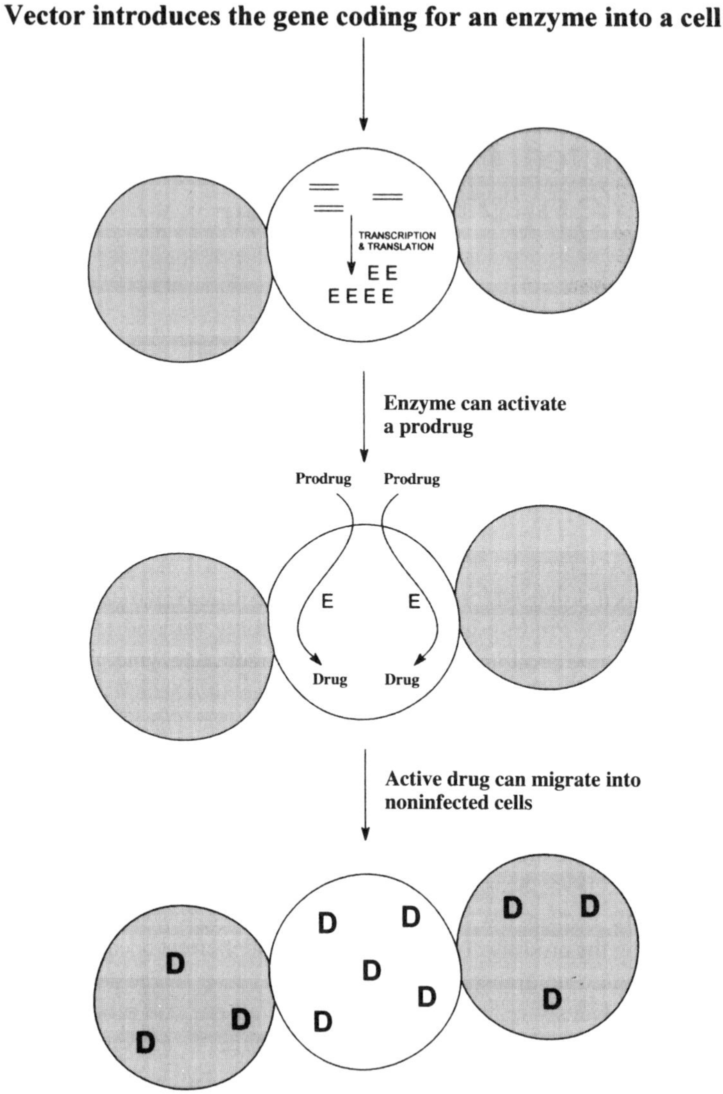

Figure 1. The generation of a cytotoxic drug by GDEPT. In the first phase the cell is transduced with a gene that codes for a prodrug activating enzyme. This enzyme (**E**) is expressed, and a prodrug is administered which is converted to an active drug (**D**) by the bound enzyme. Importantly the active drug can migrate and have cytotoxic effects on cells that have not been transduced.

cell whereas in ADEPT the enzyme is extracellular. Thus, for GDEPT the prodrug must be able to enter the cell. Therefore, enzyme prodrug systems developed for ADEPT that rely on excluding a charged prodrug (such as a glucuronide) from the cell, while the active drug is not, are not directly applicable to GDEPT. However, it should be noted that a GDEPT system that allows extracellular expression of the activating enzyme has now been described.

This chapter describes the various enzyme/prodrug systems that have been described for GDEPT. However, success for this technique requires a choice of an enzyme/prodrug system and also a delivery system by which the gene that encodes for the enzyme can be delivered efficiently and accurately to a human tumor. This aspect is common to all types of gene therapy and has been very well reviewed. At present, targeting is achieved by altering the surface components of viruses and liposomes to achieve a level of target cell recognition so that transcriptional elements can be incorporated and the incorporated gene is expressed only in the target cells. Retroviral based vectors can infect only replicating target cells, but integration occurs leading to stable expression of the transduced gene. Adenoviral vectors are more efficient at transduction and do not require cell division. But they do not integrate, so it is very difficult to regulate gene expression. However, for the long term success of GDEPT (and any gene therapy procedure) advances in these systems are required particularly to improve their targeting. It has been proposed that future gene therapy vectors will be not based on any single virus but will be synthetic vehicles, custom designed to incorporate specific targeting features relevant to the target disease and tissue (Fig. 2) (Miller and Vile, 1995).

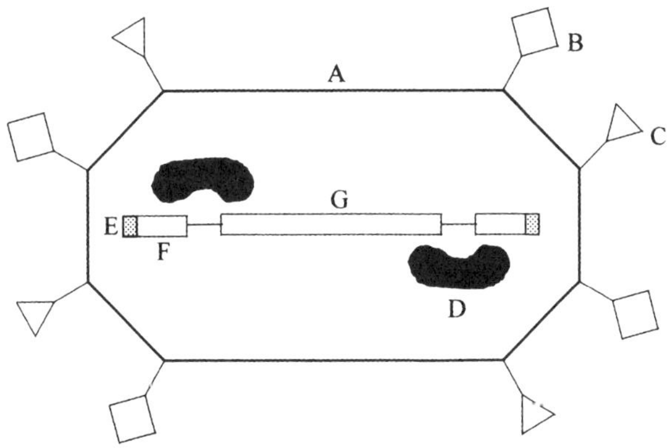

Figure 2. An ideal synthetic gene therapy vector as proposed by Miller and Vile (1995). Its features include (A) a stable nonimmunogenic lipid envelope; (B) ligands to confer a particular affinity on the vector; (C) moieties to encourage fusion with the target cell membrane; (D) site-specific recombinase for direct integration of the vector DNA; (E) sequences to enable homologous recombination of the vector DNA and a specific target loci of the genome; (F) tissue-specific promoter region to confer restricted expression on the therapeutic gene; and (G) the therapeutic DNA. Certain features of the vector, such as homologous recombination, are not required for GDEPT.

As with ADEPT, a number of different enzyme/prodrug combinations have been proposed for GDEPT. Again, as in ADEPT, an enzyme is required for which there is no endogenous activity. Bacterial or viral enzymes are particularly suited to ADEPT. First, many of these enzymes have no corresponding activity in humans. Secondly, because expression is intracellular, there should be no problems with an immune response against a "foreign" protein. Thirdly, bacterial enzymes have few requirements for posttranslational modification, and more complex proteins may not fold correctly in different species. Because activation takes place intracellularly, the prodrug should be freely diffusible, and thus should be a neutral species with an appropriate partition coefficient. Because not all of the tumor cells are transduced by the vector, a bystander effect is required. Thus, the active drug should be able to diffuse away from the site of activation. However, its half-life needs to be such that it cannot migrate and affect normal tissue. The active drug should also not be phase-specific proliferation-dependent because many tumors have a low mitotic index (Mendelsohn, 1960), and all malignant cells need to be killed, not just those that are proliferating.

2. WHY GDEPT?

Cancer gene therapy strategies have been divided into two categories. The first is correction of the genetic defect in tumor cells and the second is killing them. Cell killing can be mitigated either by the host immune system or by direct cytotoxic effects. Each of these approaches imposes different requirements on the delivery system, as summarized in Table I. Corrective gene therapy may not be a feasible approach for curing cancer because most tumor cells contain multiple genetic mutations. However, if, for example, a mutant oncogene were a realistic target for corrective therapy, what is required from the delivery system? First, it needs to deliver a correct copy to every tumor cell—it must be very efficient because gene correction works only in cells that receive the gene and will have no

Table I. The Requirements of the Delivery System for the Various Types of Gene Therapy[a]

Requirement of delivery system	Type of gene therapy			
	Gene correction	Cytotoxic	Immunotherapy	GDEPT
Accuracy	+	+++	+	++
Efficiency	+++	+++	+	++
Stability	+++	+	+	+

[a]Modified from Vile and Russell, 1994.

effect on those on do not. Secondly, expression of the corrective gene needs to be stable because the mutant phenotype will reappear once the corrective gene ceases to be expressed. However, thirdly, delivery need not be accurate and specific to tumor cells because, in theory, delivery of a therapeutic gene (such as a functional tumor suppresser gene) should not be deleterious to a normal cell. Opposite demands are made when a cytotoxic gene (coding for, for example, diphtheria toxin) is delivered. Stable expression is not required, but delivery must be accurate because expression would be deleterious to a normal cell. Immunotherapy puts the fewest constraints on the delivery system. In theory, immunotherapy attempts to activate an immune response by stimulating the immune system to recognize tumor-specific antigens as "non-self" by expressing immunomodulary genes, such as IL-2. These antigens are not present on normal cells. Thus, expression of immunomodulary genes should not be deleterious to them, so there is less requirement for accuracy. Expression and stability are also lesser requirements because, once activated, the immune system amplifies the antitumor response and kills any nontransduced tumor cells anywhere in the body. This represents an ideal situation, and it may be limited by autoimmunity. Further, human tumors have been extensively selected against immune recognition. It is possible to transduce genes coding for antigens but, in this case, systemic protection is lost and delivery would have to be accurate. However, the bystander effect mediated by the immune system can still be substantial.

GDEPT is a recognition of the limitations of the present delivery systems. Compared to cytotoxic gene therapy, less accuracy is required because some tumor specificity can be conferred by the pharmacology of the prodrug and its route of administration. Efficiency is improved because there is a bystander effect that can kill nontransduced tumor cells. Long-term stability is not required because transduced cells are killed after prodrug administration. Further, in contrast to all other types of gene therapy, GDEPT is controllable. The expressed enzyme is nontoxic, and cytotoxicity occurs only after administering the prodrug. Therefore, unexpected toxicity can be countered by lowering the dose of the prodrug.

3. ENZYMES AND PRODRUGS USED FOR GDEPT

3.1. Varicella-Zoster Virus Thymidine Kinase

Varicella-zoster virus produces a thymidine kinase (VZV-tk) that has substrate specificity distinct from normal mammalian cellular kinases. VZV-tk can selectively monophosphorylate 6-methoxypurine arabinoside (ara-M). The monophosphate is further metabolized by cellular enzymes (nucleoside diphosphate kinase, AMP kinase, AMP deaminase, adenylosuccinate synthetase lyase) to the

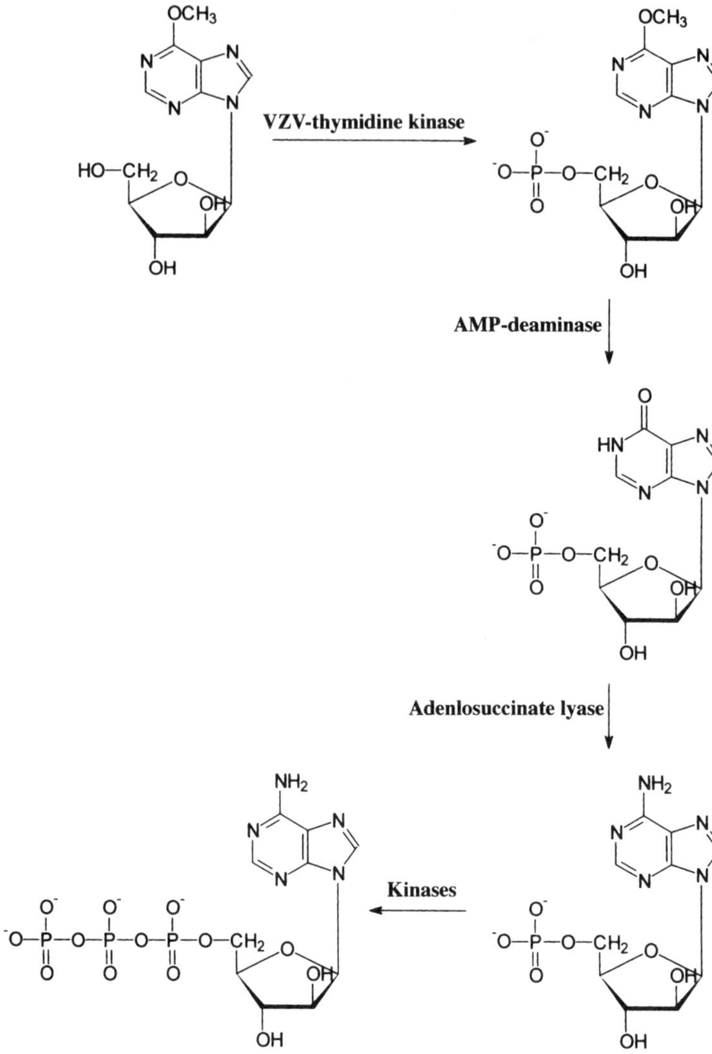

Figure 3. The activation of ara-M to a cytotoxic triphosphate, ara-ATP, by varicella-zoster virus thymidine kinase.

cytotoxic adenine arabinonucleoside triphosphate (ara-ATP) (Fig. 3), a compound that inhibits DNA synthesis. Only very low levels of the triphosphate are detected in mammalian cells in the absence of VZV-tk. It is thought that these low levels result from the mammalian enzyme adenosine deaminase that (poorly) convert araM to araH. Cellular phosphatases convert araH to the monophosphate that is on the activation pathway of araM (Huber *et al.*, 1991).

Retroviruses constructed containing the VZV-tk gene are transcriptionally regulated by either the hepatoma-associated α-fetoprotein or liver-associated albumin transcriptional regulatory sequences. Thus, subsequent to infection by the retrovirus, expression of VZV-tk protein was limited to either α-fetoprotein or albumin positive cells. Cells did selectively express the protein and thus were selectively sensitive to araM (Huber *et al.*, 1991). Untransduced cells could withstand >2000 μM araM. In transduced cells, sensitivity to the prodrug was proportional to VZV-tk activity and was less than 0.5 μM in a high expressing clone (Fig. 4) (Huber *et al.*, 1991).

This system was proposed as a model for treating hepatocellular carcinoma. The active metabolite of araM, araATP, is cytotoxic only in dividing cells. However, it was suggested that this could be an advantage. Retroviruses require dividing cells to integrate efficiently, and normal hepatocytes, which are essentially nondividing, will not be transduced efficiently. Further, even if they did integrate the gene, they would still be resistant to araM because they are not

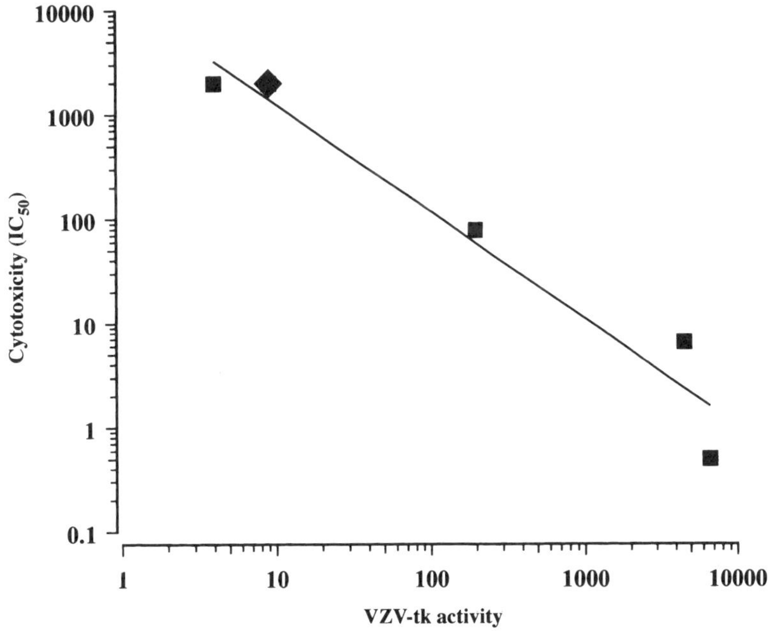

Figure 4. The correlation between VZV-tk enzyme activity and the cytotoxicity of the prodrug ara-M. Human HepG2 cells were transduced with a retrovirus containing the VZV-tk gene, transcriptionally regulated by either the hepatoma-associated α-fetoprotein or liver-associated albumin transcriptional regulatory sequences. Enzyme activity is expressed as pmoles of ara-M phosphorylated/mg protein/30 min and the IC_{50} values are for a 5 day exposure to the prodrug. Non-transduced cells (♦) have some VZV-tk but this can be ascribed to the mammalian enzyme, adenosine deaminase. Data from Huber *et al.*, 1991.

dividing (Huber *et al.*, 1991). However, many tumors have a low mitotic index, and all malignant cells need to be killed, not just those that are proliferating. No further work has been published on this system, and it appears to have been superseded by the conceptually similar HSV-tk systems described following.

3.2. Herpes Simplex Virus Thymidine Kinase

Herpes simplex virus type-1 also expresses a thymidine kinase (HSV-tk) with a substrate specificity different from cellular kinases. HSV-tk converts the purine nucleoside analogs ganciclovir (9-[2-hydroxy-1(hydroxymethyl)ethoxy]methyl guanine) (1, Fig. 5), acyclovir (9-[2-hydroxyethoxy] guanine) (2, Fig. 5) and FIAU (1[2-deoxy-2-fluoro-β-D-arabinofuranosyl]-5-iodouracil) (3, Fig. 5) into monophosphorylated compounds (Fig. 5). Mammalian kinases can convert these into nucleotide triphosphates (Fig. 5) that inhibit DNA replication and act as a chain terminator for DNA synthesis. However, the prodrugs are only poor substrates for mammalian nucleotide monophosphate kinase, and these compounds have been used clinically to treat herpes.

HSV-tk/gancilovir is the most described GDEPT system. Delivery of the gene coding for HSV-tk to animal tumors *in vivo* has been achieved using retroviruses, adenoviruses, and naked DNA. Retroviral vectors are being used to deliver the gene to intracranial and leptomeningeal tumors in rats, and the system is proposed for the first clinical trials of GDEPT (Hart and Vile, 1995; Oldfield *et al.*, 1993). Transfer of HSV-tk in tumor cells confers sensitivity to ganciclovir and its analogs both *in vitro* and *in vivo* (Fig. 6). *In vitro*, cells expressing HSV-tk are generally inhibited by 1–50 μM ganciclovir, and these levels can be achieved by patients treated with the drug (Abe *et al.*, 1993; Barba *et al.*, 1993; Borrelli *et al.*, 1988; Chen *et al.*, 1994; Moolten *et al.*, 1992; Smythe *et al.*, 1994; Tanaka *et al.*, 1996; Tong *et al.*, 1996a,b; Vile and Hart, 1993). In the absence of prodrug, expression of HSV-tk is not detrimental and expressing cells grow normally *in vivo* and *in vitro*. Transgenic mice that constitutively express HSV-tk have been produced (Borrelli *et al.*, 1988). Because of the tissue specificity of the promoter, the enzyme was expressed only in the lymphoid organs. Following a 7-day infusion of FIAU the spleen and thymus were visually atrophied, but the other organs were normal (Borrelli *et al.*, 1988). Such studies demonstrate *in vivo* the feasibility of GDEPT therapy and show that tissue specific promoters do regulate expression of the introduced gene.

A more clinically relevant model for GDEPT has been developed for brain tumors. These tumors are a good model because, apart from the tumor, the cellular elements of the brain are nondividing. Because dividing cells are more easily transduced, the tumor cells would be targeted more directly. Established gliomas

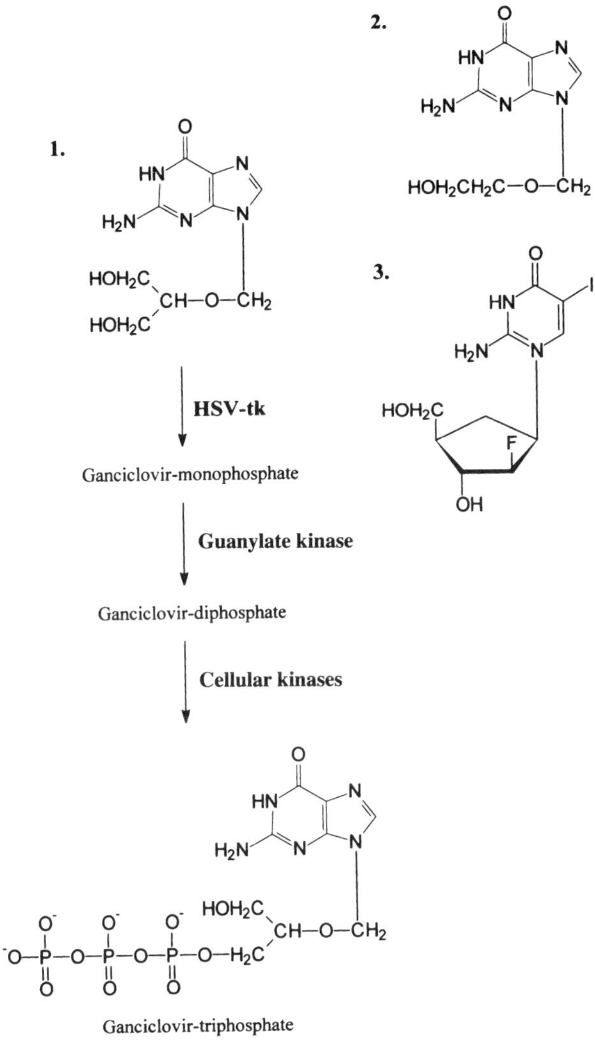

Figure 5. The activation of ganciclovir (1) by herpes simplex virus thymidine kinase (HSV-tk). Cellular kinases can convert the monophosphate into the cytotoxic triphosphate. Acyclovir (2) and FIAU (3) are activated by the same metabolic route.

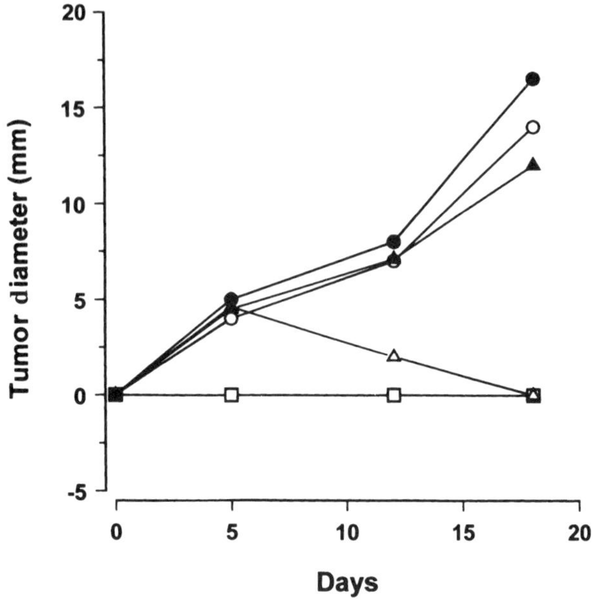

Figure 6. The antitumor effect of gancilovir on HSV-tk expressing tumor cells. Mice were inoculated s.c. with KBALD tumor cells (2×10^5) either stably transduced with the HSV-tk gene (open symbols) or not (closed symbols). Then the mice were treated with ganciclovir (150 mg/kg twice a day for 3.5 days) commencing on day 0 (□) or day 5 (△), or left untreated (○). Data from Freeman *et al.*, 1993.

in rats could be cured if they had been transfected with the HSV-tk gene and subsequently treated with gancilovir. Rats with brian tumors established by intracerebral injection of wild-type or HSV-tk modified glioma cells or by a mixture of wild-type and HSV-tk-modified cells were studied with and without ganciclovir treatments. All rats with HSV-tk expressing tumors treated with ganciclovir survived 90 days whereas all untreated rats died within 25 days. Further, tumors composed of combinations of 9L and 9L-TK cells could be eliminated by ganciclovir treatments even when only one-half of the tumor cells carried the HSV-tk gene (Barba *et al.*, 1993). Rats with a cerebral glioma were also given an intratumoral injection of murine fibroblasts that were producing a retroviral vector in which the HSV-tk had been inserted. The HSV-tk retroviral vectors that were produced transduced the neighboring proliferating glioma cells (Culver *et al.*, 1992; Ram *et al.*, 1993). The fibroblasts were necessary to attain effective retrovirus titers, but rat glioma cells could be transduced directly *in vivo* by using an adenovirus-based vector (Chen *et al.*, 1994). Replication-defective recombinant adenoviral particles (3×10^8) carrying the HSV-tk gene were injected into the tumors, and subsequently the mice were treated with ganciclovir for 6 days.

Tumor size in untreated and treated animals was compared 20 days after tumor implantation. The mean cross-sectional areas of the tumors in the treated animals were 23-fold smaller than in control animals, and the tumor volumes were reduced by >500-fold, although it was estimated that only 50% of the tumor cells were transduced by the vector (Chen *et al.*, 1994). Therapeutic results have also been obtained with tumors derived from other malignancies, such as liver (Caruso *et al.*, 1993), breast, (Manome *et al.*, 1994), head and neck squamous cell cancer (O'Malley *et al.*, 1995), lung (Osaki *et al.*, 1994; Smythe *et al.*, 1994), ovary (Tong *et al.*, 1996a,b) and melanoma (Vile and Hart, 1993; Vile *et al.*, 1994).

A bystander effect has been demonstrated for the HSV-tk system (Fig. 7). Because the active metabolite of ganciclovir, the triphosphate, is highly charged it should not diffuse out of expressing cells, and thus it would not be predicted that the bystander effect is mediated by a diffusible metabolite. *In vitro*, a mixed population containing only 10% of HSV-tk expressing cells was totally killed by a dose of ganciclovir that is not cytotoxic to non-HSV-tk expressing cells (Freeman *et al.*, 1993). When the HSV-tk expressing and nonexpressing cells are separated

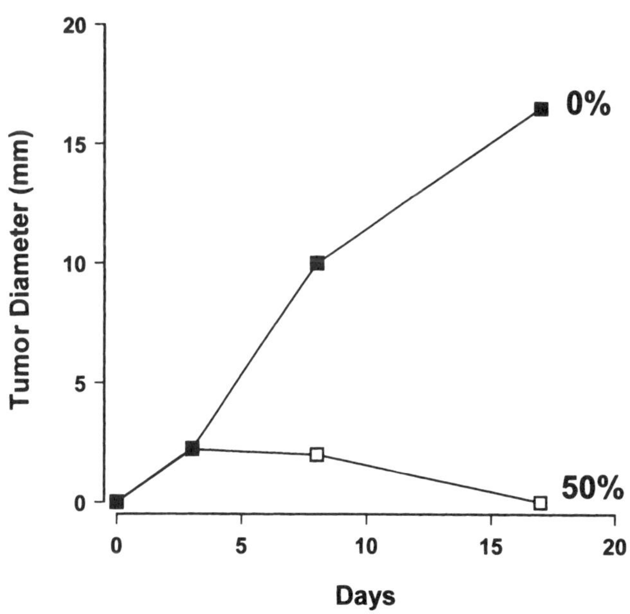

Figure 7. The antitumor effect of gancilovir on non-HSV-tk expressing tumor cells when mixed with 50% HSV-tk expressing tumor cells. Mice were inoculated s.c. with KBALD tumor cells ether alone or in a 50:50 mixture with KBALD tumor cells stably transduced with the HSV-tk gene. Then the mice were treated with ganciclovir (150 mg/kg twice a day for 5 days) commencing on day 3. Data from Freeman *et al.*, 1993.

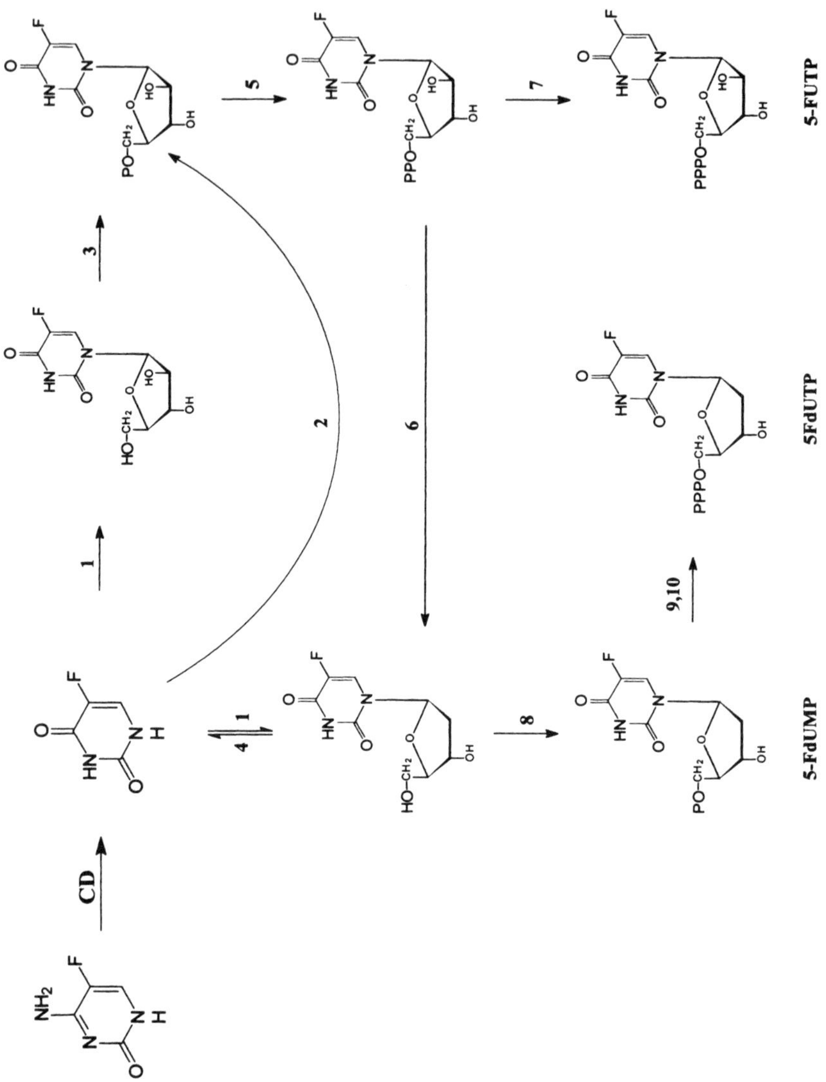
CD
1
2
3
4
5
6
7
8
9,10
5-FdUMP
5FdUTP
5-FUTP

by a 0.4 μm membrane, the bystander effect is abrogated, suggesting that a diffusible metabolite is not involved and that intimate cell to cell contact is necessary (Freeman *et al.*, 1993). Metabolic cooperation is involved, and a ganciclovir metabolic product, presumably a phosphorylated form, can pass from HSV-tk expressing cells to nonexpressing cells and mediate cytotoxicity but only as a consequence of direct contact (Bi *et al.*, 1993; Fick *et al.*, 1995). The uptake of apoptotic vessels by the nonexpressing tumor cells has also been demonstrated (Freeman *et al.*, 1993). *In vivo*, there is also a cell-mediated immune component to the bystander effect. In mixed-tumor experiments, ganciclovir resulted in almost total tumor regression in the immunocompetent BALB/c mice but not in immunocompromised athymic BALB/c mice (Gagandeep *et al.*, 1996).

3.3. Cytosine Deaminase

This enzyme, found in many fungi and bacteria, catalyzes the deamination of cytosine to uracil. It is not found in mammalian cells. Cytosine deaminase also converts the clinically used antifungal agent 5-fluorocytosine (5-FC) into the known antitumor drug 5-fluorouracil (5-FU). 5-FU is further metabolized to 5-fluorouridine 5′-triphosphate, 5-fluoro-2′-deoxyuridine 5′-triphosphate and 5-fluoro-2′-deoxyuridine 5′-monophosphate (Fig. 8). The first two compounds inhibit RNA and DNA synthesis, respectively, and the monophosphate is a potent inhibitor of thymidylate synthase that is an important enzyme in DNA biosynthesis.

Mouse 3T3 cells are sensitive to 5-FC when tranfected with the gene encoding cytosine deaminase (Mullen *et al.*, 1992). However, in mixing experiments, expressing cells could be killed without affecting nontransfected cells. Thus, there was not a bystander effect. Similar effects were seen when mouse fibrosarcomas and adencarcinomas were transduced with a retroviral vector containing the cytosine deaminase gene (Mullen *et al.*, 1994). When injected into syngeneic mice, the cytosine deaminase expressing tumors could be eliminated *in vivo* by systemic treatment with 5-fluorocytosine without significant toxicity to the host, although delaying the prodrug treatment reduces its effectiveness (Fig. 9). Interestingly, animals whose tumors were eliminated by prodrug treatment, resisted

Figure 8. The activation of 5-fluorocytosine (5-FC). The enzyme cytosine deaminase (CD) converts 5-FC to 5-fluorouracil (5-FU). This is converted by cellular enzymes to form the antimetabolites 5-fluorouridine 5′-triphosphate (5-FUTP), 5-fluoro-2′-deoxyuridine 5′-triphosphate (5-FdUTP) and 5-fluoro-2′-deoxyuridine 5′-monophosphate (5-FdUMP). (1) nucleoside phosphorylases; (2) phosphoribosyl transferase; (3) uridine kinase; (4) thymidine phosphorylase; (5) UMP kinase; (6) ribonucleoside reductase; (7) nucleotide diphosphate kinases; (8) deoxyuridine kinase; (9) nucleoside kinases; (10) dUMP kinase.

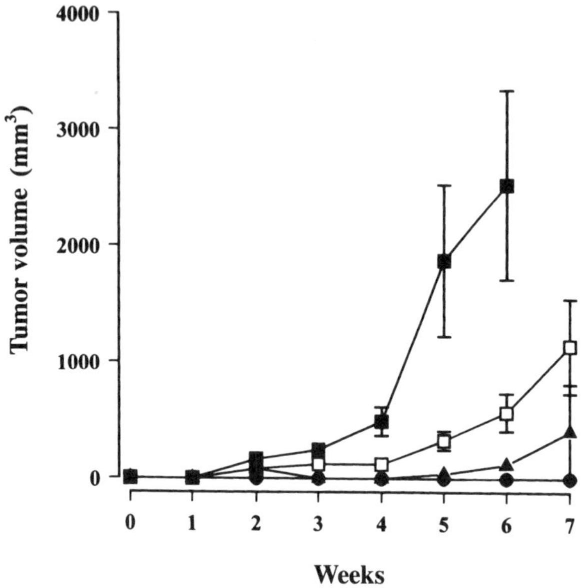

Figure 9. The effect of 5-fluorocytidine (5-FC) on the growth of a cytosine deaminase expressing mouse adenocarcinoma. Mice received 5-FC (37.5 mg) twice daily for 10 days either 4, 11, or 18 days after tumor inoculation. Data from Mullen *et al.*, 1994.

subsequent rechallenge with unmodified wild-type tumor. This posttreatment immunity is tumor-specific (Mullen *et al.*, 1994). It was postulated that cell death leads to a more effective antigen presentation or that the cytosine deaminase protein itself is immunogenic (Mullen *et al.*, 1994). Human colorectal tumors are also sensitive to 5-FC when they express cytosine deaminase. *In vitro*, WiDr cells were transfected with the cytosine deaminase gene. In parental cells, 5-FC had an IC_{50} of 26 mM but this decreases to only 27 μM in a cytosine deaminase expressing clone (Huber *et al.*, 1993). The sensitivity of these cells is retained *in vivo* when grown in nude mice. Using short courses of 5-FC treatment (500 mg/kg i.p., daily for 10 days), all tumors regressed, and a 75% cure rate was claimed when the treatment time was extended (Huber *et al.*, 1993). Importantly, a large bystander effect was observed both *in vitro* and *in vivo* (Huber *et al.*, 1993), in contrast to that observed with mouse tumor cells (Mullen *et al.*, 1992, 1994). 5FU was liberated into the surrounding environment when cytosine deaminase expressing tumor cells are treated with 5FC. This liberated 5FU kills neighboring, non-cytosine deaminase expressing tumor cells *in vitro* and *in vivo* (Huber *et al.*, 1994). When only 2% of the tumor mass contained cytosine deaminase expressing cells (98% non-cytosine deaminase expressing cells), significant regressions in

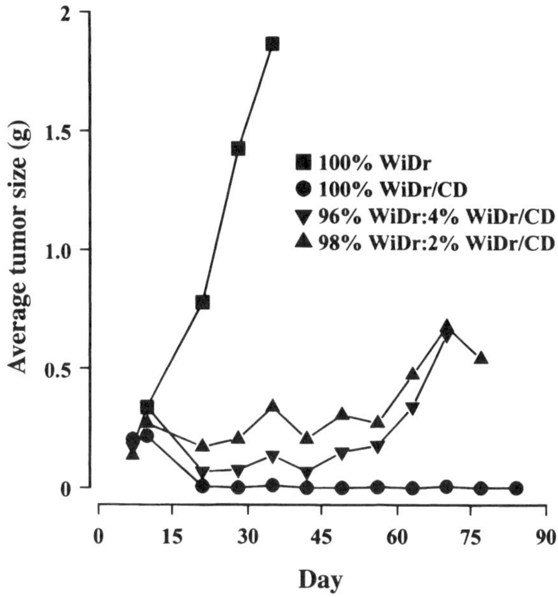

Figure 10. The antitumor effect of 5-fluorocytidine (5-FC) on mixed-cell tumor xenografts. Nude mice were implanted s.c. with mixtures of human WiDr cells, either expressing (WiDr/CD) or not expressing (WiDr) cytosine deaminase, as indicated. Eight days after implantation, mice were treated with 5-FC (500 mg/kg, daily until day 22, then three times weekly until day 63), and the tumor size determined weekly. In the absence of 5-FC, the growth of the mixed-cell tumors was identical. Data from Huber *et al.*, 1994.

tumors were observed when the host mouse is dosed with nontoxic levels of 5FC (Fig. 10) (Huber *et al.*, 1994). The lack of a bystander effect in the experiments of Mullen *et al.* (Mullen *et al.*, 1992, 1994) is surprising and suggests that much higher levels of cytosine deaminase expression were achieved in the human cells.

In a comparison of cytosine deaminase with HSV-tk transduced into human lung adenocarcinoma cell lines and driven by the CMV promoter, cytosine deaminase was superior in achieving high levels of specific enzyme activity to induce growth inhibition and to affect neighboring cell growth (Hoganson *et al.*, 1996).

3.4. Cytochrome P450

Cyclophosphamide is a commonly used antitumor agent. It is a prodrug that is activated by the microsomal cytochrome P450 mixed function oxidases to form

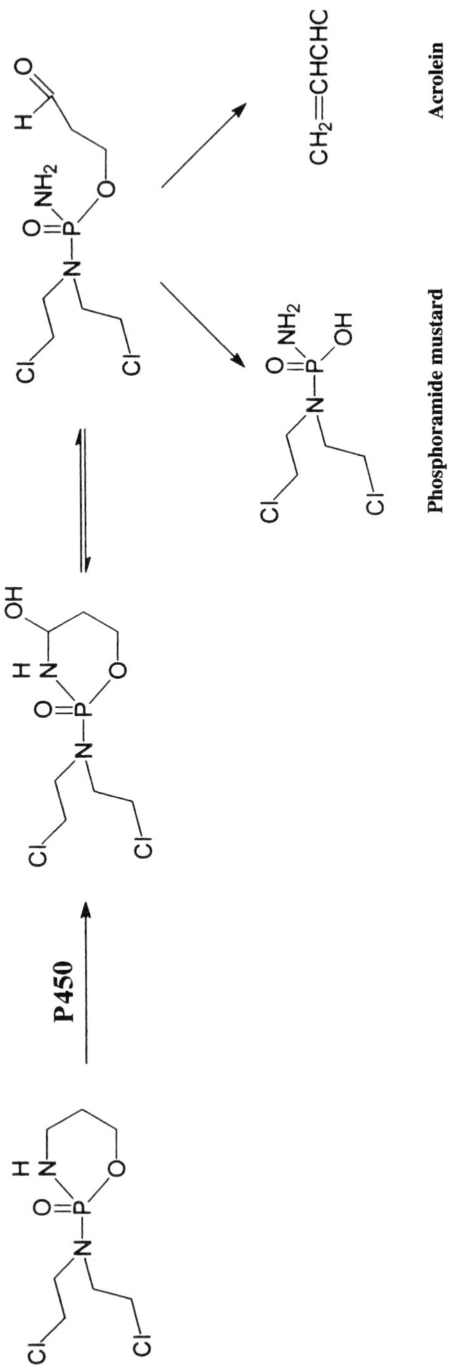

Figure 11. The activation of cyclophosphamide by cytochrome P450.

phosphoramide mustard via the 4-hydroxy intermediate that exists in equilibrium with the ring-opened tautomer, aldophosphamide (Fig. 11). Cytochrome P450 is a complex system involving many isoenzymes, not all of which activate cyclophosphamide. In humans, the isoenzymes 3A4 and 2B6 are mainly responsible for activating cyclophosphamide. These enzymes are only present in liver and not in other tissues or in tumor cells. Thus, there is a rationale for introducing a cytochrome P450 into a tumor cell. However, the 4-hydroxy/aldophosphamide intermediate is a circulating metabolite that reaches tumor cells. Therefore, this systemic activation of the prodrug might mitigate any selective activation by an introduced gene (Connors, 1995). As predicted, tumor cells can be sensitized to the cytotoxic effects of cyclophosphamide in culture by introducing cytochrome P450 2B1. Stable transfection of rat C6 glioma cells with the P450 2B1 gene rendered the cultured tumor cells sensitive to cyclophosphamide (Wei *et al.*, 1994). Further, C6 cells bearing this gene were more sensitive than parental cells to the cytotoxic action of CPA when grown subcutaneously in the flanks of athymic mice (Wei *et al.*, 1994). Similar results were seen with rat 9L gliosarcoma cells (Fig. 12) (Chen and Waxman, 1995; Manome *et al.*, 1996a) and human MCF-7 cells (Chen *et al.*, 1996). *In vitro*, a panel of human tumors was sensitized to cyclophosphamide and its analog ifosfamide when infected with the cytochrome P450 gene (Chen *et al.*, 1995).

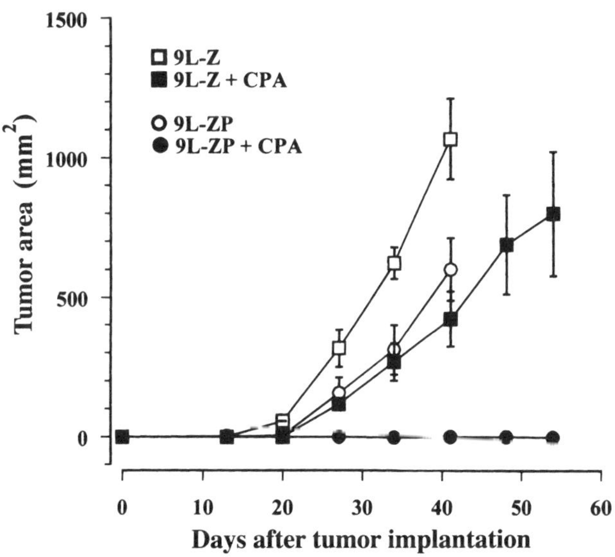

Figure 12. The antitumor effect of cyclophosphamide on rat 9L gliosarcoma cells either expressing cytochrome P450 2B1 gene (9L-ZP) or not (9L-Z). Cells (2×10^6) were inoculated by s.c. injection into the left (9L-Z) or right flank (9L-ZP) flank of a rat and treated with a single dose of cyclophosphamide (10 mg/kg) after 7 days. Data from Chen and Waxman, 1995.

Thus, it has been shown that expression of cytochrome P450 significantly increases the antitumor efficacy of cyclophosphamide without any increase in host toxicity. Therefore, there is a substantial "proximity effect" with intratumoral cyclophosphamide activation despite the high metabolic activation by the liver (Chen and Waxman, 1995). This suggests that the circulating metabolite, the 4-hydroxy/aldophosphamide intermediate, has limited access to the tumor vasculature or a low degree of cell permeability. Alternatively it has been suggested that acrolein, which is formed in equimolar amounts with phosphoramide mustard (Fig. 11), can potentiate the cytotoxicity of the mustard locally (Chen and Waxman, 1995). Acrolein is a very reactive compound and reacts rapidly with proteins and thiols. Thus it is rapidly lost from circulation. However, local generation could result in the depleting thiols, such as glutathione, and thereby modulate the cytotoxicity of phosphoramide mustard.

A bystander effect was observed with this system (Chen and Waxman, 1995; Wei *et al.*, 1995; Chen *et al.*, 1996). However, unlike the bystander effect observed with HSV-tk/ganciclovir cells, contact is not required, and the effect is mediated by diffusible, cytotoxic metabolite, presumably the 4-hydroxy/aldophosphamide intermediate. *In vitro*, the bystander effect is seen when only 10% of the cultured cells express the cytochrome P450 gene (Wei *et al.*, 1995) and is observed when contact between expressing and nonexpressing cells is prevented by a 0.45 μm membrane (Fig. 13) (Chen and Waxman, 1995).

3.5. Carboxypeptidase G2

Carboxypeptidase G2 (CPG2) was originally isolated from *Pseudomonas* sp. RS-16 and is a dimeric protein (Mr 83 kDa) consisting of two identical, zinc-containing subunits. It catalyzes the hydrolysis and cleavage of folates to pteroates and L-glutamic acid but can also cleave other substrates that are prodrugs, such as the mustard derivatives of glutamic acid (Fig. 14). The enzyme has no equivalent mammalian activity. This enzyme and potential prodrugs have been extensively studied for ADEPT in which it has shown excellent results (Chapters 4 and 8). However, the properties of both the enzyme and its prodrugs mitigate its use in GDEPT. First, because the enzyme degrade folates, expression of CPG2 inside a cell may be cytotoxic because it can deplete folate pools (Romanini *et al.*, 1989). Secondly, because the prodrugs are derivatives of glutamic acid, they are charged and would not be expected to cross a cell membrane readily. The reactivity of the mustard is reduced by the linkage of the benzoic acid moiety through an amide bond to a glutamic acid residue. However, this deactivation was not considered great enough to account for the difference in cytotoxicity. This enhanced effect has been ascribed to the fact that the prodrug is excluded from the cell. In ADEPT this

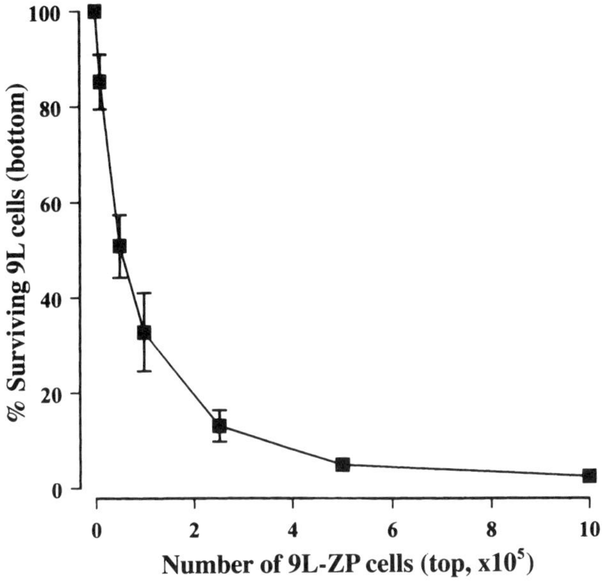

Figure 13. The bystander cytotoxicity of cyclophosphamide against non-cytochrome P450 2B1 expressing cells. Parental rat 9L cells (1 × 105) were plated into the bottom chamber of a culture plate and cytochrome P450 2B1 expressing 9L-ZP cells seeded, at the number indicated, into the upper chamber of an insert. Thus, the two populations were separated by a 0.45 μm membrane. Cells were treated with cyclophosphamide (1 mM) and the surviving 9L cells (in the bottom chamber) were counted after 5 days. Data from Chen and Waxman, 1995.

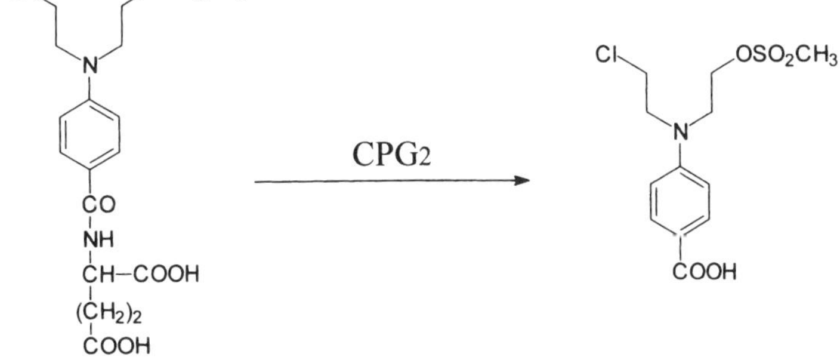

Figure 14. The activation of 4-[(2-chloroethyl)(2-mesyloxyethyl)amino]benzoyl-L-glutamic acid by the enzyme carboxypeptidase G2 (CPG2).

is ideal because the activating enzyme is localized to the cell surface. However, in GDEPT the enzyme is intracellular and would not be available to activate the prodrug.

Contrary to theoretical considerations, the expression of CPG2 in combination with the prodrug 4-[(2-chloroethyl)(2-mesyloxyethyl)amino]benzoyl-L-glutamic (CMDA) is an effective GDEPT system (Marais *et al.*, 1996). Because CPG2 is a secreted protein and the enzyme is found in the bacterial periplasm (Minton *et al.*, 1984), the bacterial sequence was engineered to remove the signal peptide from the gene. This gene was introduced into a number of cell lines by transfection under the control of a nonspecific promoter from the elongation 1α gene. Stable clones were obtained which expressed CPG2 and were sensitive to CMDA. Some clones were nearly 100-fold more sensitive than the corresponding control cell lines (Table II) (Marais *et al.*, 1996). There was no reduction in the rate of growth of the expressing cells, and therefore CPG2 was not toxic to these cells. However, normal tissue culture medium is rich in thymidine and rescues cells from folate depletion.

A large bystander effect was also observed (Marais *et al.*, 1996). When CPG2 expressing LS174T cells were mixed with nonexpressing cells, total kill of the whole population occurred when only ~5% were expressing CPG2.

Interestingly, the extracellular expression of CPG2 for GDEPT applications has been reported (Springer *et al.*, 1997). The gene was engineered so that it is expressed tethered to the outer surface of mammalian cells. It was found necessary to mutate the gene so as remove some glycosylation sites because glycosylation

Table II. The Cytotoxicity of 4-[(2-Chloroethyl)(2-mesyloxyethyl)amino]benzoyl-L-glutamic Acid (CMDA) against Human Tumor Cell Lines Transfected with the Gene Encoding for CPG2 or β-Galactosidase as a Control[a]

Cell line	Clone	CPG2 activity (units/mg protein)	IC_{50} (μM CMDA)	IC_{50} ratio (β-gal/CPG2)
A2780	β-gal	0	2150.0 ± 180	92.7
	CPG2	0.964	23.2 ± 2.9	
LS174T	β-gal	0	2770.0 ± 260	95.2
	CPG2	1.256	29.1 ± 8.9	
SKOV-3	β-gal	0	4180.0 ± 452	16.2
	CPG2	1.013	258.0 ± 14	
WiDr	β-gal	0	3230.0 ± 120	11.66
	CPG2	0.787	277.0 ± 38	

[a]The cells were treated when confluent with CMDA for 1 hr, then replaced with fresh CMDA containing medium for 18 hr. Then cells were washed, trysinized, and ~3% seeded in fresh medium. Then cell viability was assessed after four days by [^{3}H]thymidine incorporation. The IC_{50} values are relative to vehicle treated controls. Data from Marais *et al.*, 1996.

inhibits the enzyme (proteins in bacteria are not normally glycosylated). Although this reduces the efficiency of surface expression, extracellular expression is achieved. Results obtained with the CMDA prodrug, with respect to cytotoxicity and bystander effect, are similar to those obtained with intracellular expression of CPG2 (Springer *et al.*, 1997). Although glycosylation is a problem for the external expression of CPG2, this might not be the case with other enzymes and this method would allow using any of the enzyme/prodrug systems developed for ADEPT in GDEPT. However, external expression of an enzyme may stimulate an immune response against it. Although this would probably inhibit prodrug activation, it may lead to an antitumor response in its own right. This combination of immunotherapy and GDEPT may have advantages but puts extra constraints on the targeting system.

3.6. Nitroreductase

Like carboxypeptidase G2, nitroreductase (NR) was originally proposed as an enzyme for use in ADEPT (Knox *et al.*, 1993). NR was isolated and cloned from *E. coli* (Michael *et al.*, 1994), is a flavoenzyme containing FMN (flavin mononucleotide), and is a 23 kDa monomer (Anlezark *et al.*, 1995; Knox *et al.*, 1992). NR activates the prodrug CB 1954 [5-(aziridin-1-yl)-2,4-dinitrobenzamide] by reducing either of the nitro groups of CB 1954 and forms a mixture of the 2- and 4-hydroxylamines in equal proportions (Fig. 15) (Knox *et al.*, 1992). 5-(Aziridin-1-yl)-4-hydroxylamino-2-nitrobenzamide reacts chemically with cellular thioesters (such as *S*-acetylthiocholine) to form a highly reactive DNA cross-linking species (Knox *et al.*, 1988, 1991). 5-(Aziridin-1-yl)-2-hydroxylamino-4-nitrobenzamide is not as cytotoxic as the 4-hydroxylamine and is a less potent DNA cross-linking agent. However, it is still much more cytotoxic than the prodrug CB 1954. Thus CB 1954 is converted from a weak monofunctional alkylating agent to a potent difunctional agent (which can form DNA interstrand cross-links). There is a very large difference between the cytotoxicity of the prodrug and the active form (10,000) of CB 1954 (Roberts *et al.*, 1986), which makes this system attractive for GDEPT. In human cells, CB 1954 is reduced by the enzyme DT diaphorase but reduction is very slow (K_m = 1.4 mM, k_{cat} = 0.009 s^{-1}) (Boland *et al.*, 1991) compared to NR (K_m = 1.4 mM, k_{cat} = 6.0 s^{-1}) (Anlezark *et al.*, 1992), and human cells (even those expressing high levels of human DT diaphorase) are resistant to CB 1954 (Boland *et al.*, 1991). The application of nitroreductase in ADEPT was handicapped by its requirement for a cofactor (Knox *et al.*, 1993, 1995). However, this is not a problem in GDEPT because the enzyme is expressed intracellularly and thus can use the endogenous cofactors NADH or NADPH [NR can use either (Anlezark *et al.*, 1992)]. In fact, the cofactor requirement may be an advantage.

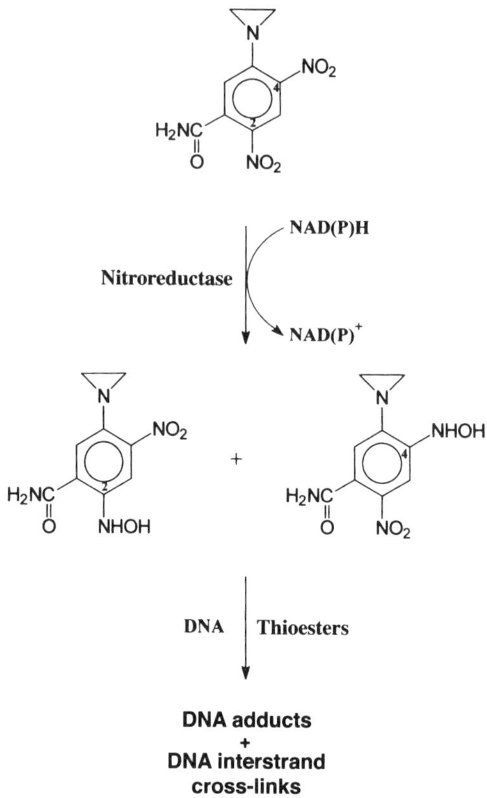

Figure 15. The activation of the prodrug CB 1954 (5-(aziridin-1-yl)-2,4-dinitrobenzamide) by the enzyme *E. coli* nitroreductase.

Any enzyme that escapes into the circulation (for example, from dying cells) is incapable of activating circulating prodrug through the lack of a cofactor because NAD(P)H is very rapidly metabolized by serum components (Friedlos and Knox, 1992).

A recombinant retrovirus that encodes NR was used to infect mammalian cells. NIH3T3 cells expressing NR were killed by CB 1954. The bulk infected, unselected cell population was about a 100-fold more sensitive to CB 1954 than the parental cells (Bridgewater *et al.*, 1995). A selected clone was even more sensitive and, using a cell count assay, was more than a 1000-fold more sensitive to CB 1954 than parental NIH3T3 cells (Fig. 16) (Bridgewater *et al.*, 1995). Optimal cytotoxicity was found after a 36 hr exposure to CB 1954, but significant inhibition of cell growth was seen after only a 4 hr exposure to the prodrug

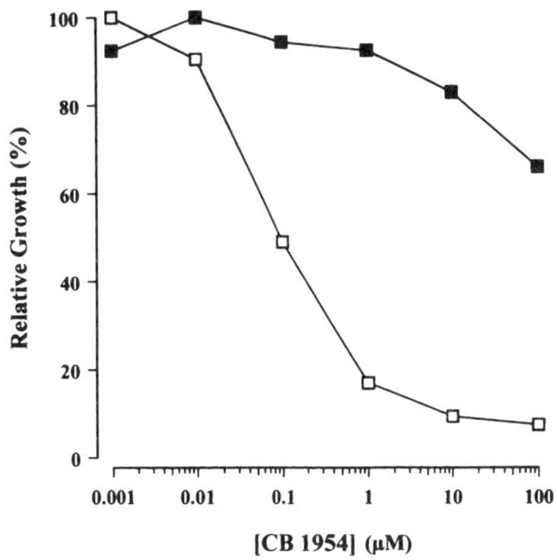

Figure 16. The cytotoxicity of CB 1954 against a nitroreductase (NR) expressing cell line. NIH3T3 cells were infected with a recombinant retrovirus containing NR and a cell clone (NIH3T3-NR) derived by limiting dilution. Parental NIH3T3 cells (■) or NIH3T3-NR cells (□) were treated with CB 1954 for 24 hr before assay. Data from Bridgewater *et al.*, 1995.

(Bridgewater *et al.*, 1995). Similar results were seen in human melanoma, ovarian carcinoma, mesothelioma (Bridgewater *et al.*, 1995), and colorectal and pancreatic cell lines (Green *et al.*, 1997). The rapid action of CB 1954 and the resulting need for a shorter exposure time may facilitate the use of this prodrug clinically (Bailey and Hart, 1997). A significant bystander effect was observed, and admixed, unmodified, NIH3T3 cells are also killed by a normally nontoxic dose of prodrug (Fig. 17) (Bridgewater *et al.*, 1995). The bystander effect is mitigated by diffusible metabolites, and both the 2- and 4-hydroxylamino derivatives of CB 1954 are released into the medium of CB 1954 treated nitroreductase-expressing NIH3T3 cells (Bridgewater *et al.*, 1997). Importantly and in contrast to the Herpes simplex virus thymidine kinase/ganciclovir enzyme/prodrug system, NR/CB1954 cell killing is cytotoxic to noncycling cells (Fig. 18) (Bridgewater *et al.*, 1995).

As with ADEPT, other potential prodrugs have been proposed for use with the nitroreductase. Chinese hamster V79 cells transfected with a nitroreductase expression vector were 770-fold more sensitive to CB 1954 than control non-expressing cells. Other prodrugs, such as nitrofurazone (1, Fig. 19) (97-fold) and the nitroimidazole compounds, misonidazole (2, Fig. 19) (21-fold) and metronidazole (3, Fig. 19) (50-fold), also exhibit increased cytotoxicity against the nitroreductase-expressing cells, and HPLC showed that they act as substrates for the purified NR enzyme (Bailey *et al.*, 1996). However, this correlation was not absolute. In particular, the quinone EO9 [3-hydroxymethyl-5-aziridinyl-1-methyl-

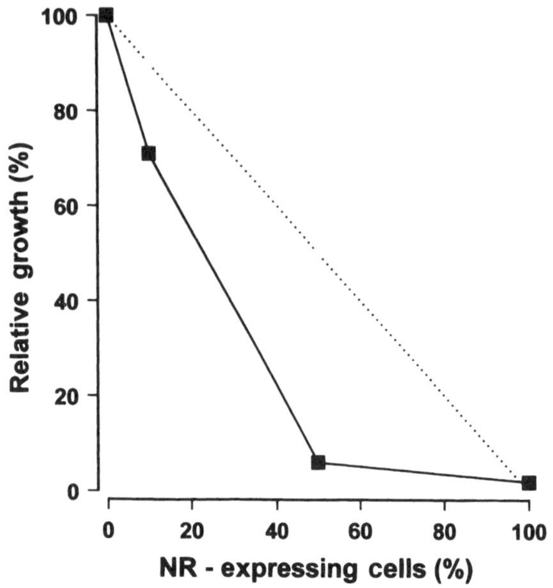

Figure 17. The bystander cytotoxicity of CB 1954 against non-nitroreductase expressing cells. NIH3T3 cells either expressing or not expressing nitroreductase were admixed in the proportions indicated and treated with CB 1954 (100 μM) for 24 hr before assay. The dashed line indicates the theoretical result if there were no bystander effect. Data from Bridgewater *et al.*, 1995.

2-(H-indole-4,7-indione)-propenol] showed only a very small differential (<3-fold) (Bailey *et al.*, 1996) probably because this compound undergoes activation by endogenous enzymes, such as NADPH: cytochrome P450 reductase (Bailey *et al.*, 1996). When misonidazole or metronidazole are activated by NR extracellularly there is little increase in cytotoxicity (Anlezark *et al.*, 1995). This suggests that the active species has a very short half-life and probably does not exhibit a large bystander effect. A series of 2,4-dinitrobenzamide mustard analogs of CB 1954 has been evaluated as potential prodrugs for GDEPT in a V79 cell line transfected with the nitroreductase (Friedlos *et al.*, 1997). The cytotoxicity (IC_{50}) of these compounds was determined in both the NR-expressing and a nonexpressing V79 cell line, and the proportion of NR-expressing cells, admixed with nonexpressing cells, required to produce 50% cytotoxicity was determined as a measure of the bystander effect. A number of compounds, more potent toward the NR-expressing cells, had higher IC_{50} ratios between expressing and nonexpressing cells and a better bystander effect than CB 1954. In this study the best prodrug was 5-[*N,N*-bis(2-bromoethyl)amino]-2,4-dinitrobenzamide (SN 24927) (4, Fig. 19) (Friedlos *et al.*, 1997). The IC_{50} ratio of this compound is 2532 and that for CB

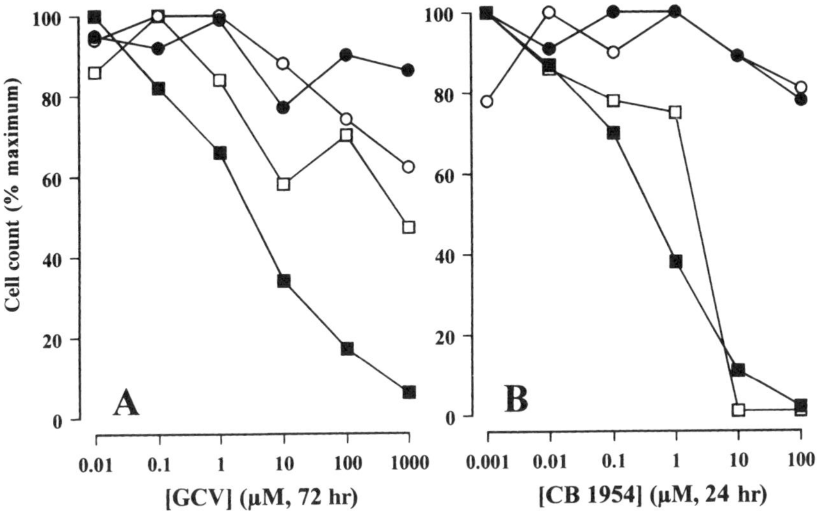

Figure 18. The effect of (A) ganciclovir (GCV) or (B) CB 1954 on NIH3T3 cells either expressing (■) or not expressing (●) either (A) herpes simplex virus thymidine kinase or (B) nitroreductase. Cells were treated either during normal growth (solid symbols) or following growth arrest in the G_0-phase of the cell cycle (open symbols). Data from Bridgewater *et al.*, 1995.

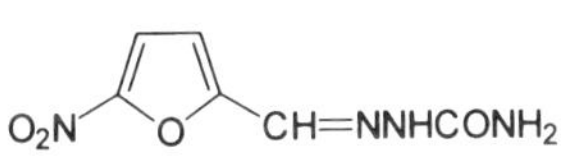

1. Nitrofurazone

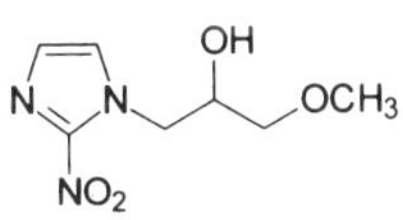

2. Misonidazole

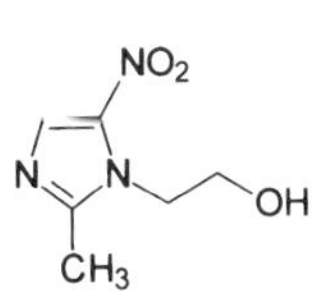

3. Metronidazole

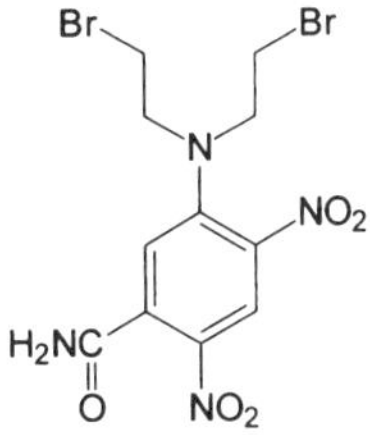

4. SN 24927

Figure 19. The chemical structures of some prodrugs activated by nitroreductase.

1954 was reported to be 177. Furthermore, only 0.1% of expressing cells were required to kill 50% of an admixed culture when treated with the mustard prodrug compared to the 1% required with CB 1954 (Friedlos *et al.*, 1997). However, the cytotoxicity ratio reported for CB 1954 is very low and, in fact, is stated to be about 2000-fold in the methods section of the paper (Friedlos *et al.*, 1997)! This higher value is agrees better with previously published data using the same cell line transfected with the same vector containing the NR gene (Bailey *et al.*, 1996). The kinetics of NR with SN 24927 have not been reported, and effects of this prodrug may be cell-line-dependent because the cytotoxicity of the prodrug alone is variable. Against Chinese hamster UV4 cells SN 24927 is 12-fold more cytotoxic than the chloroderivative, SN 23862 (Anlezark *et al.*, 1995), but is almost equipotent toward Chinese hamster V79 cells (Friedlos *et al.*, 1997). Other potential prodrugs for use in GDEPT are those activated by a self-immolative mechanism that can potentially form active drugs, such as mustards, actinomycins, mitomycin C (Mauger *et al.*, 1994), enediynes (Hay *et al.*, 1995), *seco*-CI alkylating agents (Tercel *et al.*, 1996), or tallimustine (Lee *et al.*, 1997). The prodrugs are 4-nitrobenzyloxycarbonyl derivatives of these drugs, which upon enzymatic reduction, generate the drug through self-immolation of the 4-(hydroxylamino)-benzyloxycarbonyl group. Prodrugs of this type have been synthesized and are activated by NR (Hay *et al.*, 1995; Lee *et al.*, 1997; Mauger *et al.*, 1994; Tercel *et al.*, 1996) but have not yet been tested against a NR-expressing cell line.

3.7. Other Enzymes

Two other enzymes have been considered for us in GDEPT. However, they differ from the previous examples because they are normal human enzymes. Resistance to the prodrugs they activate may be caused by low expression of these enzymes in the tumor and thus a therapeutic advantage could be gained by their selective expression.

3.7.1. THYMIDINE PHOSPHORYLASE

Human thymidine phosphorylase (dThdPase) is identical to platelet-derived endothelial cell growth factor (PD-ECGF) and catalyzes the reversible phosphorolytic cleavage of thymidine, deoxyuridine, and their analogs to the base and deoxyribose 1-phosphate. It also cleaves the glycosidic bond of the prodrug 5′-deoxy-5-fluorouridine (5′-DFUR) to form 5-fluorouracil (5-FU) (Fig. 20). Human MCF-7 breast cancer cell lines have up to 165-fold increased sensitivity to the prodrug *in vitro* when transfected with the dThdPase cDNA. Sensitivity to both 5-FU and 5-fluoro-2′-deoxyuridine are unchanged (Patterson *et al.*, 1995). A

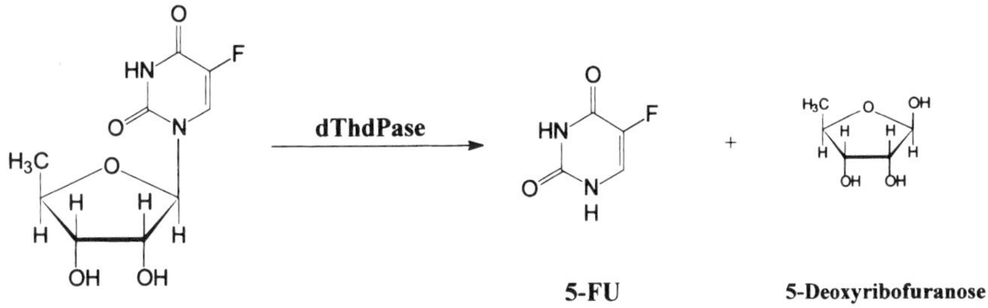

Figure 20. The activation of the prodrug 5′-deoxy-5-fluorouridine (5′-DFUR) to form 5-fluorouracil (5-FU) by the human enzyme thymidine phosphorylase (dThdPase).

substantial bystander killing effect was seen when small proportions of dThdPase-expressing cells were mixed with parental MCF-7 cells (Fig. 21) (Patterson *et al.*, 1995). Similar bystander effects have been observed when 5-FU is formed from 5-fluorocytosine by the enzyme cytosine deaminase (Huber *et al.*, 1993). It was considered that in some tumors resistance to 5′-DFUR therapy could result from low dThdPase activity and that GDEPT could raise the dThdPase levels and thus enhance sensitivity to the prodrug (Patterson *et al.*, 1995).

3.7.2. DEOXYCYTIDINE KINASE

Cytosine arabinoside (ara-C) is a cytidine analog that is incorporated into replicating DNA and induces lethal DNA strand breaks. Although ara-C is a potent antitumor agent for hematological malignancies and is the most effective agent for treating acute myeloid leukemia, it has only minimal activity against most solid tumors. The rate-limiting step in intracellular ara-C activation is phosphorylation of the prodrug by deoxycytidine kinase (dCK) which competes with the detoxification of the prodrug through deamination by cytidine deaminase (Fig. 22). Thus it was considered that solid tumors could be made more sensitive to the prodrug if the balance were shifted from the detoxification pathway to the activation pathway by introducing dCK (Manome *et al.*, 1996b). Both retroviral and adenoviral vector-mediated transduction of the dCK cDNA results in sensitizing (~10-fold after a 96 hr prodrug exposure) the 9L rat glioma cell lines to the cytotoxic effects of ara-C *in vitro* (Manome *et al.*, 1996b). The ara-C treatment of established intradermal and intracerebral rat gliomas transduced with dCK also resulted in significant antitumor effects *in vivo* (Fig. 23) (Manome *et al.*, 1996b). However, the bystander effect with this system is minimal (Manome *et al.*, 1996b) indicating rapid breakdown or inefficient transfer of the activated forms of ara-C across intracellular junctions.

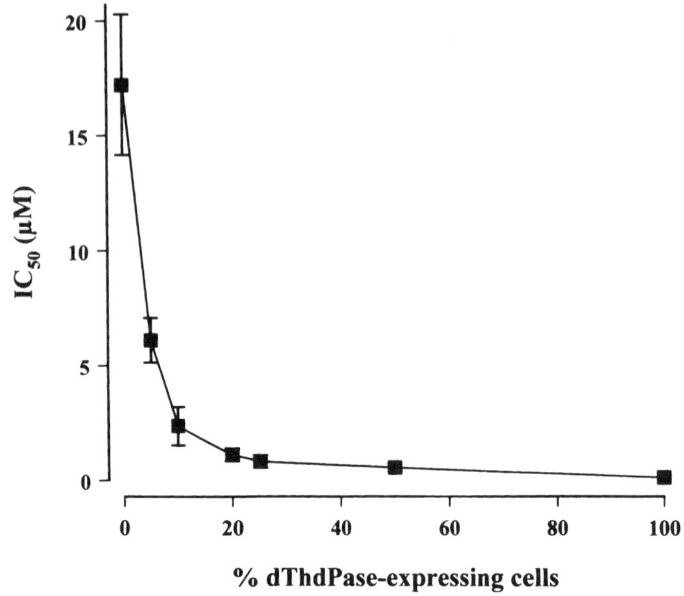

Figure 21. The effect of the proportion of cells expressing human thymidine phosphorylase (dThdPase) on the cytotoxicity of 5′-deoxy-5-fluorouridine (5′-DFUR). Human MCF-7 breast cancer cells were transfected with the dThdPase gene, and a clone expressing a 90-fold increase in dThdPase enzyme activity was selected. This clone was admixed with parental MCF-7 cells in the proportions indicated, and the mean IC_{50} for 5′-DFUR was determined by the MTT (thiazolyl blue) assay. Data from Patterson *et al.*, 1995.

4. CONCLUSIONS

Selective killing of tumor cells can be achieved by GDEPT. A diverse range of enzymes and prodrugs have been proposed for GDEPT. Most have shown encouraging results *in vitro*. There have been some excellent results using *in vivo* models, and a clinical trial has been proposed. It is highly probable that additional examples will materialize. Perhaps unfortunately, many of the active drugs formed are antimetabolites. These are cytotoxic only to dividing cells and require long exposure times for optimum effect. Alkylating agents do not suffer from these disadvantages and are also less prone to inducing drug resistance (Frei *et al.*, 1988). The generation of an alkylating agent probably offers the most potent means of killing targeted cell types. However, given the different modes of action of nitroreductase/CB 1954 and HSV-tk/ganciclovir, for example, a combination of these approaches offers a way of obtaining potentially synergistic effects. Cooper-

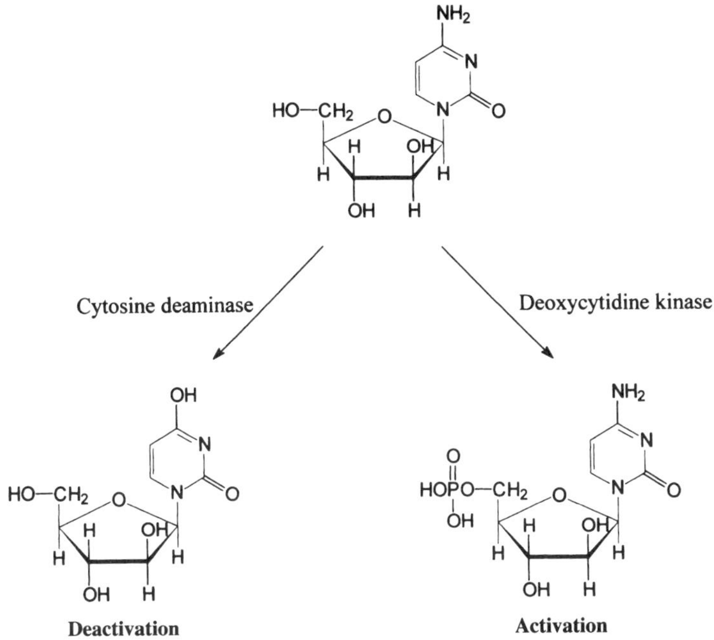

Figure 22. The competing pathways of metabolism of the prodrug Ara-C. Ara-C can either be detoxified by deamination or activated by phosphorylation. The monophosphate is further activated to a cytotoxic triphosphate analogously to ara-M (Fig. 3).

ative killing is observed when cells expressing both NR and HSV-tk are treated with a combination of CB 1954 and ganciclovir (Bridgewater *et al.*, 1995). A bystander effect is also required for the active agent because the present gene therapy vectors cannot transduce all of the target tumor cells. All the examples given, except for deoxycytidine kinase/ara-C, produce substantial bystander effects. In some models, less than 1% of genetically modified cells produce significant cytotoxicity in the bulk population (Friedlos *et al.*, 1997). In most cases, the bystander effect is caused by a diffusible cytotoxic metabolite formed by prodrug activation. In the case of the purine nucleotides, a direct cell to cell contact is involved, and gap junctions are involved in transferring the cytotoxic metabolites. Other factors may be involved, however, and it has been postulated that lytic enzymes, triggered by programmed cell death, can also be transferred (Freeman *et al.*, 1993). *In vivo*, the immune system is also involved, and there may even be a priming effect (Freeman *et al.*, 1993; Gagandeep *et al.*, 1996).

The limitations of GDEPT are not with the enzyme prodrug systems but with the delivery systems. Systemic administration of the present generation of vectors

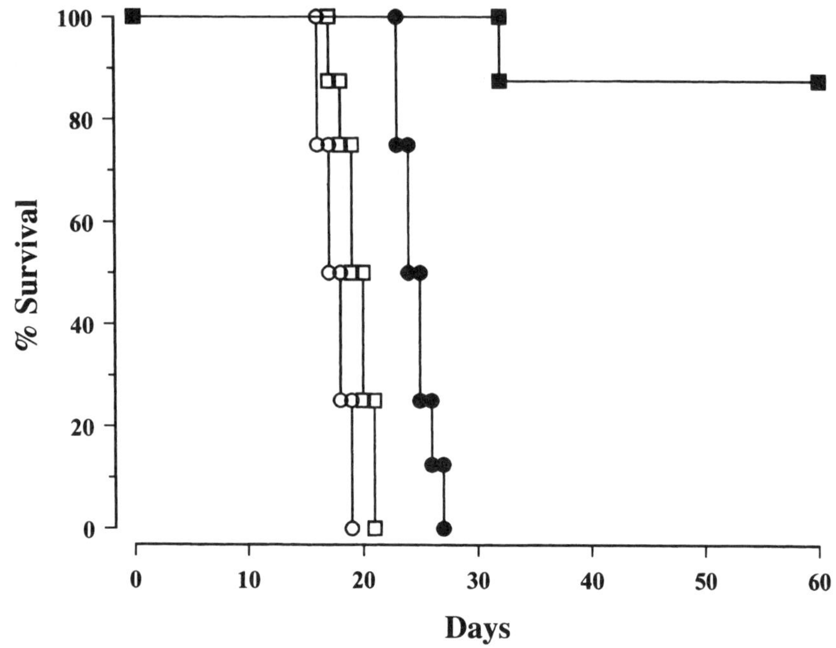

Figure 23. The antitumor effect of Ara-C on intracerebral rat 9L gliomas either expressing deoxycytidine kinase (■) or not (●). Cells (2×10^4) were stereotaxically injected into the right caudate nucleus of the brain. Then rats were treated with either saline (open symbols) or ara-C (200 mg/kg every 8 hr for 2 days) (closed symbols) on days 2–3 and 10–11. Surviving rats were alive for more than 120 days. Data from Manome *et al.*, 1996b.

is not possible, and thus GDEPT is limited to isolable tumor deposits, such as intracerebral tumors, surrounded by largely nondividing normal tissue. In such cases, GDEPT is feasible using retrovirally based vectors. Improved vectors have been proposed and are under development. However, it should be remembered that GDEPT is an answer to some of the limitations of the present gene therapy vectors, and a perfect, tumor-specific, gene delivery system would make GDEPT obsolete.

REFERENCES

Abe, A., Takeo, T., Emi, N., Tanimoto, M., Ueda, R., Yee, J. K., Friedmann, T., and Saito, H., 1993, Transduction of a drug-sensitive toxic gene into human leukemia cell lines with a novel retroviral vector. *Proc. Soc. Exp. Biol. Med.* **203:**354–359.

Anlezark, G. M., Melton, R. G., Sherwood, R. F., Coles, B., Friedlos, F., and Knox, R. J., 1992, The bioactivation of 5-(aziridin-1-yl)-2,4-dinitrobenzamide (CB 1954)—I. Purification and properties of a nitroreductase enzyme from *Escherichia coli*—a potential enzyme for antibody-directed enzyme prodrug therapy (ADEPT), *Biochem. Pharmacol.* **44:**2289–2295.

Anlezark, G. M., Melton, R. G., Sherwood, R. F., Wilson, W. R., Denny, W. A., Palmer, B. D., Knox, R. J., Friedlos, F., and Williams, A., 1995, Bioactivation of dinitrobenzamide mustards by an *E. coli* B nitroreductase, *Biochem. Pharmacol.* **50:**609–618.

Bailey, S. M., and Hart, I. R., 1997, Nitroreductase activation of CB1954—an alternative 'suicide' gene system [editorial], *Gene Ther.* **4:**80–81.

Bailey, S. M., Knox, R. J., Hobbs, S. M., Jenkins, T. C., Mauger, A. B., Melton, R. G., Burke, P. J., Connors, T. A., and Hart, I. R., 1996, Investigation of alternative prodrugs for use with *E. coli* nitroreductase in 'suicide gene' approaches to cancer therapy, *Gene Ther.* **3:**1143–1150.

Barba, D., Hardin, J., Ray, J., and Gage, F. H., 1993, Thymidine kinase-mediated killing of rat brain tumors, *J. Neurosurg.* **79:**729–735.

Bi, W. L., Parysek, L. M., Warnick, R., and Stambrook, P. J., 1993, *In vitro* evidence that metabolic cooperation is responsible for the bystander effect observed with HSV tk retroviral gene therapy, *Hum. Gene Ther.* **4:**725–731.

Boland, M. P., Knox, R. J., and Roberts, J. J., 1991, The differences in kinetics of rat and human DT diaphorase result in a differential sensitivity of derived cell lines to CB 1954 (5-(aziridin-1-yl)-2,4-dinitrobenzamide), *Biochem. Pharmacol.* **41:**867–875.

Borrelli, E., Heyman, R., Hsi, M., and Evans, R. M., 1988, Targeting of an inducible toxic phenotype in animal cells, *Proc. Natl. Acad. Sci. USA* **85:**7572–7576.

Bridgewater, J. A., Knox, R. J., Pitts, J. D., Collins, M. K., and Springer, C. J., 1997, The bystander effect of the nitroreductase CB 1954 enzyme prodrug system is due to a cell-permeable metabolite, *Hum. Gene Ther.* **8:**709–717.

Bridgewater, J. A., Springer, C. J., Knox, R. J., Minton, N. P., Michael, N. P., and Collins, M. K., 1995, Expression of the bacterial nitroreductase enzyme in mammalian cells renders them selectively sensitive to killing by the prodrug CB1954, *Eur. J. Cancer* **31a:**2362–2370.

Caruso, M., Panis, Y., Gagandeep, S., Houssin, D., Salzmann, J. L., and Klatzmann, D., 1993, Regression of established macroscopic liver metastases after in situ transduction of a suicide gene, *Proc. Natl. Acad. Sci. USA* **90:**7024–7028.

Chen, L., Chen, D., Manome, Y., Dong, Y., Fine, H. A., and Kufe, D. W., 1995, Breast cancer selective gene expression and therapy mediated by recombinant adenoviruses containing the DF3/MUC1 promoter, *J. Clin. Invest.* **96:**2775–2782.

Chen, L., and Waxman, D. J., 1995, Intratumoral activation and enhanced chemotherapeutic effect of oxazaphosphorines following cytochrome P-450 gene transfer: Development of a combined chemotherapy/cancer gene therapy strategy, *Cancer Res.* **55:**581–589.

Chen, L., Waxman, D. J., Chen, D., and Kufe, D. W., 1996, Sensitization of human breast cancer cells to cyclophosphamide and ifosfamide by transfer of a liver cytochrome P450 gene, *Cancer Res.* **56:**1331–1340.

Chen, S. H., Shine, H. D., Goodman, J. C., Grossman, R. G., and Woo, S. L., 1994, Gene therapy for brain tumors: Regression of experimental gliomas by adenovirus-mediated gene transfer *in vivo*, *Proc. Natl. Acad. Sci. USA* **91:**3054–3057.

Connors, T. A., 1995, The choice of prodrugs for gene directed enzyme prodrug therapy of cancer, *Gene Ther.* **2:**1–9.

Consalvo, M., Mullen, C. A., Modesti, A., Musiani, P., Allione, A., Cavallo, F., Giovarelli, M., and Forni, G., 1995, 5-Fluorocytosine-induced eradication of murine adenocarcinomas engineered to express the cytosine deaminase suicide gene requires host immune competence and leaves an efficient memory, *J. Immunol.* **154:**5302–5312.

Cook, D. R., Maxwell, I. H., Glode, L. M., Maxwell, F., Stevens, J. O., Purner, M. B., Wagner, E.,

Curiel, D. T., and Curiel, T. J., 1994, Gene therapy for B-cell lymphoma in a SCID mouse model using an immunoglobulin-regulated diphtheria toxin gene delivered by a novel adenovirus-polylysine conjugate, *Cancer Biother.* **9:**131–141.

Culver, K. W., Ram, Z., Walbridge, S., Ishii, H., Oldfield, E. H., and Blaese, R. M., 1992, *In vivo* gene transfer with retroviral vector-producer cells for treatment of experimental brain tumors, *Science* **256:**1550–1552.

Fick, J., Barker, F. G. II, Dazin, P., Westphale, E. M., Beyer, E. C., and Israel, M. A., 1995, The extent of heterocellular communication mediated by gap junctions is predictive of bystander tumor cytotoxicity in vitro, *Proc. Natl. Acad. Sci. USA* **92:**11071–11075.

Freeman, S. M., Abboud, C. N., Whartenby, K. A., Packman, C. H., and Koeplin, D. S., 1993, The "bystander effect": Tumor regression when a fraction of the tumor mass is genetically modified, *Cancer Res.* **53:**5274–5283.

Freeman, S. M., Whartenby, K. A., Freeman, J. L., Abboud, C. N., and Marrogi, A. J., 1996, *In situ* use of suicide genes for cancer therapy, *Semin. Oncol.* **23:**31–45.

Frei, E., Teicher, B. A., Holden, S. A., Cathcart, K. N., and Wang, Y. Y., 1988, Preclinical studies and clinical correlation of the effect of alkylating dose, *Cancer Res.* **48:**6417–6423.

Friedlos, F., Denny, W. A., Palmer, B. D., and Springer, C. J., 1997, Mustard prodrugs for activation by *Escherichia coli* nitroreductase in gene-directed enzyme prodrug therapy, *J. Med. Chem.* **40:**1270–1275.

Friedlos, F., and Knox, R. J., 1992, Metabolism of NAD(P)H by blood components. Relevance to bioreductively activated prodrugs in a targeted enzyme therapy system, *Biochem. Pharmacol.* **44:** 631–635.

Gagandeep, S., Brew, R., Green, B., Christmas, S. E., Klatzmann, D., Poston, G. J., and Kinsella, A. R., 1996, Prodrug-activated gene therapy: Involvement of an immunological component in the "bystander effect," *Cancer Gene Ther.* **3:**83–88.

Green, N. K., Youngs, D. J., Neoptolemos, J. P., Friedlos, F., Knox, R. J., Springer, C. J., Anlezark, G. M., Michael, N. P., Melton, R. G., Ford, M. J., Young, L. S., Kerr, D. J., and Searle, P. F., 1997, Sensitization of colorectal and pancreatic cell lines to the prodrug 5-(aziridin-1-yl)-2,4-dinitrobenzamide (CB 1954) by retroviral transduction and expression of the *E. coli* nitroreductase gene, *Cancer Gene Ther.* **4:**229–238.

Hart, I. R., and Vile, R. G., 1995, Targeted gene therapy, *Br. Med. Bull.* **51:**647–655.

Hay, M. P., Wilson, W. R., and Denny, W. A., 1995, A novel enediyne prodrug for antibody-directed enzyme prodrug therapy (ADEPT) using *E. coli* B nitroreductase, *Bioorg. Medicinal Chem. Lett.* **5:**2829–2834.

Hoganson, D. K., Batra, R. K., Olsen, J. C., and Boucher, R. C., 1996, Comparison of the effects of three different toxin genes and their levels of expression on cell growth and bystander effect in lung adenocarcinoma, *Cancer Res.* **56:**1315–1323.

Huber, B. E., Austin, E. A., Good, S. S., Knick, V. C., Tibbels, S., and Richards, C. A., 1993, *In vivo* antitumor activity of 5-fluorocytosine on human colorectal carcinoma cells genetically modified to express cytosine deaminase, *Cancer Res.* **53:**4619–4626.

Huber, B. E., Austin, E. A., Richards, C. A., Davis, S. T., and Good, S. S., 1994, Metabolism of 5-fluorocytosine to 5-fluorouracil in human colorectal tumor cells transduced with the cytosine deaminase gene: Significant antitumor effects when only a small percentage of tumor cells express cytosine deaminase, *Proc. Natl. Acad. Sci. USA* **91:**8302–8306.

Huber, B. E., Richards, C. A., and Krenitsky, T. A., 1991, Retroviral-mediated gene therapy for the treatment of hepatocellular carcinoma: An innovative approach for cancer therapy, *Proc. Natl. Acad. Sci. USA* **88:**8039–8043.

Knox, R. J., Friedlos, F., and Boland, M. P., 1993, The bioactivation of CB 1954 and its use as a prodrug in antibody-directed enzyme prodrug therapy (ADEPT), *Cancer Metastasis Rev.* **12:**195–212.

Knox, R. J., Friedlos, F., Jarman, M., Davies, L. C., Goddard, P., Anlezark, G. M., Melton, R. G., and

Sherwood, R. F., 1995, Virtual cofactors for an *Escherichia-coli* nitroreductase enzyme—relevance to reductively activated prodrugs in antibody-directed enzyme prodrug therapy (ADEPT), *Biochem. Pharmacol.* **49:**1641–1647.

Knox, R. J., Friedlos, F., Jarman, M., and Roberts, J. J., 1988, A new cytotoxic, DNA interstrand cross-linking agent, 5-(aziridin-1-yl)-4-hydroxylamino-2-nitrobenzamide, is formed from 5-(aziridin-1-yl)-2,4-dinitrobenzamide (CB 1954) by a nitroreductase enzyme in Walker carcinoma cells, *Biochem. Pharmacol.* **37:**4661–4669.

Knox, R. J., Friedlos, F., Marchbank, T., and Roberts, J. J., 1991, Bioactivation of CB 1954: Reaction of the active 4-hydroxylamino derivative with thioesters to form the ultimate DNA-DNA interstrand cross-linking species, *Biochem. Pharmacol.* **42:**1691–1697.

Knox, R. J., Friedlos, F., Sherwood, R. F., Melton, R. G., and Anlezark, G. M., 1992, The bioactivation of 5-(aziridin-1-yl)-2,4-dinitrobenzamide (CB1954)—II. A comparison of an *Escherichia coli* nitroreductase and Walker DT diaphorase, *Biochem. Pharmacol.* **44:**2297–2301.

Lee, M., Simpson, J. E., Woo, S., Kaenzig, C., Anlezark, G. M., EnoAmooquaye, E., and Burke, P. J., 1997, Synthesis of an aminopropyl, analog of the experimental anticancer drug tallimustine, and activation of its 4-nitrobenzylcarbamoyl prodrug by nitroreductase and NADH, *Bioorg. Medicinal Chem. Lett.* **7:**1065–1070.

Manome, Y., Abe, M., Hagen, M. F., Fine, H. A., and Kufe, D. W., 1994, Enhancer sequences of the DF3 gene regulate expression of the herpes simplex virus thymidine kinase gene and confer sensitivity of human breast cancer cells to ganciclovir, *Cancer Res.* **54:**5408–5413.

Manome, Y., Wen, P. Y., Chen, L., Tanaka, T., Dong, Y., Yamazoe, M., Hirshowitz, A., Kufe, D. W., and Fine, H. A., 1996a, Gene therapy for malignant gliomas using replication incompetent retroviral and adenoviral vectors encoding the cytochrome P450 2B1 gene together with cyclophosphamide, *Gene Ther.* **3:**513–520.

Manome, Y., Wen, P. Y., Dong, Y., Tanaka, T., Mitchell, B. S., Kufe, D. W., and Fine, H. A., 1996b, Viral vector transduction of the human deoxycytidine kinase cDNA sensitizes glioma cells to the cytotoxic effects of cytosine arabinoside *in vitro* and *in vivo*, *Nat. Med.* **2:**567–573.

Marais, R., Spooner, R. A., Light, Y., Martin, J., and Springer, C. J., 1996, Gene-directed enzyme prodrug therapy with a mustard prodrug/carboxypeptidase G2 combination, *Cancer Res.* **56:** 4735–4742.

Mauger, A. B., Burke, P. J., Somani, H. H., Friedlos, F., and Knox, R. J., 1994, Self-immolative prodrugs: Candidates for antibody-directed enzyme prodrug therapy in conjunction with a nitroreductase enzyme, *J. Med. Chem.* **37:**3452–3458.

Maxwell, I. H., Glode, L. M., and Maxwell, F., 1991, Expression of the diphtheria toxin A-chain coding sequence under the control of promoters and enhancers from immunoglobulin genes as a means of directing toxicity to B-lymphoid cells, *Cancer Res.* **51:**4299–4304.

Maxwell, I. H., Glode, L. M., and Maxwell, F., 1992, Expression of diphtheria toxin A-chain in mature B-cells: A potential approach to therapy of B-lymphoid malignancy, *Leukemia Lymphoma* **7:**457–462.

Mendelsohn, M. L., 1960, The growth fraction: A new concept applied to tumours, *Science* **132:**1496.

Michael, N. P., Brehm, J. K., Anlezark, G. M., and Minton, N. P., 1994, Physical characterisation of the *Escherichia coli* B gene encoding nitroreductase and its overexpression in *Escherichia coli* K12, *FEMS Microbiol. Lett.* **124:**195–202.

Miller, N., and Vile, R., 1995, Targeted vectors for gene therapy, *FASEB J.* **9:**190–199.

Minton, N. P., Atkinson, T., Bruton, C. J., and Sherwood, R. F., 1984, The complete nucleotide sequence of the *Pseudomonas* gene coding for carboxypeptidase G2, *Gene* **31:**31–38.

Moolten, F. S., Wells, J. M., and Mroz, P. J., 1992, Multiple transduction as a means of preserving ganciclovir chemosensitivity in sarcoma cells carrying retrovirally transduced herpes thymidine kinase genes, *Cancer Lett.* **64:**257–263.

Mullen, C. A., 1994, Metabolic suicide genes in gene therapy, *Pharmacol. Ther.* **63:**199–207.

Mullen, C. A., Coale, M. M., Lowe, R., and Blaese, R. M., 1994, Tumors expressing the cytosine deaminase suicide gene can be eliminated *in vivo* with 5-fluorocytosine and induce protective immunity to wild type tumor, *Cancer Res.* **54:**1503–1506.

Mullen, C. A., Kilstrup, M., and Blaese, R. M., 1992, Transfer of the bacterial gene for cytosine deaminase to mammalian cells confers lethal sensitivity to 5-fluorocytosine: A negative selection system, *Proc. Natl. Acad. Sci. USA* **89:**33–37.

Oldfield, E. H., Ram, Z., Culver, K. W., Blaese, R. M., DeVroom, H. L., and Anderson, W. F., 1993, Gene therapy for the treatment of brain tumors using intratumoral transduction with the thymidine kinase gene and intravenous ganciclovir, *Hum. Gene Ther.* **4:**39–69.

O'Malley, B., Jr., Chen, S. H., Schwartz, M. R., and Woo, S. L., 1995, Adenovirus-mediated gene therapy for human head and neck squamous cell cancer in a nude mouse model, *Cancer Res.* **55:** 1080–1085.

Osaki, T., Tanio, Y., Tachibana, I., Hosoe, S., Kumagai, T., Kawase, I., Oikawa, S., and Kishimoto, T., 1994, Gene therapy for carcinoembryonic antigen-producing human lung cancer cells by cell type-specific expression of herpes simplex virus thymidine kinase gene, *Cancer Res.* **54:**5258–5261.

Patterson, A. V., Zhang, H., Moghaddam, A., Bicknell, R., Talbot, D. C., Stratford, I. J., and Harris, A. L., 1995, Increased sensitivity to the prodrug 5′-deoxy-5-fluorouridine and modulation of 5-fluoro-2′-deoxyuridine sensitivity in MCF-7 cells transfected with thymidine phosphorylase, *Br. J. Cancer* **72:**669–675.

Ram, Z., Culver, K. W., Walbridge, S., Frank, J. A., Blaese, R. M., and Oldfield, E. H., 1993, Toxicity studies of retroviral-mediated gene transfer for the treatment of brain tumors, *J. Neurosurg.* **79:**400–407.

Roberts, J. J., Friedlos, F., and Knox, R. J., 1986, CB 1954 (2,4-dinitro-5-aziridinyl benzamide) becomes a DNA interstrand cross-linking agent in Walker tumour cells, *Biochem. Biophys. Res. Commun.* **140:**1073–1078.

Robinson, D. F., and Maxwell, I. H., 1995, Suppression of single and double nonsense mutations introduced into the diphtheria toxin A-chain gene: A potential binary system for toxin gene therapy, *Hum. Gene Ther.* **6:**137–143.

Romanini, A., Sobrero, A. F., Chou, T. C., Sherwood, R. F., and Bertino, J. R., 1989, Enhancement of trimetrexate cytotoxicity *in vitro* and *in vivo* by carboxypeptidase G2, *Cancer Res.* **49:**6019–6023.

Smythe, W. R., Hwang, H. C., Amin, K. M., Eck, S. L., Davidson, B. L., Wilson, J. M., Kaiser, L. R., and Albelda, S. M., 1994, Use of recombinant adenovirus to transfer the herpes simplex virus thymidine kinase (HSVtk) gene to thoracic neoplasms: An effective *in vitro* drug sensitization system, *Cancer Res.* **54:**2055–2059.

Springer, C. J., Spooner, R. A., Light, Y., Martin, J., Stribbling, S., Friedlos, F., and Marais, R., 1997, Intracellular and extracellular expression of the carboxypeptidase G2 enzyme for activation of a mustard prodrug in gene-directed enzyme prodrug therapy (GDEPT), *Br. J. Cancer* **75(Suppl 1):**13.

Tanaka, T., Kanai, F., Okabe, S., Yoshida, Y., Wakimoto, H., Hamada, H., Shiratori, Y., Lan, K., Ishitobi, M., and Omata, M., 1996, Adenovirus-mediated prodrug gene therapy for carcinoembryonic antigen-producing human gastric carcinoma cells in vitro, *Cancer Res.* **56:**1341–1345.

Tercel, M., Denny, W. A., and Wilson, W. R., 1996, A novel nitro-substituted *seco*-CI: Application as a reductively activated ADEPT prodrug, *Bioorg. Medicinal Chem. Lett.* **6:**2741–2744.

Tong, X. W., Block, A., Chen, S. H., Contant, C. F., Agoulnik, I., Blankenburg, K., Kaufman, R. H., Woo, S. L., and Kieback, D. G., 1996a, In vivo gene therapy of ovarian cancer by adenovirus-mediated thymidine kinase gene transduction and ganciclovir administration, *Gynecol. Oncol.* **1:**175–179.

Tong, X. W., Block, A., Chen, S. H., Woo, S. L., and Kieback, D. G., 1996b, Adenovirus-mediated thymidine kinase gene transduction in human epithelial ovarian cancer cell lines followed by exposure to ganciclovir, *Anticancer Res.* **16:**1611–1617.

Vile, R. G., and Hart, I. R., 1993, Use of tissue-specific expression of the herpes simplex virus thymidine kinase gene to inhibit growth of established murine melanomas following direct intratumoral injection of DNA, *Cancer Res.* **53:**3860–3864.

Vile, R. G., Nelson, J. A., Castleden, S., Chong, H., and Hart, I. R., 1994, Systemic gene therapy of murine melanoma using tissue specific expression of the HSVtk gene involves an immune component, *Cancer Res.* **54:**6228–6234.

Wei, M. X., Tamiya, T., Chase, M., Boviatsis, E. J., Chang, T. K., Kowall, N. W., Hochberg, F. H., Waxman, D. J., Breakefield, X. O., and Chiocca, E. A., 1994, Experimental tumor therapy in mice using the cyclophosphamide-activating cytochrome P450 2B1 gene, *Hum. Gene Ther.* **5:** 969–978.

Wei, M. X., Tamiya, T., Rhee, R. J., Breakefield, X. O., and Chiocca, E. A., 1995, Diffusible cytotoxic metabolites contribute to the in vitro bystander effect associated with the cyclophosphamide/cytochrome P450 2B1 cancer gene therapy paradigm, *Clin. Cancer Res.* **1:**1171–1177.

Appendix

Enzymes and Prodrugs Proposed for Cancer Therapy

The enzymes may have been targets for prodrug therapy or targeted using either antibody-directed enzymes prodrug therapy (ADEPT) or gene-directed enzyme prodrug therapy (GDEPT).

Enzyme-Prodrug Strategies for Cancer Therapy, edited by Melton and Knox. Kluwer Academic/Plenum Publishers, New York, 1999.

Enzyme	Prodrug	Drug	Application	References
Alkaline phosphatase	Phenolmustard phosphate	Phenolmustard	ADEPT	Wallace and Senter, 1991
	Doxorubicin phosphate	Doxorubicin	ADEPT	Senter, 1990
	Mitomycin phosphate	Mitomycin alcohol	ADEPT	Sahin *et al.*, 1990; Senter *et al.*, 1989
	Etoposide phosphate	Etoposide	ADEPT	Haisma *et al.*, 1992; Senter *et al.*, 1988, 1989
Aminopeptidase	2-L-Pyroglutamylmethotrexate	Methotrexate	Prodrug therapy ADEPT	Smal *et al.*, 1995
Azoreductase	Azobenzene mustards	Phenylenediamine-mustards (various)	Prodrug therapy	Austrup and Warwick, 1975; Bukhari *et al.*, 1973
Carboxypeptidase A/B	Methotrexate-alanine	Methotrexate	ADEPT	Esswein *et al.*, 1991; Haenseler *et al.*, 1992; Kuefner *et al.*, 1989; Perron and Page, 1996; Smith *et al.*, 1997
	α-Linked derivatives of quinazoline antifolates	TS inhibitors	ADEPT	Springer *et al.*, 1996
Carboxypeptidase G2	Benzoic acid mustard glutamates	Banzoic acid mustards (various)	ADEPT	Bagshawe, 1989; Bagshawe *et al.*, 1988; Dowell *et al.*, 1996; Springer, 1993; Springer *et al.*, 1990, 1994, 1995
			GDEPT	Marais *et al.*, 1996
Cytochrome P450	Cyclophosphamide Ifosfamide	Phosphamide mustard (+ acrolein?)	GDEPT	Chen and Waxman, 1995; Chen *et al.*, 1996; Manome *et al.*, 1996a; Wei *et al.*, 1994, 1995
Cytosine deaminase	5-Fluorocytosine	5-Fluorouracil	GDEPT	Hoganson *et al.*, 1996; Huber *et al.*, 1993, 1994; Mullen *et al.*, 1992, 1994
			ADEPT	Senter *et al.*, 1991; Wallace *et al.*, 1994
Deoxycytidine kinase	Ara-C	Adenine arabinonucleoside triphosphate (araATP)	GDEPT	Manome *et al.*, 1996b

DT diaphorase	5-(Aziridine-1-yl)-2,4-dinitrobenzamide (CB 1954)	5-(Aziridine-1-yl)-4-hydroxylamino-2-nitrobenzamide	Prodrug therapy	Cobb *et al.*, 1969; Knox *et al.*, 1988a,b, 1993
	3-Hydroxymethyl-5-aziridinyl-1-methyl-2-(1H indole-4,7-dione)prop-β-en-a-ol (EO9)	Reduced-EO9 products	Prodrug therapy	Robertson *et al.*, 1994
α-Galactosidase	N-[4-(α-galactopyranosyl)-benyloxycarbonyl]-daunorubicin	Daunorubicin	ADEPT	Andrianomenjanahary *et al.*, 1992
Glucose oxidase	Glucose	Hydrogen peroxide	Radical generating system	Ito *et al.*, 1989; Muzykantov *et al.*, 1988, 1990; Stanislawski *et al.*, 1989
β-Glucosidase	Amygdalin	Cyanide	ADEPT	Rowlinson-Busza *et al.*, 1992; Rowlinson-Busza and Epenetos, 1992
β-Glucuronidase	Phenolmustard-glucuronide	Phenolmustard	Prodrug therapy	Connors and Whisson, 1966; Whisson *et al.*, 1966
	Epirubicin-glucuronide	Epirubicin	ADEPT	Haisma *et al.*, 1994; Roffler *et al.*, 1991; Wang *et al.*, 1992
	Daunorubicin-glucuronides	Daunorubicin	ADEPT	Haisma *et al.*, 1994
γ-Glutamyl transferase	γ-Glutamyl *p*-phenylenediamine-mustard	Phenylenediamine-mustard	Prodrug therapy	Smith *et al.*, 1984
β-Lactamase	Vinca-cephalosporin	4-Desacetylvinblastine-3-carboxyhydrazide	ADEPT	Meyer *et al.*, 1992, 1993, 1995; Shepherd *et al.*, 1991
	Nitrogen-mustard-cephalosporin	Nitrogen mustards (various)	ADEPT	Alexander *et al.*, 1991; Svensson *et al.*, 1992
	Doxorubicin-cephalosporin	Doxorubicin	ADEPT	Vrudhula *et al.*, 1995

Enzyme	Prodrug	Drug	Application	References
Nitroreductase	5-(Aziridine-1-yl)-2,4-dinitrobenzamide (CB 1954)	5-(Aziridine-1-yl)-4-hydroxylamino-2-nitrobenzamide	ADEPT	Anlezark *et al.*, 1992; Knox *et al.*, 1992, 1993
			GDEPT	Bailey *et al.*, Bridgewater *et al.*, 1995, 1997; Green *et al.*, 1997
	2-4-dinitrobenzamide mustards	Activated mustards	ADEPT and GDEPT	Anlezark *et al.*, 1995; Friedlos *et al.*, 1997
	4-Nitrobenzyloxycarbonyl derivatives	E.g., Actinomycin D, mitomycin C, enediyenes, tallimustine	ADEPT and GDEPT	Hay *et al.*, 1995; Lee *et al.*, 1997; Mauger *et al.*, 1994; Tercel *et al.*, 1996
Penicillin amidase	Palytoxin-4-hydroxyphenylacetamide	Palytoxin	ADEPT	Bignami *et al.*, 1992
	Doxorubicin-phenoxyacetamide	Doxorubicin	ADEPT	Kerr *et al.*, 1990; Vrudhula *et al.*, 1993
	Melphalan-phenoxyacetamide	Melphalan	ADEPT	Vrudhula *et al.*, 1993
Plasmin	Peptidyl-*p*-phenylenediamine-mustard	Phenylenediamine-mustard	Prodrug therapy	Chakravarty *et al.*, 1983a
	Peptidyl-*p*-doxorubicin	Doxorubicin	Prodrug therapy	Chakravarty *et al.*, 1983b
Ribonuclease A (mutant)	Ribotide prodrug	Aniline mustard	ADEPT	Blakey *et al.*, 1997
Thymidine phosphorylase	5′-Deoxy-5-fluorouridine (5′-DFUR)	5-FU	GDEPT	Patterson *et al.*, 1995
Thymidine kinase (viral)	Ganciclovir	Ganciclovir triphosphate	GDEPT	Abe *et al.*, 1993; Barba *et al.*, 1993; Borrelli *et al.*, 1988; Chen *et al.*, 1994; Culver *et al.*, 1992; Moolten *et al.*, 1992; Ram *et al.*, 1993; Smythe *et al.*, 1994; Tanaka *et al.*, 1996; Tong *et al.*, 1996a, b; Vile and Hart, 1991
	Adenine arabinonucleoside (ara-M)	Adenine arabinonucleoside triphosphate (araATP)	GDEPT	Huber *et al.*, 1993
Xanthine oxidase	Hypoxanthine	Superoxide, hydrogen peroxide	Radical generating system	Ito *et al.*, 1990

REFERENCES

Abe, A., Takeo, T., Emi, N., Tanimoto, M., Ueda, R., Yee, J. K., Friedmann, T., and Saito, H., 1993, Transduction of a drug-sensitive toxic gene into human leukemia cell lines with a novel retroviral vector, *Proc. Soc. Exp. Biol. Med.* **203:**354–359.

Alexander, R. P., Beeley, N. R. A., O'Driscoll, M., O'Neill, F. P., Millican, T. A., Pratt, A. J., and Willenbrock, F. W., 1991, Cephalosporin nitrogen mustard carbamate prodrugs for "ADEPT", *Tetrahedron Lett.* **32:**3269–3272.

Andrianomenjanahary, S., Dong, X., Florent, J. C., Gaudel, G., Gesson, J. P., Jacquesy, J. C., Koch, M., Michel, S., and Mondon, M., 1992, Synthesis of novel targeted pro-prodrugs of anthracyclines potentially activated by a monoclonal antibody galactosidase conjugate, *Bioorg. Medicinal Chem. Lett.* **2:**1093–1096.

Anlezark, G. M., Melton, R. G., Sherwood, R. F., Coles, B., Friedlos, F., and Knox, R. J., 1992, The bioactivation of 5-(aziridin-1-yl)-2,4-dinitrobenzamide (CB 1954)—I. Purification and properties of a nitroreductase enzyme from *Escherichia coli*—a potential enzyme for antibody-directed enzyme prodrug therapy (ADEPT), *Biochem. Pharmacol.* **44:**2289–2295.

Anlezark, G. M., Melton, R. G., Sherwood, R. F., Wilson, W. R., Denny, W. A., Palmer, B. D., Knox, R. J., Friedlos, F., and Williams, A., 1995, Bioactivation of dinitrobenzamide mustards by an *E. coli* B nitroreductase, *Biochem. Pharmacol.* **50:**609–618.

Autrup, H., and Warwick, G. P., 1975, Some characteristics of two azoreductase systems in rat liver. Relevance to the activity of 2-[4′-di(2″-bromopropyl)-aminophenylazo]benzoic acid (CB10-252), a compound possessing latent cytotoxic activity, *Chem. Biol. Interactions* **11:**329–342.

Bagshawe, K. D., 1989, The first Bagshawe lecture. Towards generating cytotoxic agents at cancer sites, *Br. J. Cancer* **60:**275–281.

Bagshawe, K. D., Springer, C. J., Searle, F., Antoniw, P., Sharma, S. K., Melton, R. G., and Sherwood, R. F., 1988, A cytotoxic agent can be generated selectively at cancer sites, *Br. J. Cancer* **58:** 700–703.

Bailey, S. M., Knox, R. J., Hobbs, S. M., Jenkins, T. C., Mauger, A. B., Melton, R. G., Burke, P. J., Connors, T. A., and Hart, I. R., 1996, Investigation of alternative prodrugs for use with *E. coli* nitroreductase in 'suicide gene' approaches to cancer therapy, *Gene Ther.* **3:**1143–1150.

Barba, D., Hardin, J., Ray, J., and Gage, F. H., 1993, Thymidine kinase-mediated killing of rat brain tumors, *J. Neurosurg.* **79:**729–735.

Bignami, G. S., Senter, P. D., Grothaus, P. G., Fischer, K. J., Humphreys, T., and Wallace, P. M., 1992, *N*-(4′-hydroxyphenylacetyl)palytoxin: A palytoxin prodrug that can be activated by a monoclonal antibody-penicillin G amidase conjugate, *Cancer Res.* **52:**5759–5764.

Blakey, D. C., Boyle, F. T., Davies, D. H., Dowell, R. I., Heaton, D. W., Rose, M. S., Slater, A. M., Eggelte, H. J., Tarragona-Fiol, A., and Taylorson, C. J., 1997, Mutant human ribonuclease for use in antibody-directed enzyme prodrug therapy (ADEPT), *Br. J. Cancer* **75(Suppl 1):**36 (Abstract).

Borrelli, E., Heyman, R., Hsi, M., and Evans, R. M., 1988, Targeting of an inducible toxic phenotype in animal cells, *Proc. Natl. Acad. Sci. USA* **85:**7572–7576.

Bridgewater, J. A., Knox, R. J., Pitts, J. D., Collins, M. K., and Springer, C. J., 1997, The bystander effect of the nitroreductase CB 1954 enzyme prodrug system is due to a cell-permeable metabolite, *Hum. Gene Ther.* **8:**709–717.

Bridgewater, J. A., Springer, C. J., Knox, R. J., Minton, N. P., Michael, N. P., and Collins, M. K., 1995, Expression of the bacterial nitroreductase enzyme in mammalian cells renders them selectively sensitive to killing by the prodrug CB1954, *Eur. J. Cancer* **31a:**2362–2370.

Bukhari, A., Connors, T. A., Gilsenan, A. M., Ross, W. C., Tisdale, M. J., Warwick, G. P., and Wilman, D. E., 1973, Cytotoxic agents designed to be selective for liver cancer, *J. Natl. Cancer Inst.* **50:** 243–247.

Chakravarty, P. K., Carl, P. L., Weber, M. J., and Katzenellenbogen, J. A., 1983a, Plasmin-activated prodrugs for cancer chemotherapy. 2. Synthesis and biological activity of peptidyl derivatives of doxorubicin, *J. Med. Chem.* **26:**638–644.

Chakravarty, P. K., Carl, P. L., Weber, M. J., and Katzenellenbogen, J. A., 1983b, Plasmin-activated prodrugs for cancer chemotherapy. 1. Synthesis and biological activity of peptidylacivicin and peptidylphenylenediamine mustard, *J. Med. Chem.* **26:**633–638.

Chen, L., and Waxman, D. J., 1995, Intratumoral activation and enhanced chemotherapeutic effect of oxazaphosphorines following cytochrome P-450 gene transfer: Development of a combined chemotherapy/cancer gene therapy strategy, *Cancer Res.* **55:**581–589.

Chen, L., Waxman, D. J., Chen, D., and Kufe, D. W., 1996, Sensitization of human breast cancer cells to cyclophosphamide and ifosfamide by transfer of a liver cytochrome P450 gene, *Cancer Res.* **56:**1331–1340.

Chen, S. H., Shine, H. D., Goodman, J. C., Grossman, R. G., and Woo, S. L., 1994, Gene therapy for brain tumors: Regression of experimental gliomas by adenovirus-mediated gene transfer *in vivo*, *Proc. Natl. Acad. Sci. USA* **91:**3054–3057.

Cobb, L. M., Connors, T. A., Elson, L. A., Khan, A. H., Mitchley, B. C., Ross, W. C., and Whisson, M. E., 1969, 2,4-Dinitro-5-ethyleneiminobenzamide (CB 1954): A potent and selective inhibitor of the growth of the Walker carcinoma 256, *Biochem. Pharmacol.* **18:**1519–1527.

Connors, T. A., and Whisson, M. E., 1966, Cure of mice bearing advanced plasma cell tumours with aniline mustard: The relationship between glucuronidase activity and tumour sensitivity, *Nature* **210:**866–867.

Culver, K. W., Ram, Z., Walbridge, S., Ishii, H., Oldfield, E. H., and Blaese, R. M., 1992, *In vivo* gene transfer with retroviral vector-producer cells for treatment of experimental brain tumors, *Science* **256:**1550–1552.

Dowell, R. L., Springer, C. J., Davies, D. H., Hadley, E. M., Burke, P. J., Boyle, F. T., Melton, R. G., Connors, T. A., Blakey, D. C., and Mauger, A. B., 1996, New mustard prodrugs for antibody-directed enzyme prodrug therapy: Alternatives to the amide link, *J. Med. Chem.* **39:**1100–1105.

Esswein, A., Hanseler, E., Montejano, Y., Vitols, K. S., and Huennekens, F. M., 1991, Construction and chemotherapeutic potential of carboxypeptidase-A/monoclonal antibody conjugate, *Adv. Enzyme Regul.* **31:**3–12.

Friedlos, F., Denny, W. A., Palmer, B. D., and Springer, C. J., 1997, Mustard prodrugs for activation by *Escherichia coli* nitroreductase in gene-directed enzyme prodrug therapy, *J. Med. Chem.* **40:**1270–1275.

Green, N. K., Youngs, D. J., Neoptolemos, J. P., Friedlos, F., Knox, R. J., Springer, C. J., Anlezark, G. M., Michael, N. P., Melton, R. G., Ford, M. J., Young, L. S., Kerr, D. J., and Searle, P. F., 1997, Sensitization of colorectal and pancreatic cell lines to the prodrug 5-(aziridin-1-yl)-2,4-dinitrobenzamide (CB 1954) by retroviral transduction and expression of the *E. coli* nitroreductase gene, *Cancer Gene Ther.* **4:**229–238.

Haenseler, E., Esswein, A., Vitols, K. S., Montejano, Y., Mueller, B. M., Reisfeld, R. A., and Huennekens, F. M., 1992, Activation of methotrexate-α-alanine by carboxypeptidase-A monoclonal antibody conjugate, *Biochemistry* **31:**891–897.

Haisma, H. J., Boven, E., van Muijen, M., De Vries, R., and Pinedo, H. M., 1992, Analysis of a conjugate between anti-carcinoembryonic antigen monoclonal antibody and alkaline phosphatase for specific activation of the prodrug etoposide phosphate, *Cancer Immunol. Immunother.* **34:**343–348.

Haisma, H. J., van-Muijen, M., Pinedo, H. M., and Boven, E., 1994, Comparison of two anthracycline-based prodrugs for activation by a monoclonal antibody-β-glucuronidase conjugate in the specific treatment of cancer, *Cell Biophys.* **25:**185–192.

Hay, M. P., Wilson, W. R., and Denny, W. A., 1995, A novel enediyne prodrug for antibody-directed enzyme prodrug therapy (ADEPT) using *E. coli* B nitroreductase, *Bioorg. Medicinal Chem. Lett.* **5:**2829–2834.

Hoganson, D. K., Batra, R. K., Olsen, J. C., and Boucher, R. C., 1996, Comparison of the effects of three different toxin genes and their levels of expression on cell growth and bystander effect in lung adenocarcinoma, *Cancer Res.* **56:**1315–1323.

Houba, P. H., Leenders, R. G., Boven, E., Scheeren, J. W., Pinedo, H. M., and Haisma, H. J., 1996, Characterization of novel anthracycline prodrugs activated by human β-glucuronidase for use in antibody-directed enzyme prodrug therapy, *Biochem. Pharmacol.* **52:**455–463.

Huber, B. E., Austin, E. A., Good, S. S., Knick, V. C., Tibbels, S., and Richards, C. A., 1993, *In vivo* antitumor activity of 5-fluorocytosine on human colorectal carcinoma cells genetically modified to express cytosine deaminase, *Cancer Res.* **53:**4619–4626.

Huber, B. E., Austin, E. A., Richards, C. A., Davis, S. T., and Good, S. S., 1994, Metabolism of 5-fluorocytosine to 5-fluorouracil in human colorectal tumor cells transduced with the cytosine deaminase gene: Significant antitumor effects when only a small percentage of tumor cells express cytosine deaminase, *Proc. Natl. Acad. Sci. USA* **91:**8302–8306.

Huber, B. E., Richards, C. A., and Krenitsky, T. A., 1991, Retroviral-mediated gene therapy for the treatment of hepatocellular carcinoma: An innovative approach for cancer therapy, *Proc. Natl. Acad. Sci. USA* **88:**8039–8043.

Ito, H., Morizet, J., Coulombel, L., Goavec, M., Rousseau, V., Bernard, A., and Stanislawski, M., 1989, An immunotoxin system intended for bone marrow purging composed of glucose oxidase and lactoperoxidase coupled to monoclonal antibody 097, *Bone Marrow Transplant.* **4:**519–527.

Ito, H., Morizet, J., Coulombel, L., and Stanislawski, M., 1990, T cell depletion of human bone marrow using an oxidase-peroxidase enzyme immunotoxin, *Bone Marrow Transplant.* **6:**395–398.

Kerr, D. E., Senter, P. D., Burnett, W. V., Hirschberg, D. L., Hellstrom, I., and Hellstrom, K. E., 1990, Antibody-penicillin-V-amidase conjugates kill antigen-positive tumor cells when combined with doxorubicin phenoxyacetamide, *Cancer Immunol. Immunother.* **31:**202–206.

Knox, R. J., Boland, M. P., Friedlos, F., Coles, B., Southan, C., and Roberts, J. J., 1988a, The nitroreductase enzyme in Walker cells that activates 5-(aziridin-1-yl)-2,4-dinitrobenzamide (CB 1954) to 5-(aziridin-1-yl)-4-hydroxylamino-2-nitrobenzamide is a form of NAD(P)H dehydrogenase (quinone) (EC 1.6.99.2), *Biochem. Pharmacol.* **37:**4671–4677.

Knox, R. J., Friedlos, F., and Boland, M. P., 1993, The bioactivation of CB 1954 and its use as a prodrug in antibody-directed enzyme prodrug therapy (ADEPT), *Cancer Metastasis Rev.* **12:**195–212.

Knox, R. J., Friedlos, F., Jarman, M., and Roberts, J. J., 1988b, A new cytotoxic, DNA interstrand cross-linking agent, 5-(aziridin-1-yl)-4-hydroxylamino-2-nitrobenzamide, is formed from 5-(aziridin-1-yl)-2,4-dinitrobenzamide (CB1954) by a nitroreductase enzyme in Walker carcinoma cells, *Biochem. Pharmacol.* **37:**4661–4669.

Knox, R. J., Friedlos, F., Sherwood, R. F., Melton, R. G., and Anlezark, G. M., 1992, The bioactivation of 5-(aziridin-1-yl)-2,4-dinitrobenzamide (CB1954)—II. A comparison of an *Escherichia coli* nitroreductase and Walker DT diaphorase, *Biochem. Pharmacol.* **44:**2297–2301.

Kuefner, U., Lohrmann, U., Montejano, Y. D., Vitols, K. S., and Huennekens, F. M., 1989, Carboxypeptidase-mediated release of methotrexate from methotrexate-α-peptides, *Biochemistry* **28:**2288–2297.

Lee, M., Simpson, J. E., Woo, S., Kaenzig, C., Anlezark, G. M., EnoAmooquaye, E., and Burke, P. J., 1997, Synthesis of an aminopropyl, analog of the experimental anticancer drug tallimustine, and activation of its 4-nitrobenzylcarbamoyl prodrug by nitroreductase and NADH, *Bioorg. Medicinal Chem. Lett.* **7:**1065–1070.

Manome, Y., Wen, P. Y., Chen, L., Tanaka, T., Dong, Y., Yamazoe, M., Hirshowitz, A., Kufe, D. W., and Fine, H. A., 1996a, Gene therapy for malignant gliomas using replication incompetent retroviral and adenoviral vectors encoding the cytochrome P450 2B1 gene together with cyclophosphamide, *Gene Ther.* **3:**513–520.

Manome, Y., Wen, P. Y., Dong, Y., Tanaka, T., Mitchell, B. S., Kufe, D. W., and Fine, H. A., 1996b, Viral vector transduction of the human deoxycytidine kinase cDNA sensitizes glioma cells to the cytotoxic effects of cytosine arabinoside *in vitro* and *in vivo*, *Nat. Med.* **2:**567–573.

Marais, R., Spooner, R. A., Light, Y., Martin, J., and Springer, C. J., 1996, Gene-directed enzyme prodrug therapy with a mustard prodrug/carboxypeptidase G2 combination, *Cancer Res.* **56:** 4735–4742.

Mauger, A. B., Burke, P. J., Somani, H. H., Friedlos, F., and Knox, R. J., 1994, Self-immolative prodrugs: Candidates for antibody-directed enzyme prodrug therapy in conjunction with a nitroreductase enzyme, *J. Med. Chem.* **37:**3452–3458.

Meyer, D. L., Jungheim, L. N., Law, K. L., Mikolajczyk, S. D., Shepherd, T. A., Mackensen, D. G., Briggs, S. L., and Starling, J. J., 1993, Site-specific prodrug activation by antibody-β-lactamase conjugates: Regression and long-term growth inhibition of human colon carcinoma xenograft models, *Cancer Res.* **53:**3956–3963.

Meyer, D. L., Jungheim, L. N., Mikolajczyk, S. D., Shepherd, T. A., Starling, J. J., and Ahlem, C. N., 1992, Preparation and characterization of a β-lactamase-Fab′ conjugate for the site-specific activation of oncolytic agents, *Bioconjugate Chem.* **3:**42–48.

Meyer, D. L., Law, K. L., Payne, J. K., Mikolajczyk, S. D., Zarrinmayeh, H., Jungheim, L. N., Kling, J. K., Shepherd, T. A., and Starling, J. J., 1995, Site-specific prodrug activation by antibody-β-lactamase conjugates: Preclinical investigation of the efficacy and toxicity of doxorubicin delivered by antibody directed catalysis, *Bioconjugate Chem.* **6:**440–446.

Moolten, F. S., Wells, J. M., and Mroz, P. J., 1992, Multiple transduction as a means of preserving ganciclovir chemosensitivity in sarcoma cells carrying retrovirally transduced herpes thymidine kinase genes, *Cancer Lett.* **64:**257–263.

Mullen, C. A., Coale, M. M., Lowe, R., and Blaese, R. M., 1994, Tumors expressing the cytosine deaminase suicide gene can be eliminated *in vivo* with 5-fluorocytosine and induce protective immunity to wild type tumor, *Cancer Res.* **54:**1503–1506.

Mullen, C. A., Kilstrup, M., and Blaese, R. M., 1992, Transfer of the bacterial gene for cytosine deaminase to mammalian cells confers lethal sensitivity to 5-fluorocytosine: A negative selection system, *Proc. Natl. Acad. Sci. USA* **89:**33–37.

Muzykantov, V. R., Saharov, D. V., Sinitsyn, V. V., Domogatsky, S. P., Goncharov, N. V., and Danilov, S. M., 1988, Specific killing of human endothelial cells by antibody-conjugated glucose oxidase, *Anal. Chem.* **169:**383–389.

Muzykantov, V. R., Trubetskaya, O. V., Puchnina, E. A., Saharov, D. V., and Domogatsky, S. P., 1990, Cytotoxicity of glucose oxidase conjugated with antibodies to target cells: Killing efficiecy depends on conjugate internalization, *Biochim. Biophys. Acta* **1053:**27–31.

Patterson, A. V., Zhang, H., Moghaddam, A., Bicknell, R., Talbot, D. C., Stratford, L. J., and Harris, A. L., 1995, Increased sensitivity to the prodrug 5′-deoxy-5-fluorouridine and modulation of 5-fluoro-2′-deoxyuridine sensitivity in MCF-7 cells transfected with thymidine phosphorylase, *Br. J. Cancer* **72:**669–675.

Perron, M. J., and Page, M., 1996, Activation of methotrexate-phenylalanine by monoclonal antibody–carboxypeptidase A conjugate for the specific treatment of ovarian cancer in vitro, *Br. J. Cancer* **73:**281–287.

Ram, Z., Culver, K. W., Walbridge, S., Frank, J. A., Blaese, R. M., and Oldfield, E. H., 1993, Toxicity studies of retroviral-mediated gene transfer for the treatment of brain tumors, *J. Neurosurg.* **79:**400–407.

Robertson, N., Haigh, A., Adams, G. E., and Stratford, I. J., 1994, Factors affecting sensitivity to EO9 in rodent and human tumour cells *in vitro*: DT-diaphorase activity and hypoxia, *Eur. J. Cancer* **7:**1013–1019.

Roffler, S. R., Wang, S. M., Chern, J. W., Yeh, M. Y., and Tung, E., 1991, Anti-neoplastic glucuronide prodrug treatment of human tumor cells targeted with a monoclonal antibody-enzyme conjugate, *Biochem. Pharmacol.* **42:**2062–2065.

Rowlinson-Busza, G., Bamias, A., Krausz, T., and Epenetos, A. A., 1992, Antibody-guided enzyme nitrile therapy (Agent): *In vitro* cytotoxicity and *in vivo* tumour localisation, in: *Antibody-Guided Enzyme Nitrile Therapy (Agent): In Vitro Cytotoxicity and In Vitro Tumour Localisation* (Epenetos, A. A., ed.), Chapman and Hall, London, pp. 111–118.

Rowlinson-Busza, G., and Epenetos, A. A., 1992, Targeted delivery of biologic and other antineoplastic agents, *Curr. Opinion Oncol.* **4:**1142–1148.

Sahin, U., Hartmann, F., Senter, P., Pohl, C., Engert, A., Diehl, V., and Pfreundschuh, M., 1990, Specific activation of the prodrug mitomycin phosphate by a bispecific anti-CD30/anti-alkaline phosphatase monoclonal antibody, *Cancer Res.* **50:**6944–6948.

Senter, P. D., 1990, Activation of prodrugs by antibody-enzyme conjugates: A new approach to cancer therapy, *FASEB J.* **4:**188–193.

Senter, P. D., Saulnier, M. G., Schreiber, G. J., Hirschberg, D. L., Brown, J. P., Hellstrom, I., and Hellstrom, K. E., 1988, Anti-tumor effects of antibody-alkaline phosphatase conjugates in combination with etoposide phosphate, *Proc. Natl. Acad. Sci. USA* **85:**4842–4846.

Senter, P. D., Schreiber, G. J., Hirschberg, D. L., Ashe, S. A., Hellstrom, K. E., and Hellstrom, I., 1989, Enhancement of the *in vitro* and *in vivo* antitumor activities of phosphorylated mitomycin C and etoposide derivatives by monoclonal antibody-alkaline phosphatase conjugates, *Cancer Res.* **49:** 5789–5792.

Senter, P. D., Su, P. C. D., Katsuragi, T., Sakai, T., Cosand, W. L, Hellstrom, I., and Hellstrom, K. E., 1991, Generation of 5-fluorouracil from 5-fluorocytosine by monoclonal antibody-cytosine deaminase conjugates, *Bioconjugate Chem.* **2:**447–451.

Shepherd, T. A., Jungheim, L. N., Meyer, D. M., and Starling, J. J., 1991, A novel targeted delivery system utilizing a cephalosporin-oncolytic prodrug activated by an antibody B-lactamase conjugate for the treatment of cancer, *Bioorg. Medicinal Chem. Lett.* **1:**21–26.

Smal, M. A., Dong, Z., Cheung, H. T., Asano, Y., Escoffier, L., Costello, M., and Tattersall, M. H., 1995, Activation and cytotoxicity of 2-α-aminoacyl prodrugs of methotrexate, *Biochem. Pharmacol.* **49:**567–574.

Smith, G. D., Chakravarty, P. K., Connors, T. A., and Peters, T. J., 1984, Synthesis and preliminary characterization of a novel substrate for γ-glutamyl transferase. A potential anti-hepatoma drug, *Biochem. Pharmacol.* **33:**527–529.

Smith, G. K., Banks, S., Blumenkopf, T. A., Cory, M., Humphreys, J., Laethem, R. M., Miller, J., Moxham, C. P., Mullin, R., Ray, P. H., Walton, L. M., and Wolfe, L. A., 1997, Toward antibody-directed enzyme prodrug therapy with the T268G mutant of human carboxypeptidase A1 and novel *in vivo* stable prodrugs of methotrexate, *J. Biol. Chem.* **272:**15804–15816.

Smythe, W. R., Hwang, H. C., Amin, K. M., Eck, S. L., Davidson, B. L., Wilson, J. M., Kaiser, L. R., and Albelda, S. M., 1994, Use of recombinant adenovirus to transfer the herpes simplex virus thymidine kinase (HSVtk) gene to thoracic neoplasms: An effective *in vitro* drug sensitization system, *Cancer Res.* **54:**2055–2059.

Springer, C. J., 1993, CMDA 4-[2-chloroethyl)[2-(mesyloxy)ethyl]amino]benzoyl-L-glutamic acid. Antineoplastic prodrug, *Drugs of the Future* **18:**212–215.

Springer, C. J., Antoniw, P., Bagshawe, K. D., Searle, F., Bisset, G. M., and Jarman, M., 1990, Novel prodrugs which are activated to cytotoxic alkylating agents by carboxypeptidase G2, *J. Med. Chem.* **33:**677–681.

Springer, C. J., Bavetsias, V., Jackman, A. L., Boyle, F. T., Marshall, D., Pedley, R. B., and Bisset, G. M., 1996, Prodrugs of thymidylate synthase inhibitors: Potential for antibody directed enzyme prodrug therapy (ADEPT), *Anticancer Drug Des.* **11:**625–636.

Springer, C. J., Dowell, R., Burke, P. J., Hadley, E., Davis, D. H., Blakey, D. C., Melton, R. G., and Niculescu Duvaz, I., 1995, Optimization of alkylating agent prodrugs derived from phenol and aniline mustards: A new clinical candidate prodrug (ZD2767) for antibody-directed enzyme prodrug therapy (ADEPT), *J. Med. Chem.* **38:**5051–5065.

Springer, C. J., Niculescu-Duvaz, I., and Pedley, R. B., 1994, Novel prodrugs of alkylating agents derived from 2-fluoro- and 3-fluorobenzoic acids for antibody-directed enzyme prodrug therapy, *J. Med. Chem.* **37:**2361–2370.

Stanislawski, M., Rousseau, V., Goavec, M., and Ito, H., 1989, Immunotoxins containing glucose oxidase and lactoperoxidase with tumoricidal properties: *In vitro* killing effectiveness in a mouse plasmacytoma cell model, *Cancer Res.* **49:**5497–5504.

Svensson, H. P., Kadow, J. F., Vrudhula, V. M., Wallace, P. M., and Senter, P. D., 1992, Monoclonal antibody-β-lactamase conjugates for the activation of a cephalosporin mustard prodrug, *Bioconjugate Chem.* **3:**176–181.

Tanaka, T., Kanai, F., Okabe, S., Yoshida, Y., Wakimoto, H., Hamada, H., Shiratori, Y., Lan, K., Ishitobi, M., and Omata, M., 1996, Adenovirus-mediated prodrug gene therapy for carcinoembryonic antigen-producing human gastric carcinoma cells in vitro, *Cancer Res.* **56:**1341–1345.

Tercel, M., Denny, W. A., and Wilson, W. R., 1996, A novel nitro-substituted *seco*-CI: Application as a reductively activated ADEPT prodrug, *Bioorg. Medicinal Chem. Lett.* **6:**2741–2744.

Tong, X. W., Block, A., Chen, S. H., Contant, C. F., Agoulnik, I., Blankenburg, K., Kaufman, R. H., Woo, S. L., and Kieback, D. G., 1996a, In vivo gene therapy of ovarian cancer by adenovirus-mediated thymidine kinase gene transduction and ganciclovir administration, *Gynecol. Oncol.* **1:**175–179.

Tong, X. W., Block, A., Chen, S. H., Woo, S. L., and Kieback, D. G., 1996b, Adenovirus-mediated thymidine kinase gene transduction in human epithelial ovarian cancer cell lines followed by exposure to ganciclovir, *Anticancer Res.* **16:**1611–1617.

Vile, R. G., and Hart, I. R., 1993, Use of tissue-specific expression of the herpes simplex virus thymidine kinase gene to inhibit growth of established murine melanomas following direct intratumoral injection of DNA, *Cancer Res.* **53:**3860–3864.

Vrudhula, V. M., Senter, P. D., Fischer, K. J., and Wallace, P. M., 1993, Prodrugs of doxorubicin and melphalan and their activation by a monoclonal antibody-penicillin-G amidase conjugate, *J. Med. Chem.* **36:**919–923.

Vrudhula, V. M., Svensson, H. P., and Senter, P. D., 1995, Cephalosporin derivatives of doxorubicin as prodrugs for activation by monoclonal antibody-β-lactamase conjugates, *J. Med. Chem.* **38:** 1380–1385.

Wallace, P. M., MacMaster, J. F., Smith, V. F., Kerr, D. E., Senter, P. D., and Cosand, W. L., 1994, Intratumoral generation of 5-fluorouracil mediated by an antibody-cytosine deaminase conjugate in combination with 5-fluorocytosine, *Cancer Res.* **54:**2719–2723.

Wallace, P. M., and Senter, P. D., 1991, *In vitro* and *in vivo* activities of monoclonal antibody-alkaline phosphatase conjugates in combination with phenol mustard phosphate, *Bioconjugate Chem.* **2:**349–352.

Wang, S. M., Chern, J. W., Yeh, M. Y., Ng, J. C., Tung, E., and Roffler, S. R., 1992, Specific activation of glucuronide prodrugs by antibody-targeted enzyme conjugates for cancer therapy, *Cancer Res.* **52:**4484–4491.

Wei, M. X., Tamiya, T., Chase, M., Boviatsis, E. J., Chang, T. K., Kowall, N. W., Hochberg, F. H., Waxman, D. J., Breakefield, X. O., and Chiocca, E. A., 1994, Experimental tumor therapy in mice using the cyclophosphamide-activating cytochrome P450 2B1 gene, *Hum. Gene Ther.* **5:**969–978.

Wei, M. X., Tamiya, T., Rhee, R. J., Breakefield, X. O., and Chiocca, E. A., 1995, Diffusible cytotoxic metabolites contribute to the in vitro bystander effect associated with the cyclophosphamide/cytochrome P450 2B1 cancer gene therapy paradigm, *Clin. Cancer Res.* **1:**1171–1177.

Whisson, M. E., Connors, T. A., and Jeney, A., 1966, Mechanism of cure of large plasma cell tumors, *Arch. Immunol. Ther. Exp. Warsz.* **14:**825–831.

Index